MOLECULAR AND CELLULAR ASPECTS OF REPRODUCTION

ADVANCES IN EXPERIMENTAL MEDICINE AND BIOLOGY

Recent Volumes in this Series

A Continuation Order Plan is available for this series. A continuation order will bring delivery of each new volume immediately upon publication. Volumes are billed only upon actual shipment. For further information please contact the publisher.

MOLECULAR AND CELLULAR ASPECTS OF REPRODUCTION

Edited by

Dharam S. Dhindsa

National Institute of Health
Bethesda, Maryland

and

Om P. Bahl

State University of New York
Buffalo, New York

SPRINGER SCIENCE+BUSINESS MEDIA, LLC

Library of Congress Cataloging in Publication Data

Molecular and cellular aspects of reproduction.

(Advances in experimental medicine and biology; v. 205)
"Proceedings of a workshop on Molecular and Cellular Aspects of Reproduction
sponsored by the Reproductive Biology Study Section of the Division of Research
Grants, and National Institute of Child Health and Human Development, National In-
stitutes of Health, held October 20–21, 1985, in Bethesda, Maryland"—T.p. verso.
Includes bibliographies and index.
1. Reproduction—Congresses. 2. Cell membranes—Congresses. 3. Gene expression
—Congresses. 4. Genetic regulation—Congresses. 5. Cell interaction—Congresses. I.
Dhindsa, Dharam S. II. Bahl, Om P. III. National Institutes of Health (U.S.).
Reproductive Biology Study Section. IV. National Institutes of Child Health and
Human Development (U.S.) V. Series. [DNLM: 1. Cell Membrane—congresses. 2.
Gene Expression Regulation—congresses. 3. Reproduction—congresses. W1 AD559/
QH 471 M718 1985]
QP251.M55   1986   599'.016   86-20533
ISBN 978-1-4684-5211-2   ISBN 978-1-4684-5209-9 (eBook)
DOI 10.1007/978-1-4684-5209-9

Proceedings of workshop on Molecular and Cellular Aspects of Reproduction
sponsored by the Reproductive Biology Study Section of the Division of
Research Grants, and National Institute of Child Health and Human
Development, National Institutes of Health, held October 20–21, 1985,
in Bethesda, Maryland

© 1986 Springer Science+Business Media New York
Originally published by Plenum Press, New York 1986
Softcover reprint of the hardcover 1st edition 1986

Recently, considerable attention has been focused on studies of membrane structure and function--involvement of cell surface components in intercellular interaction, in translocation of ligands and receptors across cell membranes, and in the immunological properties of cells and gene expression and regulation. These investigations have led to the development of powerful technical tools which can be of immense value in the study of animal and human reproduction. The investigations of problems such as gamete interaction, fertilization, embryo implantation, and development have reached a stage where further meaningful progress in their understanding does not seem likely unless the conventional approaches are coupled with more modern molecular and cellular techniques. Furthermore, it is only through such basic studies that potential means of fertility regulation will emerge. The various physiological events in animal reproduction such as fertilization and implantation essentially involve an interaction between specific cell membrane components. Similarly, embryogenesis involves the expression and regulation of genes at various stages of development. Therefore, the entire Workshop was specifically devoted to two topics: 1) Structure, function, and biosynthesis of membrane components, and 2) Gene expression and regulation as related to animal reproduction. The presentations relating to each topic are presented in separate sections in this book.

The presentations on each topic of the Workshop were arranged so that the fundamental aspects of the topics were reviewed prior to the discussion of their application to specific systems. For instance, the basic concepts of glycoprotein chemistry, mechanisms of glycosylation and its regulation, and the general facts about transbilayer membrane proteins were reviewed in the first few presentations. This was followed by a discussion of specific systems such as sperm and zona pellucida components and their interaction. The first part of the Workshop concluded with presentations on the cellular and molecular aspects of membrane components in polypeptide hormone action such as receptor phosphorylation and internalization. The second part of the Workshop began with a presentation on chromatin structure and gene expression, and this was followed by reviews of structure and expression of specific genes, including placental and pituitary genes of gonadotropins, prolactin, and growth hormone. Mammalian gene transfer and gene expression, which are important in the understanding of gene regulation in development, were also reviewed. The use of cDNA probes in the characterization of Sertoli cell products for the elucidation of cell function was discussed. Finally, the subunit structure of progesterone receptor and DNA binding as related to steroid hormone action was presented.

The program of this workshop was organized by Dr. Om P. Bahl and Dr. Dharam S. Dhindsa with the assistance and advice of the Reproductive Biology Study Section members. The Reproductive Biology Study Section would like to thank the Division of Research Grants of the National

Institutes of Health for sponsoring the Workshop. This Workshop was also supported in part by the National Institute of Child Health and Human Development. Our sincere thanks are extended to Mrs. Debbi Shoaee-Tehrani for the final typing of the manuscripts and for editorial assistance.

Om P. Bahl, Ph.D.

Dharam S. Dhindsa, D.V.M., Ph.D.

CONTENTS

PART I

MEMBRANES: STRUCTURE, FUNCTION, AND BIOSYNTHESIS

CHARACTERIZATION OF GLYCOPROTEINS: CARBOHYDRATE
STRUCTURES OF GLYCOPROTEIN HORMONES

Om P. Bahl and Premanand V. Wagh

Department of Biological Sciences
State University of New York at Buffalo
Buffalo, New York 14260

INTRODUCTION

Gylcoproteins are a complex group of macromolecules that are widely distributed in nature. They are present virtually in all forms of life and are involved in important biological functions (Wagh and Bahl, 1981; Sharon and Liz, 1982; Kobata, 1984). The importance of the role that the carbohydrates play in the functions of glycoproteins has been fairly well recognized (Ashwell and Harford, 1982; Kalyan et al., 1982; Kalyan and Bahl, 1983). In recent years, considerable progress has been made toward our understanding of the structure, function, and biosynthesis of the carbohydrates (Kornfeld and Kornfeld, 1985). This has been due largely to the refinement in oligosaccharide separation and structural techniques. A vast amount of structural information gathered as a result has revealed that microheterogeneity is of general occurrence in glycoproteins. The function of this microheterogeneity is ambiguous at present. Moreover, it is difficult to reconcile the phenomenon of microheterogeneity with the fact that there is an intimate and precise relationship between structure and function in biological macromolecules. Nevertheless, the existence of microheterogeneity has complicated the structural determination of the carbohydrate prosthetic groups. In the seventies, microheterogeneity was merely considered as an artifact and was ignored during the structural determination of heterosaccharides. Consequently, the structures reported then were average structures and should be reexamined (Kobata, 1984). In this report, the problem of micro-heterogeneity the general strategy for heterosaccharide structural characterization are considered with respect to three glycoprotein hormones: human choriogonadotropin (hCG), ovine luteinizing hormone (oLH), and equine choriogonadotropin (eCG).

1

GENERAL STRUCTURAL FEATURES OF CARBOHYDRATES IN GLYCOPROTEINS

The nature of covalent bond that links the carbohydrate to the peptide chains is a major aspect of the structural uniqueness of glycoproteins. The glycosylation of amino acids is not a random event because it is well documented that glycosylation occurs at specific sites on the protein molecule. For instance, in human chorionic gonadotropin with the known structure (Bellisario et al., 1973; Carlsen et al., 1973; Mise and Bahl, 1980; Mise and Bahl, 1981; Morgan et al., 1975), the specific sites are at 4 of the 7 asparaginyl residues in the α- and β-subunits (i.e., 52 and 78 in the α-subunit and 13 and 30 in the β-subunit) and 4 of the 13 serine residues (i.e., 121, 127, 132, and 138) in the β-subunit.

Three major types of glycopeptide bonds have been found to occur among known glycoproteins. These can be differentiated strictly on the basis of the attachment of a given sugar to a specific functional group of an amino acid. First, when the attachment of the carbohydrate is via the amide group of asparagine, a glycosylamine or N-glycosidic bond is formed. Second, whenever the carbohydrate is linked through the hydroxyl group of serine, threonine, hydroxylysine, or hydroxyproline, the glycosidic bond is of the O-glycosidic type. Third is the S-glycosidic bond which is formed when the carbohydrate is attached through the thiol group of cysteine. The S-glycosidic bond has been found only in a few cases (Wagh and Bahl, 1981).

In animal glycoproteins, the predominant protein-carbohydrate linkages are Asn- and Ser/Thr-linked, N- and O-glycosidic linkages, respectively. The N-glycosidic linkage [2-acetamido-1-(β-L-aspartamido)-1,2-dideoxy-β-D-glucose] is stable to mild alkaline conditions, a criterion which distinguishes this linkage from the O-glycosidic linkage.

The asparaginyl glycopeptides known thus far exhibit three different types of structural patterns (Fig. 1):
1. High mannose type, commonly known as "simple" carbohydrate units made up of two monosaccharides, N-acetyl-glucosamine and mannose.
2. Complex type heteropolysaccharides containing more than two types of sugars.
3. Hybrid type carbohydrates sharing some of the features of both the high mannose and complex type structures.

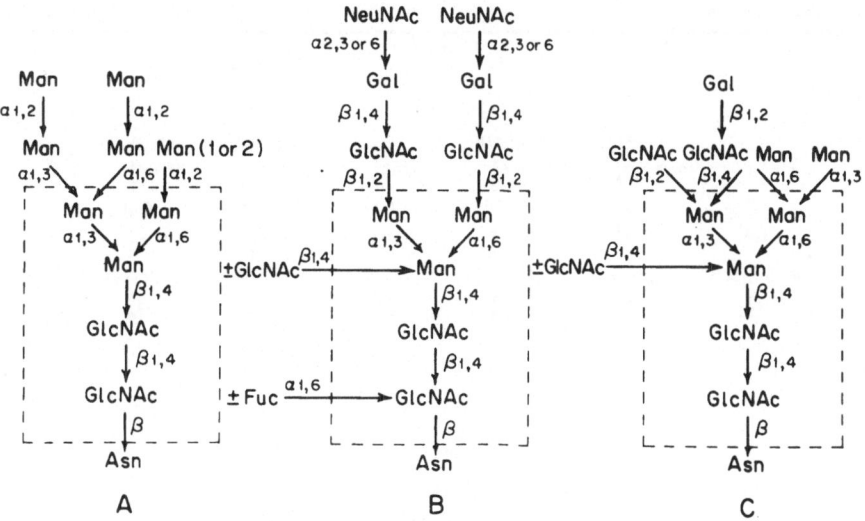

Fig. 1. Major types of asparagine-linked carbohydrate structures. Sugar residues within the boxed areas represent the common pentasaccharide core. (A) High mannose or "simple" type. (B) Complex type. (C) Hybrid type.

Most of the Asn-linked carbohydrates have a common inner core: Manα1,3(Manα1,6)Manβ1,4GlcNAcβ1,4GlcNAc and the variation in their structures, therefore, resides in the number and arrangement of the monosaccharides in the outer chains. The number of branches generally varies from 2 to 4 forming bi-, tri-, and tetra-antennary structures. In rare cases more highly branched structures have also been found (Anumula and Bahl, 1985a). Figure 2 shows the variations in the substituents which are attached to the outer chain N-acetylglucosamine residues of complex oligosaccharides.

Another outer branch structure consists of repeating N-acetyllactosamine disaccharides (Galβ1,4,GlcNAcβ1,3), first discovered in erythrocyte membrane glycoproteins called erythroglycans (Jarnefelt et al., 1978; Krusius et al., 1978). Recently, these have been found in fibronectin (Takasaki et al., 1979) and a soluble serum glycoprotein, eCG (Anumula and Bahl, 1985a, 1985b).

The Ser/Thr-linked O-glycosidic bonds involving N-acetylgalactosamine are quite common in animal glycoproteins. The size of the O-glycosidic carbohydrates varies from a disaccharide to a polysaccharide with as many as 40 to 50 sugar residues present as linear or multibranched chains (Fig. 3). The O-glycosylation sites can vary from a single site in

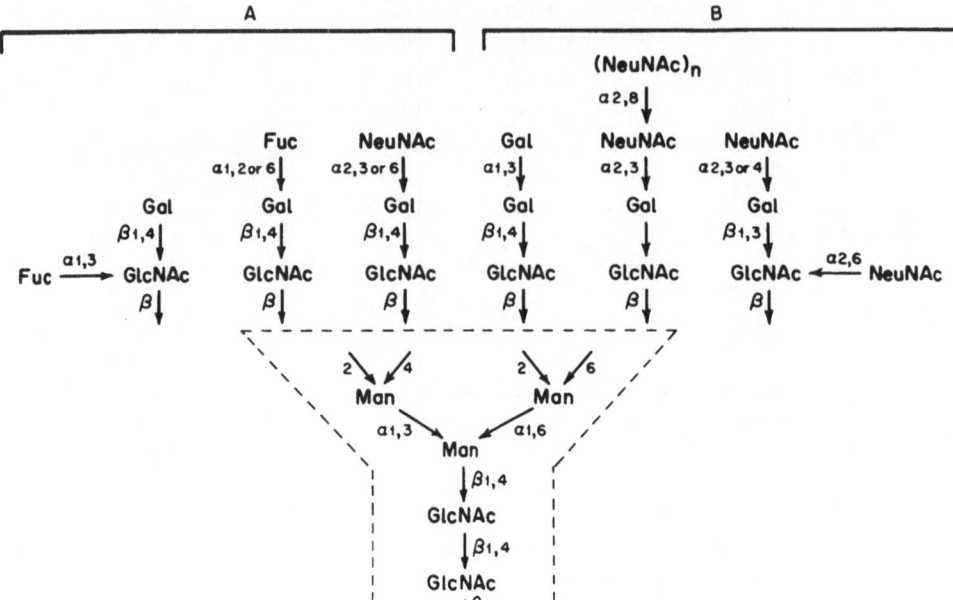

Fig. 2. Variations in the substituents attached to the outer chain N-acetylglucosamine residues of complex oligosaccharides. Sugars within the dashed lines indicate the pentasaccharide core. (A) Very frequently occurring outer branches. (B) Rarely occurring sugar chains. (Adapted from Kornfeld and Kornfeld, 1985.)

immunoglobulin A to several in fetuin and hCG containing 3 and 4 carbohydrate chains, respectively, and to 800 chains in ovine submaxillary mucins (Wagh and Bahl, 1981).

MICROHETEROGENEITY OF GLYCOPROTEINS

Microheterogeneity is referred to the carbohydrate structural variants present on a single glycosylation site in a glycoprotein (Montgomery, 1972). Initially, microheterogeneity was ascribed to minor variations of sugars at the non-reducing termini of the carbohydrate units. More recently, major variations due to heterogeneity in the degree and location of branching have been found. In this regard, although the term "microheterogeneity" becomes a misnomer, we have used both terms interchangeably. Since Asn-linked glycoproteins have a common pentasaccharide inner core ($Man_3GlcNAc_2$), the source of microheterogeneity is due to the variation in the number and arrangement of monosaccharides

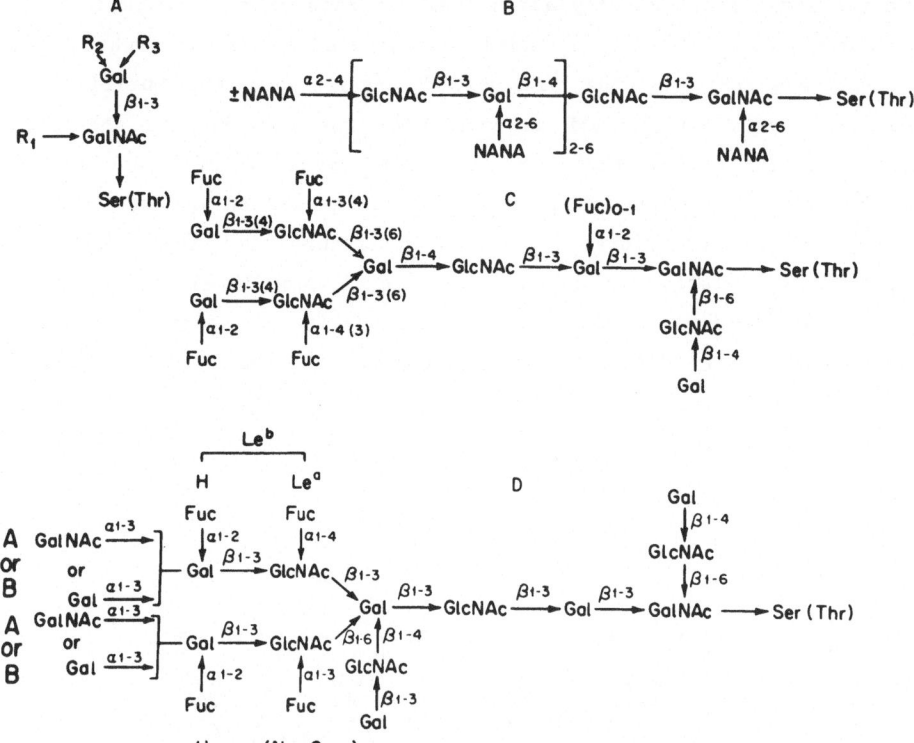

Fig. 3. The diversity in the O-linked oligosaccharides with linkages between N-acetylgalactosamine and Serine/Threonine. (A) R_1, R_2, and R_3 = H: antifreeze glycoprotein; R_1 and R_2 = NeuNAc and R_3 = H: human chorionic gonadotropin; R_1 = NeuNAc, R_2 = Fuc, and R_3 = GalNAc: porcine submaxillary glycoprotein A. (B) Rat sublingual glycoprotein (acidic oligosaccharide). (C) Human salivary glycoprotein (neutral oligosaccharide). (D) Composite heteropolysaccharide structure of blood group substances. (Taken from Wagh and Bahl, 1981.)

in the outer chains and in the location and degree of branching. The changes in the non-reducing peripheral monosaccharides such as sialic acid and fucose are quite frequent and were considered earlier to be the result of incomplete glycan synthesis, partial degradation, or an artifact introduced during the isolation (Hatton et al., 1983). It has been shown that this type of microheterogeneity is not due to partial degradation in vivo or introduced during the isolation process. A sialic acid deficient form of human transferrin has been isolated from the plasma of normal individuals as well as that from chronic alcoholics. Similarly, α_1-acid glycoprotein with decreased sialic acid has been found in the sera of patients suffering from other chronic illnesses. Wong et al. (1974) showed that this lack of sialic acid was not due to in vivo desialylation since fully sialylated radiolabeled rabbit and bovine transferrins did not

5

undergo significant desialylation over a period of eight days in rabbits. When Regoeczi et al (1977) added radiolabeled homogenous subfraction of human transferrin to the plasma and carried out the purification on DEAE-cellulose, no charged heterogeneity was observed in the isolated fraction, indicating that the microheterogeneity of transferrin was neither due to its degradation in vivo nor caused during purification in vitro. It is, therefore, reasonable to assume that the presence of a partially sialylated glycoprotein in vivo is more likely to be a feature of biosynthesis rather than degradation. Another type of microheterogeneity discovered in glycoproteins is the one in which there

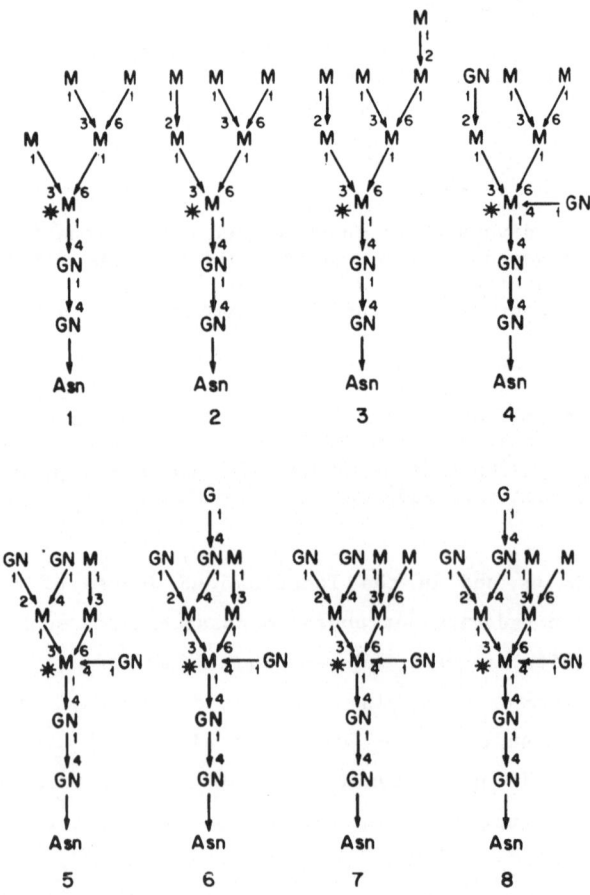

Fig. 4. Oligosaccharide variants found in the asparagine-linked carbohydrate of ovalbumin. GN = GlcNAcβ; M = Manα; and *M = Manβ. (Adapted from Kobata, 1984.)

are major changes in the number and location of outer branches. It was
first noted by Cunningham (1971), who showed the existence of several
carbohydrate variants on a single glycosylation site in ovalbumin. Later,
these variants were separated and analyzed by Kobata (1984; Fig. 4). A
number of other examples can be cited for this type of microheterogeneity,
e.g., fetuin has three glycosylation sites and has triantennary structures
of two different types (Fig. 5). Human transferrin has on two sites,
biantennary, triantennary, and a small amount of tetra-antennary
structures (Marz et al., 1982). In disease state, microheterogeneity is
further enhanced. For example, human chorionic gonadotropin has four
biantennary structures but in choriocarcinoma or hyditidiform mole
(Mizuochi et al., 1983) triantennary structures are also formed. Human
thyroglobulin from malignant tissues not only contained less sialic acid
and had triantennary and high molecular weight carbohydrates with
repeating N-acetyllactosamine, but also had carbohydrate units which were
phosphorylated (Osawa et al., 1985). Briefly, carbohydrates show
microheterogeneity due to variation in the peripheral sugars and in the
degree of branching and further modification by phosphorylation or
sulfation. Microheterogeneity in the O-glycosidic carbohydrates has also
been found but the cause of this microheterogeneity is not clear at
present and the detailed information on the structural changes in disease

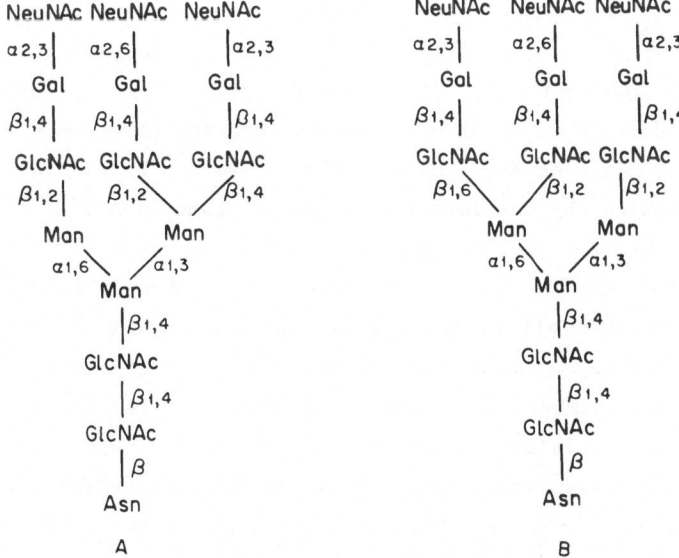

Fig. 5. Heterogeneity of asparagine-linked carbohydrate
units of bovine fetuin. Structures described by:
(A) Baenziger and Fiete, 1979; (B) Nilsson et al.,
1979.

states in such carbohydrates is rather limited (Carlstedt et al., 1985; Anumula and Bahl, 1985b).

STRUCTURAL CHARACTERIZATION OF CARBOHYDRATES

As stated above, the carbohydrate structural analysis is complicated because of its microheterogeneity and branching. Elucidation of carbohydrate structure involves determination of its 1) molecular size, 2) carbohydrate composition, 3) intersugar and anomeric linkages, 4) the number and location of the branching, 5) sequences of monosaccharides, and, finally 6) the protein carbohydrate-linkage. Although great strides have been made in the methodology for the structural characterization of carbohydrates, it is still not as well developed as that for proteins and nucleic acids.

Before embarking on the structural studies, it is necessary to cleave the carbohydrate from the peptide and fractionate the released carbohydrate into its various heterogeneous forms. The cleavage of the carbohydrate from the polypeptide can be achieved chemically by basic hydrolysis using either hydrazine, i.e., hydrazinolysis (Takasaki et al., 1982) or a mixture of NaOH and $NaBH_4$ (Bedi et al., 1982) or enzymatically by endoglycosidases such as endo-β-N-acetylglucosaminidase F or H from Flavobacterium meningosepticum (Elder and Alexander, 1982) and Streptomyces plicatus (Trimble et al., 1982), respectively, and N- and O-glycanases, Peptide: N- (Tarentino et al., 1985) or O-glycosidase (Umemoto et al., 1977). N-glycosidic linkages are relatively more stable to alkali than O-glycosidic linkages; the latter can be cleaved under mild alkaline conditions by β-elimination (Wagh and Bahl, 1981).

Endo-H is specific for high mannose or hybrid type oligosaccharides while Endo-F is for all types of N-linked carbohydrates. Therefore, in most cases Endo-F would be much more useful. N-glycanases have been isolated from a number of sources, hen oviduct (Tarentino and Maley, 1976), almond emulsion (Taga et al., 1984), and Flavobacterium meningosepticum (Tarentino et al., 1985), differing in specificities. The N-glycanase from the bacterial source is the most promising enzyme since it can hydrolyze protein-carbohydrate linkages in intact glycoproteins. The O-glycanase has been isolated from Diplococcus pneumoniae (Umemoto et al., 1977). The enzyme can hydrolyze O-N-acetylgalactosaminidic linkages in glycoproteins containing Galβ1,3GalNAc-sugar chains.

After cleavage, radiolabel can be introduced in the carbohydrate by reduction with NaB^3H_4 to facilitate monitoring the fractionation procedure and the exoglycosidase degradation products. In the cleavage of O-glycosidic linkages, the NaB^3H_4 is included in the reaction mixture. While the chemical methods for the carbohydrate-peptide cleavage have been widely and successfully used, the enzymatic methods are much more convenient and reliable giving a high recovery of the carbohydrates. The chemical methods introduce some artifactual heterogeneity at the reducing end of the oligosaccharide chains (Lloyd and Kabat, 1969).

A large number of effective methods for fractionation of the carbo-hydrate variants have been developed. These include paper chromatography (French et al., 1984), gel filtration (Yamashita et al., 1982) using Bio-Gel P-4, thin-layer chromatography (Holmes and O'Brian, 1979; Anumula and Bahl, 1985a, 1985b), and HPLC (Mellis and Baenziger, 1981; Turco, 1981). In special situations, immobilized lectin columns such as Con-A Sepharose have also been applied, particularly in the separation of biantennary from multiantennary structures (Bayard and Kerckaert, 1980). Availability of malto-oligosaccharides including the standard oligosaccharides of known structures, bi-, tri-, and tetra-antennary and high mannose type and various others with intermediate structures, have proven useful as molecular weight-markers in structural studies. Glucose malto-oligosaccharides are prepared by controlled hydrolysis of amylose while bi-, tri-, and tetra-antennary structures and the intermediates of well characterized glycoproteins are obtained by hydrolysis with Endo-F or N-glycanase followed by controlled digestions of the carbohydrate with exoglycosidases (Anumula and Bahl, 1985a, 1985b). Correlation between the glucose residue number in the malto-oligosaccharides and the number of sugar residues in a glycoprotein carbohydrate can be determined in any of the fractionation systems. Thus, the contribution of the various sugars to the molecular weight in terms of glucose residues in each fractionation system can be determined (Table I).

Whereas the above approach yields information on the saccharide variants in a glycoprotein, it does not indicate the glycosylation site(s) to which a particular variant(s) is located. In order to obtain this information it is imperative to isolate homogeneous glycopeptides using specific proteolytic enzymes such as trypsin and Staphylococcus aureus protease. Each glycopeptide is then subjected to chemical or preferably enzymatic cleavage as described above. The drawback of this approach, however, is that in the process of glycopeptide purification, one may lose

Table I. Molecular sizes of sugar residues in complex type oligosaccharides expressed in glucose units of maltooligosaccharides in Bio-Gel P-4.[a]

Oligosaccharide Structure	Glucose Units
M $\xrightarrow{\alpha 1,6}$ M $\xrightarrow{\beta 1,4}$ GN $\xrightarrow{\beta 1,4}$ GNH$_2$; M $\xrightarrow{\alpha 1,3}$	7·28
M $\xrightarrow{\alpha 1,6}$ M $\xrightarrow{\beta 1,4}$ GN $\xrightarrow{\beta 1,4}$ GNH$_2$ (Fuc $\xrightarrow{\alpha 1,6}$) ; M $\xrightarrow{\alpha 1,3}$	8·20
GN $\xrightarrow{\beta 1,2}$ M $\xrightarrow{\alpha 1,6}$ M $\xrightarrow{\beta 1,4}$ GN $\xrightarrow{\beta 1,4}$ GNH$_2$ (Fuc $\xrightarrow{\alpha 1,6}$) ; GN $\xrightarrow{\beta 1,2}$ M $\xrightarrow{\alpha 1,3}$	12·25
G $\xrightarrow{\beta 1,4}$ GN $\xrightarrow{\beta 1,2}$ M $\xrightarrow{\alpha 1,6}$ M $\xrightarrow{\beta 1,4}$ GN $\xrightarrow{\beta 1,4}$ GNH$_2$; G $\xrightarrow{\beta 1,4}$ GN $\xrightarrow{\beta 1,2}$ M $\xrightarrow{\alpha 1,3}$	13·50
G $\xrightarrow{\beta 1,4}$ GN $\xrightarrow{\beta 1,2}$ M $\xrightarrow{\alpha 1,6}$ M $\xrightarrow{\beta 1,4}$ GN $\xrightarrow{\beta 1,4}$ GNH$_2$ (Fuc $\xrightarrow{\alpha 1,6}$) ; G $\xrightarrow{\beta 1,4}$ GN $\xrightarrow{\beta 1,2}$ M $\xrightarrow{\alpha 1,3}$	14·45
G $\xrightarrow{\beta 1,4}$ GN $\xrightarrow{\beta 1,2}$ M (GN $\xrightarrow{\alpha 1,6 / \beta 1,4}$) $\xrightarrow{\beta 1,4}$ M $\xrightarrow{\beta 1,4}$ GN $\xrightarrow{\beta 1,4}$ GNH$_2$ (Fuc $\xrightarrow{\alpha 1,6}$) ; G $\xrightarrow{\beta 1,4}$ GN $\xrightarrow{\beta 1,2}$ M $\xrightarrow{\alpha 1,3}$	15·00
G $\xrightarrow{\beta 1,4}$ GN $\xrightarrow{\beta 1,4}$ M ; G $\xrightarrow{\beta 1,4}$ GN $\xrightarrow{\beta 1,2}$ M $\xrightarrow{\alpha 1,6}$ M $\xrightarrow{\beta 1,4}$ GN $\xrightarrow{\beta 1,4}$ GNH$_2$ (Fuc $\xrightarrow{\alpha 1,6}$) ; G $\xrightarrow{\beta 1,4}$ GN $\xrightarrow{\beta 1,2}$ M $\xrightarrow{\alpha 1,3}$	17·10
G $\xrightarrow{\beta 1,4}$ GN $\xrightarrow{\beta 1,4}$ M ; G $\xrightarrow{\beta 1,4}$ GN $\xrightarrow{\beta 1,2}$ M $\xrightarrow{\alpha 1,6}$ M $\xrightarrow{\beta 1,4}$ GN $\xrightarrow{\beta 1,4}$ GNH$_2$ (Fuc $\xrightarrow{\alpha 1,6}$) ; G $\xrightarrow{\beta 1,4}$ GN $\xrightarrow{\beta 1,6}$ M ; G $\xrightarrow{\beta 1,4}$ GN $\xrightarrow{\beta 1,2}$ M $\xrightarrow{\alpha 1,3}$	19·00

---

[a] Abbreviations: G = Galactose; GN = N-acetylglucosamine; GNH$_2$ = N-acetylglucosaminitol; M = Mannose; and Fuc = Fucose (Yamashita et al., 1982).

some heterogeneous forms. The glycopeptides can be purified by the above procedures used for carbohydrates including ion exchange chromatography and paper electrophoresis (Kessler et al., 1979a, 1979b).

The methods which have been generally used for carbohydrate structural characterization are listed below:

1. Methylation combined with GC-MS analysis.
2. Periodate oxidation oxidation and Smith degradation combined with methylation analysis.
3. Treatment with endo- and exoglycosidases.
4. Acetolysis.
5. Nitrous acid deamination.
6. Nuclear magnetic resonance (NMR).

These techniques, with the exception of NMR, are discussed here with respect to their applications to the analysis of the three glycoprotein hormones. The availability of high-resolution proton NMR instruments has also provided a powerful method for the structural and conformational analysis of carbohydrates (Hounsell et al., 1984).

STRUCTURAL ANALYSIS OF CARBOHYDRATES OF GLYCOPROTEIN HORMONES

We previously reported the structural characterization of carbohydrate units of hCG (Kessler et al., 1979a, 1979b) and oLH (Bedi et al., 1982) and more recently, we have completed the analysis of eCG carbohydrates (Anumula and Bahl, 1985a, 1985b). The studies of hCG and oLH carbohydrates were carried out on the purified and electrophoretically homogeneous glycopeptides obtained from homogeneous preparations of the subunits, and no attempt was made to isolate the heterogeneous forms of the heterosaccharides. Accordingly, the structures assigned to them were average structures. Structural investigation of the eCG carbohydrate, on the other hand, has addressed the problem of carbohydrate microheterogeneity.

Human Choriogonadotropin

The two noncovalently-linked subunits of hCG (38,000 d), hCG-α (15,000 d) and hCG-β (23,000 d) contain four Asn-linked carbohydrates at asparagine residues 52 and 78 in the α-subunit and 13 and 30 in the β-subunit. The β-subunit also has four Ser-linked oligosaccharide chains at positions 121, 127, 132, and 138 in the carboxyterminal glycopeptide BC-19 (residues 109-145). The structural analysis of the Asn-linked carbohydrates was carried out on four electrophoretically homogeneous tryptic glycopeptides αT-8 (residues 51-53), αT-11a (residues 76-91), βT-2 (residues 11-20), and βT-3 (residues 21-43).

Methods for the structural analysis of the four N-linked carbohydrate units included techniques such as methylation combined with GC-MS, periodate oxidation of Smith degradation, and sequential exo-glycosidase treatment. All the N-glycosidic carbohydrates in hCG were found to have the same basic structure except that fucose was present in only the hCG-β carbohydrates. The sugar compositions of the glycopeptides summarized in Table II suggest similarity in their structures.

Methylation Analysis. Among the chemical methods for structural determination, the methylation analysis is still the most important technique for determining the intersugar linkages and the location of branching. It does not allow deduction of the sequence of sugars nor their anomeric linkages.

The four asparaginyl glycopeptides were digested with Pronase prior to methylation by the Hakamori (1964) procedure in order to remove nearly all the amino acids except the one linked to asparagine to facilitate the extraction of the permethylated glycopeptides by chloroform. The Pronase-digested, methylated glycopeptides were hydrolyzed, reduced, and acetylated according to the procedure of Stellner et al. (1973). The resultant, partially methylated alditol and hexosaminitol acetates were identified by comparison of their retention times relative to that of 1,5-di-O-acetyl-2,3,4,6-tetra-O-methyl-D-glucitol with the reported values (Lindberg, 1972) as well as by comparison with those of authentic standards. Confirmatory evidence was obtained by GC-MS analysis. Fragmentation patterns were compared with those reported for partially methylated alditol acetate standards (Jansson et al., 1976) and for partially methylated hexosaminitol acetate standards (Stellner et al., 1973). The methylated sugar derivatives obtained from all four asparaginyl glycopeptides are shown in Table III. The only hexitol acetates found in each case were those corresponding to the following: 2,3,4,6-tetra-O-methyl-D-galactose; 3,4,6-tri-O-methyl-D-mannose; 2,4-di-O-methyl-D-mannose; and 2-deoxy-3,6-di-O-methyl-2-N-methylacetamido-D-glucose. However, the glycopeptide βT-3 also showed the presence of

2,3,4-tri-O-methyl-L-fucose and 3-deoxy-3-O-methyl-2-N-methylacetamido-D-glucose. Neither mannose nor N-acetylglucosamine was found to be present at the non-reducing terminus.

Table II. Amino acid and carbohydrate compositions of asparaginyl-oligosaccharide-containing glycopeptides of desialyzed hCGα and hCG-β*.

Component	αT-8	αT-11a	βT-2	βT-3
		Residues		
Aspartic acid	1.00 (1)	1.00 (1)	1.00 (1)	1.00 (1)
Threonine	1.76 (2)	1.72 (2)	0.70 (1)	4.76 (5)
Serine	2.46 (2)	1.23 (1)		
Glutamic acid	1.12 (1)	1.16 (1)	0.80 (1)	0.98 (1)
Proline			2.56 (1)	0.92 (2)
Glycine				1.18 (2)
Alanine	1.35 (1)	1.00 (1)	1.38 (2)	0.89 (1)
S-Carboxamido-methylcysteine	n.d. (2)	n.d. (3)	n.d. (1)	n.d. (4)
Valine	1.61 (2)	1.21 (1)	0.46 (1)	1.21 (2)
Methionine				0.30 (1)
Isoleucine			0.68 (1)	1.21 (2)
Leucine			1.03 (1)	
Tyrosine		1.80 (2)		0.90 (1)
Lysine	1.01 (1)		1.10 (1)	
Histidine		1.50 (2)		
Arginine			0.90 (1)	0.95 (1)
N-Acetyl-glucosamine	3.9 (4)	4.0 (4)	3.6 (4)	4.0 (4)
Fucose			0.2	0.2
Galactose	1.1 (2)	1.9 (2)	1.9 (2)	2.4 (2)
Mannose	2.6 (3)	2.7 (3)	2.6 (3)	2.6 (3)

* (Kessler et al., 1979a).

Table III. Linkage analysis-relative retention times[a] of methylated sugars as their alditol acetates obtained from methylated N-glycopeptides of hCG-α and hCG-β.

Methylated Sugar	αT-8	αT-11a	βT-2	βT-3	Literature Value[b]	Values from Reference Compounds
2,3,4-tri-O-methyl-fucose	-	-	-	0.60	0.58	0.61[c]
2,3,4,6-tetra-O-methyl-galactose	1.19	1.19	1.17	1.19	1.19	1.14[d]
3,4,6-tri-O-methyl-mannose	1.80	1.83	1.81	1.82	1.82	1.85[e]
2,4-di-O-methyl-mannose	4.42	4.41	4.33	4.26	4.51	4.44[f]
2-deoxy-3,6-di-O-methyl-2-N-methyl-acetamido-glucose	1.47	1.48	1.50	1.52	-	1.50[g,h]
2-deoxy-3-O-methyl-2-N-methyl-acetamido-glucose	-	-	-	2.58	-	-

[a] For neutral sugar, with respect to 2,3,4,6-tetra-O-methyl glucose and for amino sugars with respect to 2-deoxy-3,4,6-tri-O-methyl-2-N-methylacetamido-glucose (Kessler et al., 1979a).

[b] Jansson et al. (1976).

[c-h] Relative retention times obtained from the following reference compounds: c, Fucl,2αGal; d, ovine submaxillary oligosaccharide; e, Manα1,2Man; f, α₁-acid glycopeptide; g, chito-tetraose; h, Manα1,4GlcNAc.

Table IV. Smith degradation of asparaginyl-oligosaccharide-containing glycopeptides of desialyzed hCG-α and hCG-β.

| Sugar | Native | OR-1 | Smith Degradation[a] | | |
			ORH-1	OR-2	OR-3
			Residues per mol of aspartic acid		
αT-8:					
Galactose	1.1 (2)[b]	0.0	0.0	0.0	0.0
Mannose	2.6 (3)	1.7	0.9	0.3	0.0
N-acetyl-glucosamine	3.9 (4)	3.5	2.2	2.0	1.0
αT-11a:					
Galactose	1.9 (2)	0.0	0.0	0.0	0.0
Mannose	2.7 (3)	1.4	0.9	0.3	0.0
N-acetyl-glucosamine	4.0 (4)	3.2	2.1	2.1	1.1
βT-2:					
Galactose	1.9 (2)	0.0	0.0	0.0	0.0
Mannose	2.6 (3)	1.6	1.3	0.2	0.0
Fucose	0.2 (1)	0.0	0.0	0.0	0.0
N-acetyl-glucosamine	3.6 (4)	3.4	2.3	2.1	0.9
βT-3:					
Galactose	2.4 (2)	0.0	0.5	0.0	0.0
Mannose	2.6 (3)	1.0	0.9	0.1	0.0
Fucose	0.2 (1)	0.0	0.0	0.0	0.0
N-acetyl-glucosamine	4.0 (4)	3.4	1.9	2.0	0.8

[a] Abbreviations: OR, periodate oxidized-borohydride reduced; ORH-1, periodate oxidized-borohydride reduced-acid hydrolyzed. Numerals refer to the number of Smith degradation cycles, e.g., OR-1 refers to the first cycle and OR-2 refers to the second cycle, etc.

[b] Number of residues in completed chains (Kessler et al., 1979a).

Periodate Oxidation. One of the approaches to study the structure of the carbohydrate component of glycoproteins is the sequential oxidation of polysaccharide with periodic acid. This technique (Goldstein et al., 1965), commonly called the "Smith degradation," provides in most cases valuable structural information concerning the intersugar linkages with a polysaccharide. In this procedure the polysaccharide is oxidized with periodate, and the resulting polyaldehydic structure is reduced to polyalcohol with sodium borohydride. Whereas the non-cyclic acetal linkages derived from the reduced periodate oxidized polysaccharide are susceptible to hydrolysis by acid under mild conditions, the glycosidic linkage of a residue that is not vulnerable to oxidation by periodate is comparatively stable to acid. This difference in the stability of the linkages yields specific glycosides of oligosaccharide characteristic of the parent polysaccharide.

Each of the four desialylated asparaginyl glycopeptides was subjected to Smith degradation in three oxidation-reduction-hydrolysis cycles. The results are presented in Table IV. In the first cycle (OR-1) periodate caused the destruction of all galactose and fucose residues, indicating that they were terminally located. Furthermore, two residues of mannose but none of the N-acetylglucosamine residues were destroyed in this step. However, upon mild acid hydrolysis of OR-1, two residues of N-acetylglucosamine were lost from the glycopeptide, leaving one residue of mannose and two residues of N-acetylglucosamine per mole of aspartic acid (ORH-1). This observation suggested that the two mannose residues destroyed were located internally to the two N-acetylglucosamine residues lost during acid hydrolysis. This was later confirmed by exoglycosidase treatment studies. That the two mannose residues by periodate suggests that they were not substituted at C-3. The C-4 oxygen of these mannoses was ruled out as a possible site by isolation of N-acetylglucosaminyl glycerol from all four glycopeptides after the first Smith cycle. This was done by passing the ORH-1 materials though a G-25 Sephadex column. Fractions subsequent to the elution of core-glycopeptide were analyzed for glucosamine and those showing its presence were pooled. Paper chromatography of the products, after hydrolysis, showed the presence of glucosamine and glycerol, indicating GlcNAc to Man linkage as 1,2- or 1,6-. The substitution at C-2 of the mannoses destroyed by periodate during the first Smith cycle was later established by methylation studies.

The oxidation of the core glycopeptide with periodate destroyed the remaining mannose residue, leaving glycopeptides with two residues of N-acetylglucosamine per mole of aspartic acid (OR-2). The third treatment with periodate destroyed one of the two remaining N-acetylglucosamine residues and left glycopeptides which contained just the N-acetylglucos-amine residue involved in the linkage with asparagine (OR-3). Methylation of the core glycopeptides showed the presence of terminal mannose at the non-reducing end and 1,4-di-O- substituted N-acetylglucosamine. Thus, the sequence of monosaccharides in the core glycopeptides was established as Man-GlcNAc-GlcNAc-Asn.

Exoglycosidases Treatment. Whereas methylation analysis and periodate oxidation techniques provide information concerning the intersugar linkages and limited knowledge of the sequence of sugars, the anomeric configuration and complete sequence of sugars cannot be deduced from these techniques.

Exoglycosidases have proven to be extremely useful in the determination of the sequence of sugars and their anomeric linkages. A number of exoglycosidases have been purified to homogeneity and some of these have been obtained in crystalline forms. Several of these enzymes which hydrolyze a wide variety of glycosyl linkages have been effectively used in the elucidation of the carbohydrate structure of various glycoproteins (Wagh and Bahl, 1981).

The sequence of peripheral mannosaccharides in glycopeptides and their anomeric configurations were determined by individual and sequential treatment of hCG-α, hCG-β, and their asparaginyl glycopeptides with specific exoglycosidases. These data are summarized in Tables V and VI.

Table V. Sequential treatment of desialyzed hCG-α and hCG-β with glycosidases.

Glycosidase	Asialo-hCG-α			Asialo-hCG-β		
	Gal	GlcNAc	Man	Gal	GlcNAc	Man
	Residues Released[a]					
A. Individual Treatment						
1. β-N-acetyl-glucosaminidase	–	0.9	–	0	1.9	–
2. α-mannosidase	–	–	0	–	–	0
B. Sequential Treatment						
1. β-galactosidase	2.3	–	–	7.5	–	–
2. Then β-N-acetyl-glucosaminidase	–	2.6	–	–	5.5	–
3. Then α-mannosidase	–	–	4.0	–	–	4.0

[a]Values calculated on the basis of four α-linked mannose residues per subunit (Kessler et al., 1979a).

17

α-Mannosidase failed to release any mannose from either the subunits or glycopeptides, confirming the internal location of this sugar. Approximately one-half of the terminal galactose residues were released by treatment with β-galactosidase. Subsequent treatment with β-N-acetyl-hexosaminidase hydrolyzed 55 to 100% of the now terminally exposed N-acetylglucosamine residues. Finally, mannose was released by α-mannosidase establishing the sequence of the peripheral carbohydrate chains as Galβ-GlcNAcβ-Manα. The release of N-acetylglucosamine by β-acetylglucosaminidase, without prior treatment with β-galactosidase was due to the presence of the latter enzyme in the hexosaminidase preparation causing the removal of galactose and exposing the N-acetylglucosamine residues.

When the glycopeptide from αT-8 was treated with _Aspergillus niger_ β-mannosidase, about 68% of the terminal mannose was released. Thus, the core mannose is joined to the adjoining internal N-acetylglucosamine by a β-linkage. It was not possible to carry out this study further with other core glycopeptides because of the lack of material. However, since all four asperaginyl glycopeptides show identity in most aspects of structure, it is highly unlikely that they will differ with respect to the anomeric configuration of the core mannose. The β-configuration of the penultimate N-acetylglucosamine in the core was established by treatment of the glycopeptides with β-galactosidase, β-N-acetylglucosaminidase, and α- and β-mannosidases, which resulted in the removal of all sugar residues except the one involved in the carbohydrate-peptide linkage. On the basis of the above evidence, the structure of the core was determined as Manβ1,4GlcNAcβ1,4GlcNAc-Asn.

The Nature and Location of Sialic Acid. Sialic acid was identified as N-acetylneuraminic acid by paper chromatography, no evidence being found for the presence of the N-glycolyl analog. When the native hCG subunits were oxidized with periodate, 70 to 80% of galactose residues were found to be invulnerable to oxidation. On the other hand, under similar conditions, the destruction of galactose was complete in the asialo subunits, indicating the C-3 position of the penultimate galactose residues as the site where N-acetylneuraminic acid was linked through its C-2 oxygen.

The Structures of N-Glycosidic Units of hCG. The complete structure for the asparaginyl carbohydrate units of hCG as depicted in Figure 6 was deduced employing Smith degradation, methylation, and by use of

Table VI. Sequential treatment of glycopeptides with specific glycosidases.

Glycosidase	αT-8			αT-11a			β-T-2		
	Gal	GlcNAc	Man	Gal	GlcNAc	Man	Gal	GlcNAc	Man
A. Individual Treatment:									
1. β-N-acetylglucosaminidase	–	0.9	–	–	0	–	–	0	–
2. α-mannosidase	–	–	0	–	–	0	–	–	–
B. Sequential Treatment:									
1. β-galactosidase	1.0	–	–	1.0	–	–	2.4	0	0
2. Then β-N-acetylglucosaminidase	–	1.4	–	–	1.1	–	–	2.2	–
3. Then α-mannosidase	–	–	2.0	–	–	2.0	–	–	2.0

[a]Values calculated on the basis of the release of two α-linked mannose residues per glycoprotein (Kessler et al., 1979a).

19

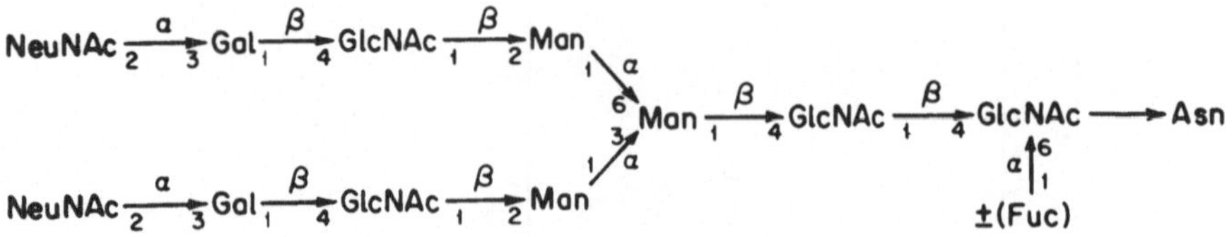

NeuNAc $\xrightarrow[3]{\alpha_2}$ Gal $\xrightarrow[1]{\beta}{}_4$ GlcNAc $\xrightarrow[1]{\beta}{}_2$ Man $\searrow{}^{\alpha}_{1}{}_6$

Man $\xrightarrow[1]{\beta}{}_4$ GlcNAc $\xrightarrow[1]{\beta}{}_4$ GlcNAc $\longrightarrow$ Asn

NeuNAc $\xrightarrow[3]{\alpha_2}$ Gal $\xrightarrow[1]{\beta}{}_4$ GlcNAc $\xrightarrow[1]{\beta}{}_2$ Man $\nearrow{}^{\alpha}_{1}{}_3$

$\alpha\vert^{6}_{1}$

±(Fuc)

Fig. 6. Structure of asparagine-linked carbohydrate units of human chorionic gonadotropins (Kessler et al., 1979a).

exoglycosidases. The anomeric configurations, intersugar linkages, and the carbohydrate-peptide linkage are consistent with the data obtained using these three techniques. Since detectable amounts of fucose were invariably present in the glycopeptides derived from hCG-β subunit as compared to those from the α-subunit, some inferences can be drawn with regard to fucosylation of the asparaginyl carbohydrate units of the hormone. First, only the β-subunit contains fucosyl residues either in one or both of the asparagine-linked carbohydrate units and that the fractional residue figure for fucose may be a reflection of microheterogeneity. The phenomenon of microheterogeneity may not be restrictive only to fucose but is also ascribed to sialic acid residues, in the biantennary structure of hCG. Thus, if the structure shown above containing fucose is considered as the complete biosynthetic product and if fucose and/or sialic acid residue at the non-reducing terminus of the ManαMan chain is found to be absent, three incomplete sugar chain patterns can be envisaged. Indeed, the three incomplete biosynthetic products of glycosylation, i.e., absence of fucose, absence of sialic acid, and absence of fucose and sialic acid have been reported for asparagine-linked sugar chains of hCG (Kobata, 1984). A fourth unusual mono-antennary structural pattern depicting the absence of the trisaccharide sialyl-N-acetyllactosamine has also been reported by the same author. However, it appears from our methylation and exoglycosidase studies of the glycopeptides that such a pattern may have resulted as an artifact due to the hormone preparation used and therefore, may not have any functional significance.

Structure and Location of the O-Glycosidic Carbohydrate Units. The serine-linked oligosaccharide was prepared from the intact hormone by β-elimination in the presence of sodium borohydride. Approximately 86 percent of the O-glycosidic chains were cleaved as indicated by the amount of N-acetylgalactosaminitol formed. The oligosaccharide was further purified by Dowex-50 [H+] chromatography and gel-filtration on Sephadex G-25. The final produce showed a high degree of purity when examined by high-voltage paper electrophoresis and paper chromatography with a 50% recovery. The ratio of galactose, N-acetylgalactosaminitol, and sialic acid in the oligosaccharide was found to be 1:1:2. All of the sialic acid was present as N-acetylneuraminic acid.

Methylation of the oligosaccharide and examination of the partially methylated alditol acetates derived therefrom by GC-MS revealed the formation of 2,4,6-tri-O-methyl-D-galactose and 4-O-methyl-2-N-methylacetamido-2-deoxy-galactosaminitol, indicating the presence of 3-O-substituted galactose and 3,6-di-O-substituted N-acetylgalactosaminitol.

Oxidation of the oligosaccharide with periodate for 60 h followed by reduction with borohydride resulted in almost complete loss of sialic acid. Approximately 80% of the galactose residues were undegraded during oxidation, indicating the presence of 1,3-linked galactose residues. Virtually all of the N-acetylgalactosaminitol was converted to N-acetyl-threosaminitol, indicating that N-acetylgalactosaminitol was either substituted at C-3 or at both C-3 and C-6 positions. Two moles of formic acid per oligosaccharide were produced during oxidation. It was clear from these data that the sialic acid and galactose residues were joined by 2,3-linkages and that the observed 20% loss of galactose residues was due to either the hydrolysis of sialic acid during oxidation, or the fact that the original oligosaccharide was a mixture of tetra- and trisaccharide in a ratio of 4:1, showing inherent microheterogeneity of the oligosaccharide. When the carboxyterminal desialylated βC-19 glycopeptide instead of the oligosaccharide was treated with periodate, the galactosyl residues were found to be invulnerable to oxidation, indicating that galactose residues must be substituted at C-3 of the hexosamines.

The sequence and the anomeric configuration of the monosaccharides of the O-linked carbohydrate units were studied employing exoglycosidases. The βC-19 glycopeptide devoid of sialic acid was used as the substrate. When A. niger α-N-acetylgalactosaminidase and β-galactosidase were used separately, N-acetylgalactosamine was not released by the former enzyme. However, β-galactosidase released approximately 75% of the galactose residues, indicating that they were located at the non-reducing end of the carbohydrate units. On the other hand, three-fourths of the N-acetylgalactosamine residues were cleaved only after prior treatment of the glycopeptide with galactosidase. Furthermore, after treatment with β-galactosidase, N-acetylgalactosamine residues were hydrolyzed by an A. niger preparation of α-N-acetylgalactosaminidase. Therefore, the sequence

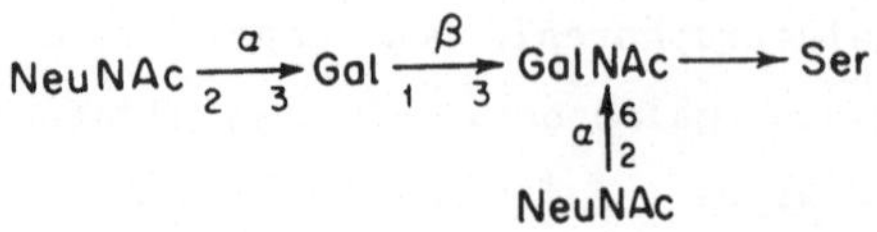

$$\text{NeuNAc} \xrightarrow[2\ \ \ 3]{\alpha} \text{Gal} \xrightarrow[1\ \ \ 3]{\beta} \text{GalNAc} \longrightarrow \text{Ser}$$
$$\alpha \big\uparrow \begin{smallmatrix}6\\2\end{smallmatrix}$$
$$\text{NeuNAc}$$

Fig. 7. Structure of the serine-linked oligosaccharides of human chorionic gonadotropin (Kessler et al., 1979b).

of sugars in the oligosaccharide was GalβGalNAcα.

From the methylation analysis, periodate oxidation results, and exoglycosidase studies, the structure for serine-linked carbohydrate units of hCGβ was established (Fig. 7).

In order to further confirm the precise location of the O-linked oligosaccharides on the polypeptide chain derived from the amino acid sequence analysis of hCG-β, the carboxyterminal peptide BC-19 containing all four O-linked oligosaccharides was isolated. It was further fragmented by trypsin into fractions BC-19T$_2$ (residues 115-122), BC-19T$_3$ (residues 123-133), and BC-19T$_4$ (residues 134-145). The amino acid and carbohydrate compositions and the amino acid sequence analysis of these fragments yielded the location of the four O-linked sugar chains on serine residues at positions 121, 127, 132, and 138.

Ovine Luteinizing Hormone

While hCG has both N- and O-glycosidically-linked carbohydrates, the pituitary glycoprotein hormones, lutropin, follitropin, and thyrotropin, contain only the N-glycosidically-linked carbohydrates (Sairam et al., 1972a, b; Liu et al., 1972a, b; Shome and Parlow, 1974; Saxena and Rathnam, 1976). In addition, the pituitary glycoprotein hormones contain N-acetylgalactosamine in their carbohydrate units. Furthermore, the presence of sulfate in the carbohydrate moieties of bLH (Parsons and Pierce, 1980) and oLH (Bedi et al., 1982) has been established by direct chemical analysis and indirectly from the observations that [^{35}S]-sulfate can be incorporated into the oligosaccharides of these two hormones either metabolically (Hortin et al., 1981; Anumula and Bahl, 1983; Green et al., 1984) or in cell-free systems (Green et al., 1985a). BLH, bTSH, hTSH, and bFSH are also sulfated in the cell-free system (Green et al., 1985b). The presence of N-acetylgalactosamine and sulfate in asparagine-linked carbohydrates, so far found only in these hormones, renders them unique from other known glycoproteins.

22

oLH contains three N-glycosidically-linked carbohydrates, two in the α-subunit and one in the β-subunit. The structural analysis was carried out on each of the three glycopeptides and the oligosaccharide obtained by alkali-sodium borohydride treatment of oLH and its individual subunits (Bedi et al., 1982).

Although the principal methods, i.e., methylation, periodate oxidation, and the use of exoglycosidases, as described above for hCG, are generally adequate for the determination of carbohydrate structures of glycoproteins, it becomes essential to employ either auxiliary techniques, such as acetolysis and nitrous acid deamination in probing the finer details of the sugar chain structure when unusual carbohydrate composition is indicated. These two techniques, in conjunction with those used for hCG, were employed in the elucidation of the structure of carbohydrate units of oLH (Bedi et al., 1982).

Structure of Carbohydrate Units of oLH. For structural studies of carbohydrate, two glycopeptides from oLH-α (αGP-1 and αGP-2) and one from oLH-β (βGP-3) were isolated following tryptic and Pronase digestion of the subunits. The oligosaccharide preparation was obtained by alkaline-borohydride treatment of oLH followed by its purification by paper chromatography. The carbohydrate compositions of these preparations are shown in Table VII. Apparently, the compositions of all three glycopeptides were nearly identical with the average molar ratio of N-acetylglucosamine, N-acetylgalactosamine, mannose, galactose, and fucose per aspartic acid residue being 3.2:1.0:2.9:0.4:0.5. The carbohydrate compositions of oLH, oLH-α, oLH-β, and the oLH oligosaccharide were also similar to the three glycopeptides. The value for N-acetylglucosaminitol in the oligosaccharide was lower than expected although the total amount of N-acetylglucosamine and its corresponding alcohol was 3.1, similar to that present in the glycopeptides.

In order to identify the sugar at the reducing terminus, purified oligosaccharide preparations from either intact hormone or its individual subunits were analyzed for hexosamines and hexosaminitols following acid hydrolysis. Since the hydrolysates contained only glucosamine, galactosamine, and glucosaminitol, it was concluded that the N-acetylglucosamine residue was at the reducing termini of the oligosaccharides.

Methylation Analysis. The three glycopeptides and the oligosaccharides were methylated and the partially methylated alditol and hexosaminitol acetates were identified by comparison of their retention

Table VII. Carbohydrate compositions of oLH, oLH-α, and oLH-β and their glycopeptides and oligosaccharides.[a]

Sugar	oLH	Oligo-saccharide	oLH-α	αGP-1	αGP-2	oLH-β	βGP-3
N-acetylglucosamine	10.1	2.5	7.1	3.2	3.0	3.4	3.1
N-acetylglucosaminitol	n.d.[b]	0.5	n.d.	n.d.	n.d.	n.d.	n.d.
N-acetylgalactosamine	2.9	1.0	1.8	0.9	1.0	1.4	1.3
Mannose	7.3	2.8	6.4	3.1	3.0	2.7	2.6
Galactose	1.1	0.4	0.5	0.5	0.2	0.3	0.3
Fucose	1.4	0.5	0.7	0.5	0.1	0.5	0.5
Sialic Acid	undetect.[c]	n.d.	undetect.	n.d.	n.d.	undetect.	n.d.

Number of sugar residues were calculated on the basis of aspartic acid residues 11 for oLH, 6 for oLH-α, 5 for oLH-β (Liu et al., 1972a, b) and 1 for each glycopeptide. For oligosaccharide, the number of residues were calculated on the basis of N-acetylgalactosamine taken as one.
[a](Bedi et al., 1982)
[b]Not determined.
[c]Undetectable.

time relative to authentic standards. The results were further confirmed by GC-MS analysis. The various methylated derivatives were quantitated using the bar spectrum and the relative abundance of pertinent mass ions for each derivative. The relative number of residues of various methylated derivatives in each glycopeptide are presented in Table VIII.

When methylation analysis was performed following partial acid hydrolysis of the glycopeptides, there was a loss in 2-deoxy-3-O-methyl-2-N-methylacetamido-D-glucose; 2,3,4-tri-O-methyl L-fucose; and 2-deoxy-3,6-di-O-methyl-2-N-methylacetamido-D-galactose with the appearance of 2-deoxy-3,4,6-tri-O-methyl-2-methylacetamido-D-galactose. These data clearly indicated that N-acetylglucosamine and N-acetylgalactosamine were substituted at C-1 and C-6 or C-1 and C-4 positions by fucose or some other labile group.

Smith Degradation. Periodate oxidation of the intact hormone or individual subunits resulted in the complete destruction of galactose and fucose and two-thirds of the mannosyl residues (Table IX). The oxidized and reduced (OR) derivatives of the hormone and its subunits were hydrolyzed with mild acid and the products were separated into a low molecular weight oligosaccharide fraction and a glycopeptide fraction by chromatography on SP-Sephadex. The oligosaccharide fraction was found to be N-acetylglucosaminyl glyceraldehyde by thin-layer chromatography (t.l.c.). Upon reduction of the oligosaccharide fraction with NaB^3H_4/ $NaBH_4$ and subsequent acid hydrolysis, the products were analyzed by paper chromatography. Only glycerol was detected as a radioactive component. Sugar analysis of the reduced oligosaccharide showed the presence of glycerol, N-acetylglucosamine, and N-acetylgalactosamine in an equimolar ratio. Oxidation of the oligosaccharide with periodate resulted in the destruction of 74% N-acetylglucosamine and 66% N-acetylgalactosamine, indicating that the small molecular weight oligosaccharide was a mixture of glyceraldehyde derivatives of the N-acetylhexosamines. It seemed from these data that the substituent at C-4 position of N-acetylgalactosamine was mostly removed during hydrolysis of the oxidized oligosaccharide.

Each of the residual glycopeptides obtained from oLH, oLH-α, and oLH-β following oxidation-reduction-hydrolysis was composed of two residues of N-acetylglucosamine per residue of mannose. Treatment of this material with A. niger β-mannosidase resulted in the release of about 50% mannose. Further digestion with T. aceti β-N-acetylglucosaminidase liberated 40% N-acetylglucosamine. From these data, the structure of the core glycopeptide was found to be ManβGlcNAcβGlcNAc-Asn.

Table VIII. Molar ratios of partially methylated alditol acetates obtained from the permethylated N-glycopeptides of oLH-α and oLH-β and oLH-oligosaccharide.

Methylated Sugar	Position Substituted	Molar Ratio[a]			
		αGP-1	αGP-2	βGP-3	Oligo-saccharide
Fucose					
2,3,4-tri-O-methyl	1-O-	0.2	0.1	0.3	0.3
Galactose					
2,3,4-tetra-O-methyl	1-O-	0.3	0.1	0.3	0.2
Mannose					
3,4,6-tri-O-methyl	1,2-di-O-	2.0	1.8	2.1	1.8
2,4-di-O-methyl	1,3,6-tri-O-	1.0	1.0	1.0	1.0
2-Deoxy-2-N-methyl-acetamidoglucose					
3,4,6-tri-O-methyl	1-O-	1.0	1.0	1.1	0.9
3,6-di-O-methyl	1,4-di-O-	1.6	1.9	1.6	1.8
3-O-methyl	1,4,6-tri-O-	0.4	0.1	0.3	0.3
2-Deoxy-2-N-methyl-acetamidogalactose					
3,6-di-O-methyl	1,4-di-O	0.4	0.7	0.5	0.9

[a]Calculated for "neutral" sugars by considering 2,4-di-O-methyl-D-mannitol as one and for amino sugars assuming all methylated derivatives of N-acetyl-glucosamine to be equivalent to three residues (Bedi et al., 1982).

Table IX. Smith degradation of asparaginyl oligosaccharide containing glycopeptides of oLH-α and oLH-β.

Sugar[a]	αGP-1		αGP-2		βGP-3	
	Native	After Periodate Oxidation	Native	After Periodate Oxidation	Native	After Periodate Oxidation
Mannose	3.1	1.2	3.0	1.2	2.6	1.1
Galactose	0.5	0	0.2	0	0.3	0
Fucose	0.5	0	0.1	0	0.5	0
N-acetyl-glucosamine	3.2	3.2	3.1	3.1	3.1	3.2
N-acetyl-galactosamine	0.9	0.9	1.0	1.1	1.3	1.3

[a]Moles of monosaccharide per mol of aspartic acid (Bedi et al., 1982).

Determination of oLH Oligosaccharide. Deamination with nitrous acid is generally carried out on the oligosaccharides obtained by hydrazinolysis during which the N-acetamidohexosamines are deacetylated to form hexosamines. N-deacetylation is a prerequisite for deamination. Treatment with nitrous acid cleaves the oligosaccharide chain at the hexosamine forming small oligosaccharides with anhydromannitol and anhydrotalitol from glucosamine and galactosamine, respectively (Hase and Matsushima, 1969; Horton et al., 1972). These oligosaccharides can be characterized by the various carbohydrate structural methods.

The oLH oligosaccharide obtained by alkaline sodium borohydride hydrolysis was deaminated with nitrous acid, reduced with sodium borohydride, and the products of the deamination reaction were separated by paper chromatography according to their molecular size. Five fractions were obtained which were analyzed for hexosamine by an amino acid analyzer and for sugars and anhydromannitol and anhydrotalitol by G.C. The structures of the five fragments were found to be as follows:

Fragment	Structure
I	Manα1,3(Manα1,6)Manβ1,4-2,5 Anhydromannitol
II	ManαManβ-2,5 Anhydromannitol
III	Gal-2-5 Anhydromannitol
	+ X-2,5 Anhydrotalitol
IV	X-2,5 Anhydrotalitol
V	2,5 Anhydromannitol

Subsequent analysis of fragment IV indicated that it contained equimolar amounts of sulfate and 2,5-anhydrotalitol.

Acetolysis of oLH Óligosaccharide. Acetolysis with a mixture of acetic and sulfuric acids under controlled conditions cleaves preferentially 1,6-α-mannosidic linkages and has been used successfully in the characterization of high-mannose type carbohydrates (Kocourek and Ballou, 1969). The cleavage of the same linkage in the case of complex carbohydrates has not been as specific as that in the high-mannose type carbohydrates. The separation and analysis of the various smaller fragments resulting from acetolysis can be useful in further confirming the fine details of the structure.

In order to determine whether N-acetylgalactosamine was present at the α1,2- or α1,6-linked mannosyl branch, the oLH-oligosaccharide was

Table X. Molar ratio of partially methylated alditol acetates derived from permethylated acetolysis products of oLH-oligosaccharides.

Methylated Sugar	Position Substituted	Molar Ratio[a]		
		Fraction A-1	Fraction A-2	Fraction A-3
Mannose				
2,3,4,6-tetra-O-methyl	1-O-	0.7	0.3	0.6
3,4,6-tri-O-methyl	1,2-di-O-	0.9	0.4	0.6
2,4,6-tri-O-methyl	1,3-di-O-	0.3	0.7	1.0
2,4-di-O-methyl	1,3,6-tri-O-	0.7	0.3	
2-deoxy-2-N-methyl-acetamido-D-glucose				
3,4,6-tri-O-methyl	1-O-	0.6	0.8	0.5
3,6-di-O-methyl	1,4-di-O-	1.0	1.0	1.0
2-deoxy-2-N-methyl-acetamido-D-glucitol				
1,3,5,6-tetra-O-methyl	4-O-	n.d.[b]	n.d.	n.d.

[a] Calculated for neutral sugars on the basis of the total peak areas by GC of 1,3- and 1,3,6-substituted mannose derivatives as 1.0, and for aminosugars on the basis of 1,4-di-substituted N-acetylglucosamine derivative as 1.0 (Bedi et al., 1982).

[b] Not determined.

Table XI. Sequential treatment of glycopeptides with specific glycosidases[a].

Glycosidases GalNAc	αGP-1				αGP-2				βGP-3		
	Gal	Man	GlcNAc	GalNAc	Gal	Man	GlcNAc	GalNAc	Gal	Man	GlcNAc
1. β-galactosidase	0.3				0.2				0.3		
2. β-galactosidase + α-mannosidase	0.2	0			0.2	0			0.3		
3. β-galactosidase + β-acetylglucos-aminidase	0.2		0.8	0.1	0.2		0.23	0.05	0.3	0.4	0.1
4. β-galactosidase	0.2				0.2				0.3		
then β-N-acetyl-glucosaminidase			0.8	0.1			0.30	0.05		0.4	0.1
then α-mannosidase		0.2				n.d.[b]				n.d.	

[a] Bedi et al., 1982.
[b] Not determined.

subjected to acetolysis and the fragments were purified by thin-layer and paper chromatography. Among the various fragments obtained by preparative paper chromatography, the major fraction designated A contained about 70% of the total N-acetylglucosamine. This major fragment was further separated into three fractions designated A-1, A-2, and A-3 by thin-layer chromatography and following methylation they were analyzed by GC-MS. The data shown in Table X allowed the assignment of the following structure for the fragment A-1, A-2, and A-3: GlcNAc1,2Man1,3(Man1,6)Man1,4GlcNAc-1,4GlcNAcH$_2$, indicating that N-acetylgalactosamine was linked to the α-1,6 mannosyl branch.

Treatment of oLH Glycopeptides and Oligosaccharides with Exoglyco-sidases. The results of exoglycosidase treatment of the three oLH glycopeptides are summarized in Table XI. Treatment of the glycopeptides directly with A. niger galactosidase resulted in the release of 0.2 and 0.3 residues of galactose per mole of glycopeptide. None of the mannose present in either of the three glycopeptides could be released by α-mannosidase alone or along with β-galactosidase. However, 0.2 and 0.8 residues of N-acetylglucosamine were released from these glycopeptides after treatment with either a mixture of β-galactosidase and β-N-acetyl-glucosaminidase from A. niger or T. aceti or when the latter enzymes were used alone. A small amount of N-acetylgalactosamine was also released by β-N-acetylglucosaminidase which was probably due to intrinsic β-N-acetyl-galactosaminidase activity in the enzyme preparations. These results indicated that D-galactose and N-acetylglucosamine were located externally.

More information on the anomeric linkages was obtained from the structural characterization of the tetrasaccharide Man1,6(Man1,3)Man1,4-2,5 Anhydromannitol. When this tetrasaccharide was incubated with Jack bean α-mannosidase, two mannosyl residues were removed from the tetrasaccharide. Further digestion of the residual disaccharide with β-mannosidase resulted in the release of 2,5 Anhydromannitol. Thus, the external two mannosyl residues were linked by α-linkage while the internal mannose was β-linked to 2,5 Anhydromannitol.

The Structure of Carbohydrate Units of oLH. From the data accrued from the results of methylation, periodate oxidation, deamination, acetolysis, and enzymatic studies with exoglycosidases, it was possible to arrive at the detailed structures of the three asparagine carbohydrate units of oLH (Fig. 8).

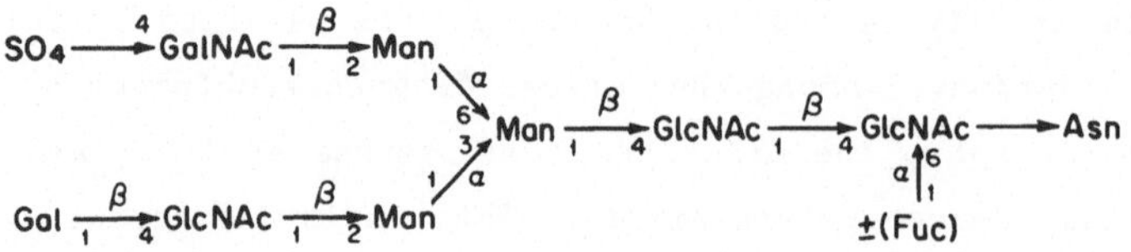

SO₄ → GalNAc →(β,1→2) Man ... (structure diagram)

$$\text{SO}_4 \xrightarrow{4} \text{GalNAc} \xrightarrow[1]{\beta}_{2} \text{Man}$$

SO4 ———4→ GalNAc ——β—→ Man
                    1   2   \α
                                 6
                                 Man ——β——→ GlcNAc ——β——→ GlcNAc ——→ Asn
                                 3   1   4   1   4
                    Gal ——β——→ GlcNAc ——β——→ Man /α   α⁶
                        1   4   1   2   1
                                                              ±(Fuc)

Fig. 8. Structure of the N-linked carbohydrate units of ovine luteinizing hormone (Bedi et al., 1982).

Recently, Anumula and Bahl (1983) purified in vitro metabolically [35]SO_4-labeled oLH from ovine pituitary slices. It was found that both N-acetylglucosamine and N-acetylgalactosamine located at the non-reducing termini were sulfated. Based on these results, a structure for oLH similar to that for bLH (Parsons and Pierce, 1980) with two sulfate groups per carbohydrate unit was proposed. More recently, detailed N-glycosidic carbohydrate structure has been reported for bLH (Fig. 9) by Green et al., (1985b) which differs from that of oLH in the following respects: Whereas the internal pentasaccharide core structure is essentially identical in oLH and bLH asparagine-linked carbohydrates, bLH contains two identical biantennary structures linked to the two mannoses. The oLH biantennary structures, on the other hand, contain two dissimilar carbohydrate chains. It is possible that the observed dissimilarity in these structures may be related to animal species variation or a result of microheterogeneity.

Equine Choriogonadotropin

Preparation of N- and O-linked Oligosaccharides. Equine CG is much larger in size (60,000 d) than hCG (38,000 d) and oLH (29,000 d). Like other glycoprotein hormones, it is also made up of two subunits, eCG-α, 17,000 d, and eCGβ, 43,000 d (Christakos and Bahl, 1979). Among the known glycoprotein hormones, eCG has the largest percentage of carbohydrate,

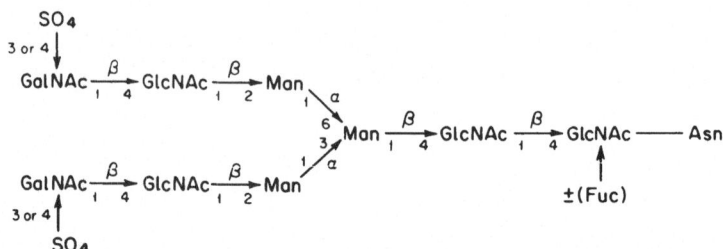

Fig. 9. Structure of N-linked carbohydrate units of bovine luteinizing hormone (Green et al., 1985b).

about 50% as compared with hCG and oLH, which contain 33% and 22% carbohydrates, respectively. The carbohydrate of eCG is distributed in three N-linked, two in eCG-α and one in eGC-β. Since eCG-β forms almost 72% of the hormone weight, the relative amount of the carbohydrate in it is much larger than eCG-α.

Unlike hCG and oLH, the structural analysis of the N-linked oligosaccharides was carried out on the Endo-F cleaved oligosaccharides rather than glycopeptides. The oligosaccharides were obtained from the Endo-F hydrolysis of Pronase glycopeptides, αGP-I, αGP-II, and βGP followed by separation of the released oligosaccharides by gel filtration on Sephadex G-50. The oligosaccharides were labeled with [^{3}H] by reduction with NaB^{3}H$_4$. αGP-I and αGP-II were prepared from the carboxymethyl eCG-α by Pronase digestion and gel filtration.

The separation of the N-linked from the O-linked carbohydrate of eCG-β, on the other hand, presented mammoth problems. First, the βGP obtained by Pronase digestion of eCG-β contained both types of carbohydrate units. Treatment of the βGP with Endo-F resulted in a mixture of N-linked oligosaccharides and the peptide containing O-glycosidic carbohydrate. Second, attempts to isolate the O-linked oligosaccharide from βGP by β-elimination also resulted in a mixture of O-linked carbohydrate and the N-linked glycopeptide and peptides due to partial alkaline degradation of the polypeptide chain. In either case, by these procedures, it was not possible to separate the released oligosaccharides.

Carboxymethyl eCG-β was digested with Endo Lys-C proteinase yielding a limited number of peptides. The peptide fragments containing Asn-linked and O-linked carbohydrates were separated by gel filtration on Sephadex G-50. βGP was obtained from the asparaginyl glycopeptides following exhaustive Pronase digestion and gel filtration of the digested materials on Sephadex G-50. The Endo Lys-C peptide fragment containing the Ser/Thr-linked carbohydrate was subjected to β-elimination, and the O-linked oligosaccharides were further fractionated by gel filtration on Bio-Gel P-30 and paper chromatography into nine fractions OS-1 to OS-9.

In addition to the problem of the N- and O-linked oligosaccharides in the β-subunit of eCG, a much more serious difficulty was encountered due to considerable heterogeneity of carbohydrates in terms of size, degree of branching (bi- to penta-antennary) and the variations in the number of

Table XII. Sugar and amino acid composition of the eCG
            N-linked glycopeptides.

Sugar	βGP	αGP-I	αGP-II
Mannose[a]	3.0	3.0	3.0
Galactose	2.5	2.4	5.1
Glucosamine	4.8	4.7	7.0
Fucose	0.3	0.1	0.2
Sialic Acids	2.3	2.2	4.5
Amino Acids[b]			
Aspartic Acid	1.0	1.1	1.1
Alanine	1.0	–	–
Histidine	–	0.6	0.8
Glutamic Acid	–	0.6	0.6
Percent Yield (%)[c]	73	70	13

[a]Number of sugar residues were calculated on the basis of
mannose content set at 3.0 (Anumula and Bahl, 1985a).

[b]Other amino acids were also detected. Ser/Thr were present
to a maximum of 0.2 residues.

[c]Final yield of the N-linked glycopeptides.

Table XIII. Sugar composition of the O-linked oligosaccha-
             rides[a]: OS-7, OS-8, and OS-9.

Sugar[b]	OS-7	OS-8	OS-9
Galactose	7.3	12.1	25.6
Glucosamine	5.5	10.5	25.0
Galactosamine	0.6	1.1	1.9
Galactosaminitol	1.0	1.0	1.0
Sialic Acids	3.1	4.9	9.2

[a]Anumula and Bahl, 1985.

[b]Moles of sugars per mole of galactosaminitol.

N-acetyllactosamine units in both Asn- and Ser/Thr-linked carbohydrates. The structural approach as used for hCG and oLH, therefore, had to be modified. This entailed cleavage of Asn- and Ser/Thr-linked oligosaccharides with Endo-F and β-elimination, respectively, followed by the separation of the various oligosaccharides. The arrangement of sugars in the outer chains and the degree of branching was determined by sequential and combined degradation with exo- and endoglycosidases followed by the analysis of the degradation products by t.l.c. using standards of known carbohydrate structures as described below.

Characterization of the N- and O-linked Oligosaccharides. The carbohydrate compositions of the N- and O-linked oligosaccharides are summarized in Tables XII and XIII. It was obvious from the compositions that both N- and O-linked Endo-F oligosaccharides were quite heterogeneous. This was further confirmed by t.l.c. as shown in Figures 10 and 11. Methylation, periodate oxidation, and deamination studies were carried out on the Endo-F oligosaccharides from αGP-I, αGP-II, and βGP and the O-linked oligosaccharides obtained by β-elimination. Since the application of the methylation and periodate oxidation techniques to the analysis of complex carbohydrates has already been illustrated in relation to hCG and oLH, these will not be discussed here. The data on the methylation, periodate oxidation, and deamination analysis of the N- and O-linked oligosaccharides of eCG have been described elsewhere. As previously stated, these techniques could only provide average structures.

In order to determine the structures of the various heterogeneous forms, the N- and O-linked oligosaccharides were sequentially degraded with exoglycosidases and the degradation products were analyzed by t.l.c. using known oligosaccharide standards obtained by the N-glycosidase digestions of transferrin, fetuin, and α_1-acid glycoproteins containing bi-, tri-, and tetra-antennary structures, respectively. The maltooligosaccharides obtained by the controlled acid hydrolysis of amylose were also used as standards (Behrens and Tabora, 1978). The number of sugar residues removed at each step of glycosidase treatment was obtained by comparing the mobility of the oligosaccharide before and after exoglycosidase treatment with the standards. Endo-F generated N-linked oligosaccharides and their glycosidase degradation products were analyzed by t.l.c. using the solvent systems of Holmes and O'Brien (1979). For the O-linked oligosaccharides, we developed a new solvent system which was able to resolve oligomers containing up to about 20 monosaccharide units.

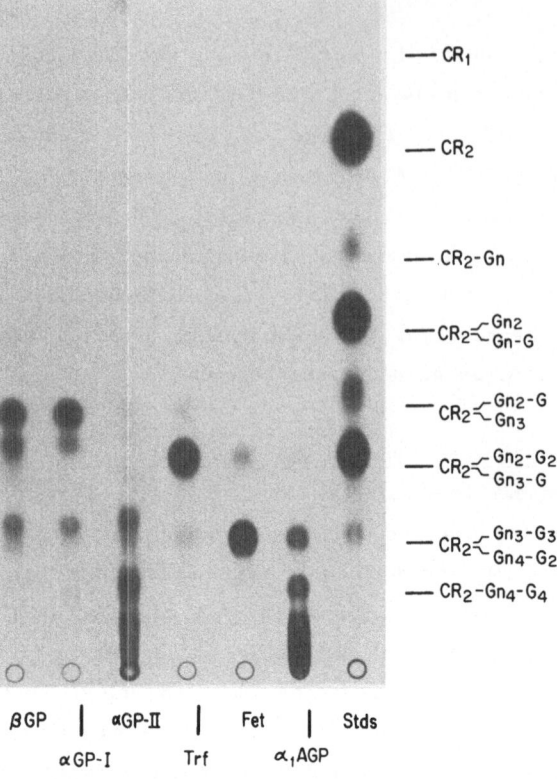

$$\begin{array}{ll}
& -\ CR_1\\
& -\ CR_2\\
& -\ CR_2\text{-}Gn\\
& -\ CR_2\!\!<_{Gn\text{-}G}^{Gn2}\\
& -\ CR_2\!\!<_{Gn3}^{Gn2\text{-}G}\\
& -\ CR_2\!\!<_{Gn3\text{-}G}^{Gn2\text{-}G_2}\\
& -\ CR_2\!\!<_{Gn4\text{-}G_2}^{Gn3\text{-}G_3}\\
& -\ CR_2\text{-}Gn_4\text{-}G_4
\end{array}$$

βGP | αGP-II | Fet | Stds
αGP-I   Trf α_1AGP

Fig. 10. Thin-layer chromatography of Endo-F derived ^{3}H-labeled oligosaccharides from βGP, αGP-I, and αGP-II. Thin-layer chromatography was developed for 24 h. Trf, Fet, and α_1AGP are the neutral oligosaccharides obtained by N-glycosidases F (in Endo-F preparation) hydrolysis of asialo transferrin (human), asialo fetuin, and asialo α_1 acid glycoprotein, respectively. Major Trf, Fet, and α_1AGP oligosaccharides contain a biantennary, a triantennary, and a tetra-antennary structure, respectively. Various standards of intermediate sizes were generated by either complete or partial digestion of these oligosaccharides. Abbreviations for standards are CR_1, $Man_3GlcNAcH_2$; CR_2, $Man_3GlcNAc\text{-}GlcNAcH_2$. G and Gn stand for galactose and N-acetylglucosamine, respectively present in the outer chains of Trf, Fet, and α_1AGP (Anumula and Bahl, 1985a).

Analysis of N-linked Oligosaccharides after Treatment with Exoglycosidases by T.L.C. It was obvious from the t.l.c. results (Fig. 10) that the N-linked Endo-F oligosaccharides from asilao αGP-I and βGP were almost identical and were different from those obtained similarly from asialo αGP-II.

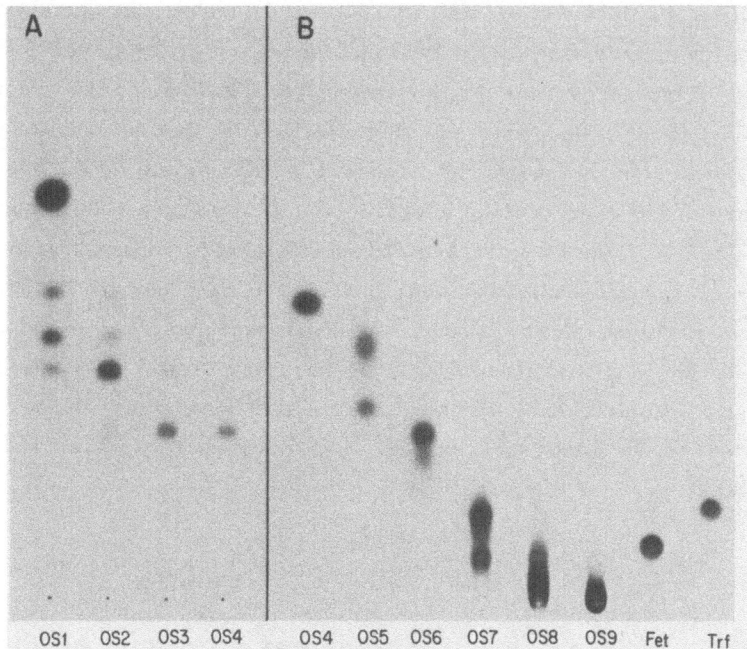

Fig. 11. Thin-layer chromatography of the O-linked oligosaccharide of eCGβ. Oligosaccharides were treated
with mild acid (0.2 N TFA at 80°C, 1 h) and
chromatographed on Silica Gel 60 coated plastic
plates in solvent system described by Anumula and
Bahl (1985b). Fet and Trf present asialo N-linked
oligosaccharides obtained by the N-glycosidase F
action on fetuin and transferrin, respectively.

Characterization of the Endo-F oligosaccharides from asialo βGP,
αGP-I, and αGP-II was based on their sequential or combined degradation
with β-galactosidase and β-N-acetylglucosaminidase directly or after
Endo-β-galactosidase (Fukuda, 1985) digestion followed by an examination
of the resulting oligosaccharides by t.l.c. using appropriate oligosaccharide standards. The treatment with exoglycosidases hydrolyzed the
outer chains leaving the native oligosaccharide to the tetraitol core.
The number of sugar residues removed at each step of enzymatic treatment
was determined by comparing the mobilities of the oligosaccharides before
and after hydrolysis with the known standards. The Endo-F oligosaccharides from asialo-glycopeptides βGP, αGP-I, and αGP-II had a single
GlcNAc (reduced) in the core ($Man_3GlcNAc_2^3H_2$ CR_1; Fig. 10) while the
standard oligosaccharides prepared from the glycoproteins had two GlcNAc
residues in the core ($Man_3GlcNAc_2^3H_2$, CR_2) as a result of the
N-glycosidase action. Therefore, it should be realized that when
identical mobilities of the standard and the Endo-F oligosaccharide of eCG
are indicated by t.l.c., the latter contains an additional sugar residue
in its outer chain.

Oligosaccharides from asialo βGP and αGP-I on treatment with β-galactosidase gave four major radioactive spots A, B, C, and D (Fig. 12, lane 1). Oligosaccharides A, B, and D lost two galactose residues to yield A_1, B_1, and D_1, whereas oligosaccharide C lost three galactose residues when compared with the standards, indicating A, B, and D being biantennary and C representing a triantennary structure. Oligosaccharides A_1, B_1, and C_1 were converted to $Man_3GlcNAcH_2$ upon treatment with β-N-acetylglucosaminidase with the loss of two residues of GlcNAc from A_1 and three residues from B_1 and C_1. Oligosaccharides B contains the "bisect" GlcNAc and migrates usually slower than $GlcNAc_2-Man_3-GlcNAcH_2$ but faster than $GlcNAc_3-Man_3-GlcNAc_3H_2$ without bisecting GlcNAc. Oligosaccharide D also lost two residues of GlcNAc and migrated slower than

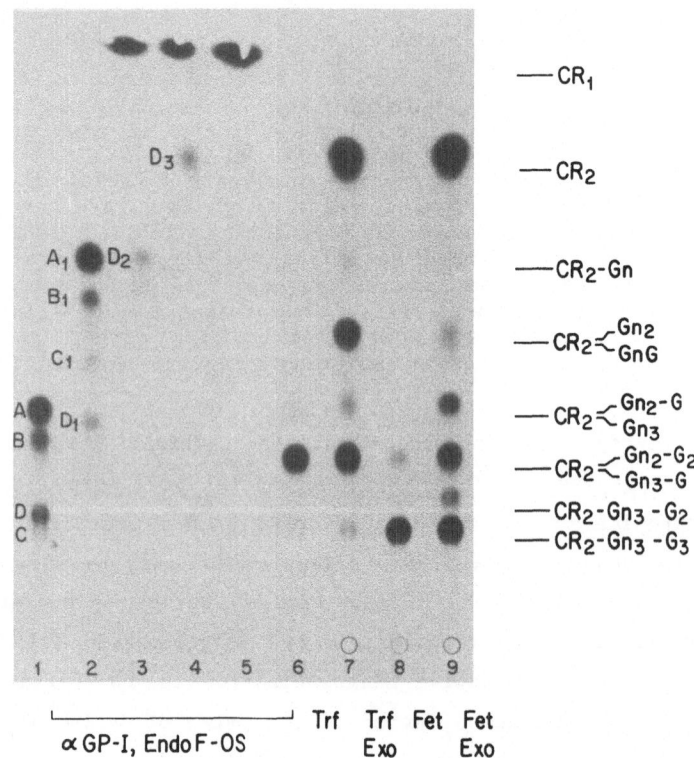

Fig. 12. Sequential treatment of Endo-F derived ^{3}H-labeled oligosaccharides from αGP-1 with exoglycosidases. Neutral oligosaccharides (lane 1) either treated alternately with β-galactosidase (lanes 2 and 4) and β-N-acetylglucosaminidase (lane 3) or with a mixture of both the enzymes (lane 5). Standards Trf and Fet represent oligosaccharides from asialo transferrin and asialo fetuin, respectively, and Trf Exo- and Fet Exo- correspondingly represent their exoglycosidase digests. For details see the legend to Figure 10.

Man_2-GlcNAc-GlcNAcH$_2$ and GlcNAc-Man$_3$-GlcNAc-GlcNAcH$_2$ from transferrin. Since the oligosaccharide is a cleavage produce of Endo-F, it would appear that D contains two sugar residues in excess of the "common core." Indeed, D was converted to Man_3-GlcNAcH$_2$ by an additional sequential treatment with β-galactosidase and β-N-acetylglucosaminidase (Fig. 12, lanes 4 and 5). It is obvious from the results that these oligosaccharides contain the trimannosyl-chitobiose "common core" structure. Relative molar proportions of the oligosaccharides, A, B, C, and D were determined from their radioactivity following B-galactosidase treatment (lane 2). Distribution of the radioactivity in the oligosaccharides from βGP and αGP-I, respectively, was 51% and 60% in A, 24% and 18% in B, 9% and 10% in C, and 16% and 12% in D. From the results obtained by the sequential exoglycosidase treatments along with the intersugar linkages from the methylation and the periodate oxidation data, the following structures from the oligosaccharides from asialo βGP and αGP-I were proposed (Fig. 13). Sialyl residues were assigned to the terminal galactoses since methylation of the glycopeptides showed only a small amount of tetra-O-methyl galactose.

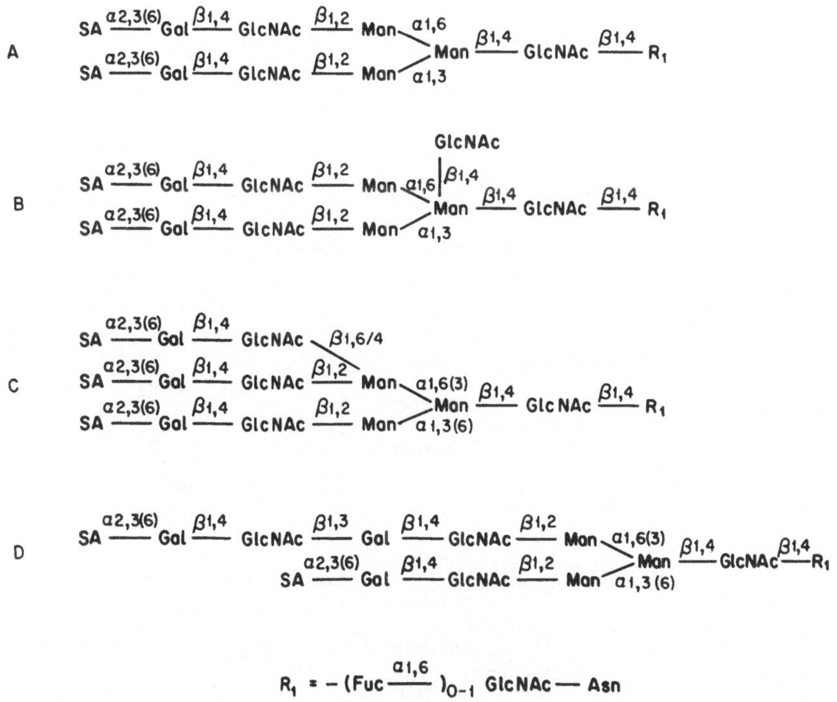

Fig. 13. Proposed structures of the Asn-linked carbohydrates in glycopeptides αGP-I and βGP.

Endo-F oligosaccharides obtained from asialo-αGP-II were much more complex in nature (Fig. 10) than those derived from asialo βGP and αGP-I. The complexity was mainly due to the variations in the structures of their outer chains rather than the core because the oligosaccharides were completely converted to $Man_3GlcNAcH_2$ on treatment with a mixture of endo-β- galactosidase, β-galactosidase, and β-N-acetylglucosaminidase. Unlike αGP-I, all oligosaccharides present in αGP-II contained repeating N-acetyllactosamine units, the number of such units attached to the core ranged from three to nine. All the αGP-II oligosaccharides were susceptible to hydrolysis with endo-β-galactosidase and decreased in size at least by three sugar residues. The details of these studies are described elsewhere (Anumula and Bahl, 1985a, b). Their structures are given in Figures 14 and 15.

Characterization of O-linked Oligosaccharides. All six O-glycosylation sites reside in the C-terminus of the β-subunit of eCG. The O-linked carbohydrates constitute 75% of the carbohydrate of the β-subunit, indicating that the size of the oligosaccharides is much larger than the usual Ser-linked oligosaccharides present in hCG. This is obvious from

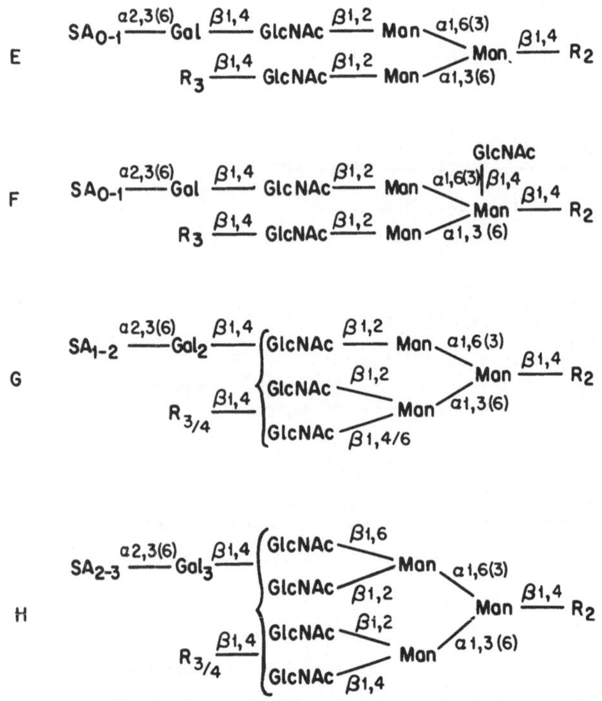

Fig. 14. Proposed structures of the Asn-linked carbohydrates in glycopeptides αGP-II.

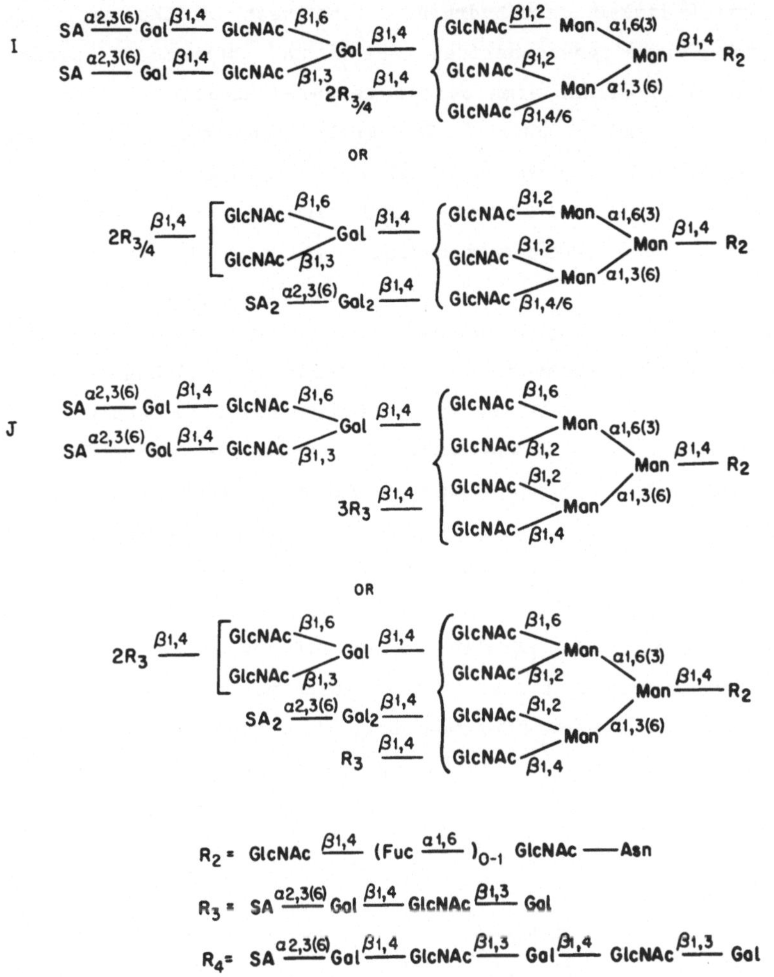

I

$$SA \xrightarrow{\alpha 2,3(6)} Gal \xrightarrow{\beta 1,4} GlcNAc \xrightarrow{\beta 1,6}$$
$$SA \xrightarrow{\alpha 2,3(6)} Gal \xrightarrow{\beta 1,4} GlcNAc \xrightarrow{\beta 1,3} Gal \xrightarrow{\beta 1,4}$$

with branch $2R_{3/4} \xrightarrow{\beta 1,4}$

$$\begin{cases} GlcNAc \xrightarrow{\beta 1,2} Man \xrightarrow{\alpha 1,6(3)} \\ GlcNAc \xrightarrow{\beta 1,2} Man \xrightarrow{\alpha 1,3(6)} Man \xrightarrow{\beta 1,4} R_2 \\ GlcNAc \xrightarrow{\beta 1,4/6} \end{cases}$$

OR

$$2R_{3/4} \xrightarrow{\beta 1,4} \begin{bmatrix} GlcNAc \xrightarrow{\beta 1,6} \\ GlcNAc \xrightarrow{\beta 1,3} Gal \xrightarrow{\beta 1,4} \\ SA_2 \xrightarrow{\alpha 2,3(6)} Gal_2 \xrightarrow{\beta 1,4} \end{bmatrix} \begin{cases} GlcNAc \xrightarrow{\beta 1,2} Man \xrightarrow{\alpha 1,6(3)} \\ GlcNAc \xrightarrow{\beta 1,2} Man \xrightarrow{\alpha 1,3(6)} Man \xrightarrow{\beta 1,4} R_2 \\ GlcNAc \xrightarrow{\beta 1,4/6} \end{cases}$$

J

$$SA \xrightarrow{\alpha 2,3(6)} Gal \xrightarrow{\beta 1,4} GlcNAc \xrightarrow{\beta 1,6}$$
$$SA \xrightarrow{\alpha 2,3(6)} Gal \xrightarrow{\beta 1,4} GlcNAc \xrightarrow{\beta 1,3} Gal \xrightarrow{\beta 1,4}$$

with branch $3R_3 \xrightarrow{\beta 1,4}$

$$\begin{cases} GlcNAc \xrightarrow{\beta 1,6} \\ GlcNAc \xrightarrow{\beta 1,2} Man \xrightarrow{\alpha 1,6(3)} \\ GlcNAc \xrightarrow{\beta 1,2} Man \xrightarrow{\alpha 1,3(6)} Man \xrightarrow{\beta 1,4} R_2 \\ GlcNAc \xrightarrow{\beta 1,4} \end{cases}$$

OR

$$2R_3 \xrightarrow{\beta 1,4} \begin{bmatrix} GlcNAc \xrightarrow{\beta 1,6} \\ GlcNAc \xrightarrow{\beta 1,3} Gal \xrightarrow{\beta 1,4} \\ SA_2 \xrightarrow{\alpha 2,3(6)} Gal_2 \xrightarrow{\beta 1,4} \\ R_3 \xrightarrow{\beta 1,4} \end{bmatrix} \begin{cases} GlcNAc \xrightarrow{\beta 1,6} \\ GlcNAc \xrightarrow{\beta 1,2} Man \xrightarrow{\alpha 1,6(3)} \\ GlcNAc \xrightarrow{\beta 1,2} Man \xrightarrow{\alpha 1,3(6)} Man \xrightarrow{\beta 1,4} R_2 \\ GlcNAc \xrightarrow{\beta 1,4} \end{cases}$$

$$R_2 = GlcNAc \xrightarrow{\beta 1,4} (Fuc \xrightarrow{\alpha 1,6})_{0-1} GlcNAc \longrightarrow Asn$$

$$R_3 = SA \xrightarrow{\alpha 2,3(6)} Gal \xrightarrow{\beta 1,4} GlcNAc \xrightarrow{\beta 1,3} Gal$$

$$R_4 = SA \xrightarrow{\alpha 2,3(6)} Gal \xrightarrow{\beta 1,4} GlcNAc \xrightarrow{\beta 1,3} Gal \xrightarrow{\beta 1,4} GlcNAc \xrightarrow{\beta 1,3} Gal$$

Fig. 15. Proposed structures of the Asn-linked carbohydrates
of αGP-II.

t.l.c. of the oligosaccharides (Fig. 11). The O-linked oligosaccharides
were prepared by β-elimination of the Endo Lys-C fragment with NaOH/NaBH$_4$
and were fractionated by gel filtration on Bio-Gel P-30 and paper
chromatography. A spectrum of nine neutral oligosaccharides, designated
OS-1 to OS-9, was obtained (after acid hydrolysis). Their sizes varied
from a disaccharide to a megalosaccharide with about 50 sugar residues.
The disaccharide constitutes only 3% of the total O-linked carbohydrates.
Thin-layer chromatographic examination of the products from endo-β-
galactosidase and exo-glycosidases, β-galactosidase and β-N-acetylglucosaminidase and chemical treatments such as nitrous acid deamination, Smith
degradation, and methylation revealed the innermost core structure
Galβ1,4-GlcNAcβ1,6(Gal1,3)GalNAcH$_2$ in all oligosaccharides. This
structure SAα2,3Galβ1,4GlcNAcβ1,6(SAα2,3Galβ1,3)GalNAcH$_2$ accounted for 25%

of all the O-linked carbohydrate. The rest of the carbohydrate was distributed in oligosaccharides which had structures preferentially extended on β1,6 GlcNAc arms by the repeated additions of varying number of N-acetyllactosamine units. The methylation analysis of the larger oligosaccharides also revealed peripheral 1,3,6-galactosyl branches giving rise to tri-, tetra-, penta-, and multiantennary structures. The structures of the O-linked carbohydrates are given in Figures 16 and 17.

The approach to the structural analysis was similar to the one used for the N-linked carbohydrates. The intersugar linkages and branching were determined by methylation and periodate oxidation analysis. The size of the oligosaccharides was estimated from the number of all sugar residues per mole of N-acetylgalactosamine as well as by comparison of their mobility by t.l.c. with those of malto-oligosaccharides (Fig. 11). The application of t.l.c. to the exoglycosidase treated nine oligo-saccharide fractions OS-1 to OS-9 for structural determination is illustrated here by the results of analysis of fraction OS-4. Fraction OS-4 was homogeneous by t.l.c. The relative molar proportion of $GalNACH_2$, GlcNAc, Gal, and sialic acid was 1:1:2:2. Methylation analysis showed $GalNAc_2$ in 1,3,6 linkage; GlcNAc in 1,4 linkage; and each Gal in β1,3 linkage. Since methylation of the mildly acid-hydrolyzed oligosaccharide

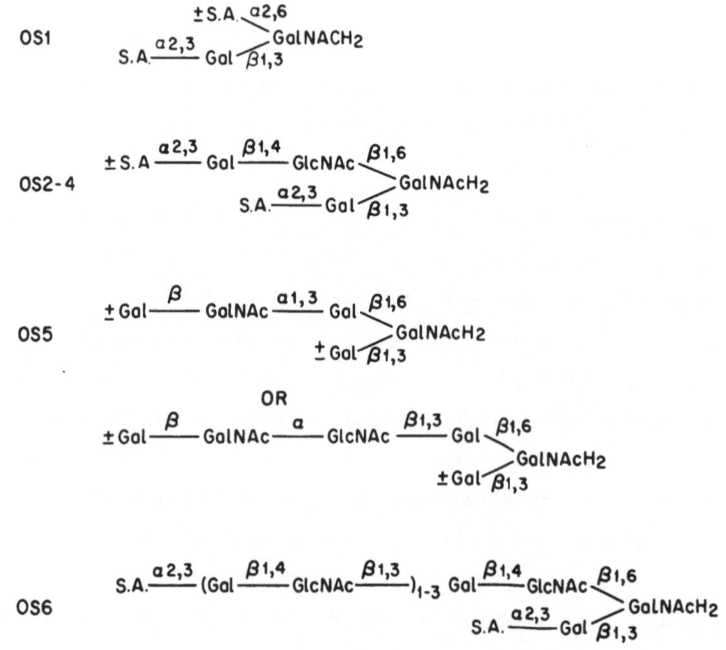

Fig. 16. Proposed structures of the O-linked carbohydrates of eCGβ.

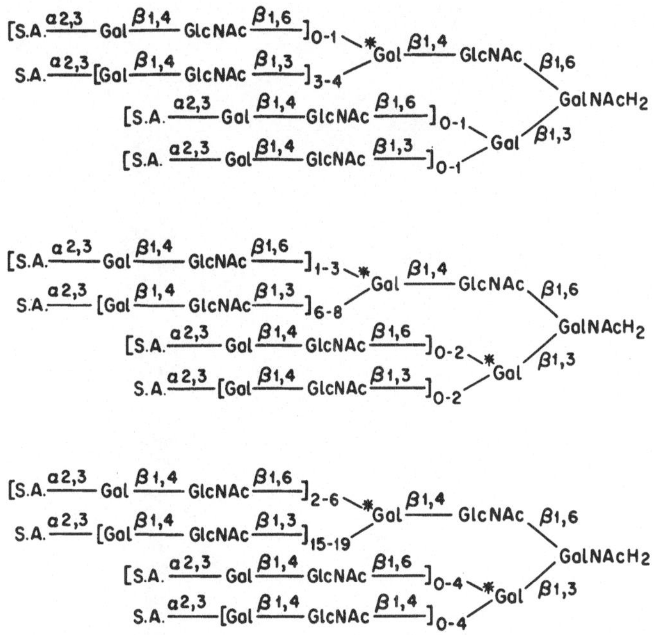

Fig. 17. Proposed structures of the high molecular weight
O-linked carbohydrates of eCGβ.

eliminated galactosyl residues and gave rise to tetra-O-methyl galactosyl
residues, it indicated that the non-reducing terminal sialic acid was
attached to the galactosyl residues by β1,3-linkages. Sequential
β-galactosidase and β-N-acetylglucosaminidase treatment resulted in a
trisaccharide and disaccharide having the structure Galβ1,3GalNAcH$_2$. It
may be noted that the β1,3-galactosyl residue was found resistant to
Aspergillus niger β-galactosidase (Fig. 18).

The structures of the high molecular weight O-oligosaccharide fractions OS-7 to OS-9 were much more complex. However, by the application of
t.l.c. for the examination of exo- and endoglycosidase products as well as
the deamination cleavage products combined with methylation and periodate
oxidation analyses, it was possible to assign them tentative structures
(Fig. 17).

CONCLUDING REMARKS

Although the amino acid sequences of the glycoprotein hormones have
been known for sometime, the structural aspects of their carbohydrates
have not been as well understood. As described above, we have determined

43

A
OS4

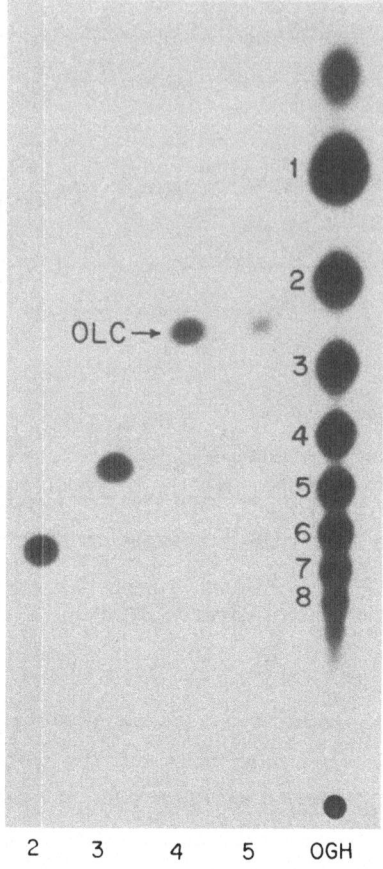

OLC →

2   3   4   5   OGH

Fig. 18. Effect of exoglycosidase treatment on the O-linked oligosaccharide, OS-4. The oligosaccharide after mild acid treatment, lane 2 (0.1 N TFA 80°, 1 h), was digested with β-galactosidase (lane 3) and β-N-acetylglucosaminidase (lane 4). OGH are the NaB^3H_4 reduced glucose oligomers. The number of glucose residues in the oligomer is also shown.

the structures of carbohydrates of three glycoprotein hormones, hCG, oLH, and eCG. It is remarkable that all three contain entirely diverse carbohydrate structures. HCG contains a commonly occurring N-linked biantennary structure and simple O-linked oligosaccharides, oLH a peripherally sulfated N-acetylgalactosamine containing biantennary structure, and eCG highly complex multibranched poly-N-acetylactosamine containing N- and O-linked carbohydrates. This is further confirmed by the presence in eCG of the blood group I and i activities which are associated with poly-N-acetyllactosamine and 3,6-branching galactosyl units.

The presence of sulfated N-acetylgalactosamine in o/bLH and bTSH is a novel carbohydrate structural feature discovered recently in N-linked oligosaccharides. The role of sulfation in the regulation of biosynthesis of bLH has been recently reported by Green et al. (1985a). Although the function of sulfate is not yet clear, it may be similar to that of sialic acid, i.e. to regulate the plasma half-life of the hormones. The lack of sulfate will expose the non-reducing N-acetylgalactosamine residues which will result in its rapid removal from circulation by the Ashwell Gal/GalNAc receptor. The availability of an oligosaccharyl sulfatase and a sulfotransferase should certainly be useful in studying the role of sulfate in the function of these hormones.

Like other glycoproteins, microheterogeneity has also been observed in glycoprotein hormones. The heterogeneity in hCG and o/bLH is not as complex and diverse as that found in eCG. The diversity of carbohydrate variants in eCG raises some interesting questions as to the functional significance of heterosaccharide variation. It is anticipated that the significance of heterogeneity will unravel when additional information on the structures of the carbohydrate variants in glycoproteins in normal and disease states becomes available. In this regard, the methodology developed for elucidation of carbohydrate structural heterogeneity should be potentially useful in the case of other equally complex glycoprotein systems.

Finally, in order to understand the biological role of heterogeneity, the availability of the various heterogeneous forms of glycoprotein with specific carbohydrate would be essential. The conventional protein purification techniques will not be suitable for the preparation of such heterogeneous forms. However, some of the following methods should provide potential alternatives. One should be able to use immobilized lectins with differing carbohydrate specificities or a single lectin with differing affinity for various carbohydrates for the separation of a glycoprotein into its heterogeneous forms with regard to heterosaccharide variants. The monoclonal antibodies against specific carbohydrate structures is another potential method for the separation of a glycoprotein into its heterogeneous forms differing in carbohydrate. The separation of the various heterogeneous forms with differing charge in the carbohydrate can be readily effected by ion exchange chromatography. Another way to obtain various forms of glycoprotein with different carbohydrates will be to synthesize them metabolically in the presence of specific inhibitors of carbohydrate-trimming glycosidases. In short, this is a challenging area of glycoprotein research and certainly will continue to attract the attention of many investigators.

ACKNOWLEDGEMENTS

The work pertaining to the carbohydrate structural elucidation of the glycoprotein hormones was supported by USPHS grants HD08766 and HD12581. The authors wish to acknowledge Ursula Brunn and Jim Stamos for their assistance in typing the manuscript and preparation of illustrations, respectively.

REFERENCES

Anumula, K. R., and Bahl, O. P., 1985a, Equine choriogonadotropin - Heterogeneity of asparagine-linked carbohydrates in the α and β subunits, J. Biol. Chem., submitted.

Anumula, K. R., and Bahl, O. P., 1985b, Equine choriogonadotropin - Unusual multiantennary Ser/Thr-linked carbohydrates in the β subunit, J. Biol. Chem., submitted.

Anumula, K. R., and Bahl, O. P., 1983, Biosynthesis of ovine lutropin in pituitary slices: Incorporation of [^{35}S]-sulfate, Arch. Biochem. Biophys., 220:645-651.

Ashwell, G., and Harford, J., 1982, Carbohydrate-specific receptors of the liver, Ann. Rev. Biochem., 51:531-554.

Bayard, B., and Kerckaert, J.-P., 1980, Evidence for uniformity of the carbohydrate chains in individual glycoprotein molecular variants, Biochem. Biophys. Res. Commun., 95:777-784.

Behrens, W. H., and Tabora, E., 1978, Dolichol intermediates in the glycosylation of proteins, in: "Methods in Enzymology," V. Ginsburg, ed., Academic Press, New York, vol. 50C, pp. 402-435.

Bedi, G., French, W. C., and Bahl, O. P., 1982, Structure of carbohydrate units of ovine luteinizing hormone, J. Biol. Chem., 257:4345-4355.

Bellisario, R., Carlsen, R. B., and Bahl, O. P., 1973, Human chorionic gonadotropin: Linear amino acid sequence of the α subunit, J. Biol. Chem., 248:6796-6809.

Carlsen, R. B., Bahl, O. P., and Swaminathan, N., 1973, Human chorionic gonadotropin: Linear amino acid sequence of the β subunit, J. Biol. Chem., 248:6810-6827.

Carlstedt, I., Sheehan, J. K., Corfield, A. P., and Gallagher, J. T., 1985, Mucous glycoproteins: A gel of a problem, Essays in Biochemistry, 20:40-76.

Christakos, S., and Bahl, O. P., 1979, Pregnant mare serum gonadotropin: Purification and physicochemical, biological and immunological characterization, J. Biol. Chem., 254:4253-4261.

Cunningham, L., 1975, Microheterogeneity and functions of glycoproteins, in: "Glycoproteins of Blood Cells and Plasma," G. A. Jamieson, and T. J. Greenwalt, eds., J. B. Lippincott Company, Philadelphia, pp. 16-34.

Elder, J. H., and Alexander, S., 1982, Endo-β-acetylglucosaminidase F: Endoglycosidase from Flavobacterium meningosepticum that cleaves both high-mannose and complex glycoproteins, Proc. Natl. Acad. Sci. USA, 79:4540-4544.

French, W. C., Henner, J. A., and Bahl, O. P., 1984, Biosynthesis of glycoproteins in human placenta: Differential labeling of mannose and heterogeneity of oligosaccharide lipid intermediates, Arch. Biochem. Biophys., 230:560-579.

Fukuda, M. N., 1985, Isolation and characterization of a new Endo-β-galactosidase from Diplococcus pneumoniae, Biochemistry, 24:2154-2163.

Goldstein, I. J., Hay, G. W., Lewis, B. A., and Smith, F., 1965, Controlled degradation of polysaccharides by periodate oxidation, reduction and hydrolysis, Methods Carbohydr. Chem., 5:361-371.

Green, E. D., Gruenebaum, J., Bielinska, M., Baenziger, J. H., and Boime, I., 1984, Sulfation of lutropin oligosaccharides with a cell-free system, Proc. Natl. Acad. Sci., 81:5320-5324.

Green, E. D., Baenziger, J. U., and Boime, I., 1985a, Cell-free sulfation of human and bovine pituitary hormones. Comparison of the sulfated oligosaccharides of lutropin, follitropin and thyrotropin, J. Biol. Chem., 260:15631-15638.

Green, E. D., Van Halbeck, H., Boime, I., and Baenziger, J. U., 1985b, Structural elucidation of the disulfated oligosaccharide from bovine lutropin, J. Biol. Chem., 260:15623-15630.

Hakamori, S., 1964, Rapid permethylation of glycolipids and polysaccharides catalyzed by methylsulfinyl carbanion in dimethyl sulfoxide, J. Biochem., (Tokyo), 55:205-208.

Hase, S., and Matsushima, Y., 1969, Aminosugar analysis by gas-liquid chromatography, J. Biochem., (Tokyo), 66:57-62.

Hatton, M. W. C., Marz, L., and Regoeczi, E., 1983, On the significance of heterogeneity of plasma glycoproteins possessing N-glycans of the complex type: A perspective, Trends Biochem. Sci., 8:287-291.

Holmes, E. W., and O'Brien, J. S., 1979, Separation of glycoprotein-derived oligosaccharides by thin-layer chromatography, Analyt. Biochem., 93:167-170.

Hortin, G., Natowicz, M., Pierce, J., Baenziger, J. U., Parsons, T., and Boime, I., 1981, Metabolic labeling of lutropin with [^{35}S]sulfate, Proc. Natl. Acad. Sci. USA, 78:7468-7472.

Horton, D., Phillips, K. D., and Defaye, J., 1972, The nitrous acid deamination of 2-amino-2-deoxy-D-mannose hydrochloride to D-glucose, Carbohydr. Res., 21:417-419.

Hounsell, E. F., Wright, D. J., Donald, A. S. R., and Feeney, J., 1984, A computerized approach to the analysis of oligosaccharide structure by high-resolution proton n.m.r., Biochem. J., 223:129-143.

Jansson, P., Kenne, Liedgren, H., Lindberg, B., and Lonngren, J., 1976, A practical guide to the methylation analysis of carbohydrates, Chem. Commun. Univ. Stockholm, 8:1-75.

Jarnefelt, J., Rush, J., Li, Y.-T., and Laine, R., 1978, Erythroglycan, a high molecular weight glycopeptide with the repeating structure (galactosyl-(1-4)-2-deoxy-2-acetamido glucosyl (1-3) comprising more than one-third of the protein-bound carbohydrate of human erythrocyte stroma, J. Biol. Chem., 253:8006-8009.

Kalyan, N. K., Lippes, H. A., and Bahl, O. P., 1982, Role of carbohydrate in human choriogonadotropin: Effect of periodate oxidation and reduction on its in vitro and in vivo biological properties, J. Biol. Chem., 257:12624-12631.

Kalyan, N. K., and Bahl, O. P., 1983, Role of carbohydrate in human chorionic gonadotropin: Effect of deglycosylation on the subunit interaction and on its in vitro and in vivo biological properties, J. Biol. Chem., 258:67-74.

Kessler, M. J., Reddy, M. S., Shah, R. H., and Bahl, O. P., 1979a, Structures of N-glycosidic carbohydrate units of human chorionic gonadotropin, J. Biol. Chem., 254:7901-7908.

Kessler, M. J., Mise, T., Ghai, R. D., and Bahl, O. P., 1979b, Structure and location of the O-glycosidic carbohydrate units of human chorionic gonadotropin, J. Biol. Chem., 254:7909-7914.

Kobata, A., 1984, The carbohydrates of glycoproteins, in: "Biology of Carbohydrates," V. Ginsburg, and P. W. Robbins, eds., John Wiley and Sons, New York, vol. 2, pp. 87-161.

Kocourek, J., and Ballou, C. E., 1969, Method for fingerprinting yeast cell wall mannans, J. Bacteriol., 100:1175-1181.

Kornfeld, R., and Kornfeld, S., 1985, Assembly of asparagine-linked oligosaccharides, Ann. Rev. Biochem., 54:631-664.

Krusius, T., Finna, J., and Rauvala, H., 1978, The poly(glycosyl) chains of glycoproteins. Characterization of a novel type of glycoprotein saccharides from human erythrocyte membrane, Eur. J. Biochem., 92:289-300.

Lindberg, B., 1972, Methylation analyses of polysaccharides, in: "Methods in Enzymology," V. Ginsburg, ed., Academic Press, New York, Vol. 28B, pp. 178-195.

Liu, W.-K., Nahm, H. S., Sweeney, C. M., Lamkin, W. M., Baker, H. N., and Ward, D. N., 1972a, The primary structure of ovine luteinizing hormone. I. The amino acid sequence of the reduced and S-aminothylated S-subunit (LHα), J. Biol. Chem., 247:4351-4364.

Liu, W.-K., Nahm, H. S., Sweeney, C. M., Holcomb, G. N., and Ward, D. N., 1972b, The primary structure of ovine luteinizing hormone. II. The amino acid sequence of the reduced, S-carboxymethylated A-subunit (LHβ), J. Biol. Chem., 247:4365-4381.

Lloyd, K. O., and Kabat, E. A., 1969, Immunological studies on blood groups. XI. Scission of oligosaccharides by sodium hydroxide in the presence of sodium borohydride: A model for the degradation of blood-group substances, Carbohydr. Res., 9:41-48.

Marz, L., Hatton, M. W. C., Berry, L. R., and Regoeczi, E., 1982, The structural heterogeneity of the carbohydrate moiety of desialylated human transferrin, Can. J. Biochem., 60:624-630.

Mellis, S. J., and Baenziger, J. U., 1981, Separation of neutral oligosaccharides by high-performance liquid chromatography, Analy. Biochem., 114:276-280.

Mise, T., and Bahl, O. P., 1980, Assignment of disulfide bonds in the α-subunit of human chorionic gonadotropin, J. Biol. Chem., 255:8516-8522.

Mise, T., and Bahl, O. P., 1981, Assignment of disulfide bonds in the β-subunit of human chorionic gonadotropin, J. Biol. Chem., 256:6587-6592.

Mizuochi, T., Nishimura, R., Derappe, C., Taniguchi, T., Hamamoto, T., Mochizuki, M., and Kobata, A., 1983, Structures of the asparagine-linked sugar chains of human chorionic gonadotropin produced in choriocarcinoma: Appearance of tetra-antennary sugar chains and unique biantennary sugar chains, J. Biol. Chem., 258:14126-14129.

Montgomery, R., 1972, Heterogeneity of the carbohydrate groups of glyco-
proteins, in: "Glycoproteins: Their Composition, Structure and
Function," Part A., A. Gottschalk, ed., Elsevier Publishing
Company, New York, pp. 518-528.

Morgan, F. J., Birken, S., and Canfield, R. E., 1975, The amino acid
sequence of human chorionic gonadotropin: The α subunit and β
subunit, J. Biol. Chem., 250:5247-5258.

Osawa, T., Yamamoto, K., Katagiri, Y., Tsuji, T., and Tarutani, O., 1985,
in: "Glycoconjugates - Proceedings of the VIIIth International
Symposium," E. A. Davidson, J. C. Williams, and N. M. DiFerrante,
eds., Praeger Scientific, New York, vol. 2., p. 418.

Parsons, T. C., and Pierce, J. G., 1980, Oligosaccharide moieties of
glycoprotein hormones: Bovine lutropin resists enzymatic deglycosylation because of terminal O-sulfated N-acetylhexosamines, Proc.
Natl. Acad. Sci., 77:7089-7093.

Regoeczi, E., Wong, K.-L., Ali, M., and Hatton, M. W. C., 1977, The
molecular components of human transferrin type C, Intl. J. Peptide
Protein Res., 10:17-26.

Sairam, M. R., Papkoff, H., and Li, C.-H., 1972a, The primary structure of
ovine interstitial cell-stimulating hormone. I. The α-subunit,
Arch. Biochem. Biophys., 153:554-571.

Sairam, M. R., Samy, T. S. A., Papkoff, H., and Li, C.-H., 1972b, The
primary structure of ovine interstitial cell-stimulating hormone.
I. The β-subunit, Arch. Biochem. Biophys., 153:572-586.

Saxena, B. B., and Rathnam, P., 1976, Amino acid sequence of the β-subunit
of follicle-stimulating hormone from human pituitary glands, J.
Biol. Chem., 251:993-1005.

Sharon, N., and Liz, H., 1982, Glycoproteins, in: "The Proteins,"
H. Neurath, and R. L. Hill, eds., Academic Press, New York, vol. V,
pp. 1-144.

Shome, B., and Parlow, A. F., 1974, Human follicle stimulating hormone
(hFSH): First proposal for the amino acid sequence of the
α-subunit (hFSHα) and first demonstration of its identity with the
α-subunit of human luteinizing hormone (hLHα), J. Clin. Endocrinol.
Metab., 39:199-205.

Stellner, K., Saito, H., and Hakamori, S., 1972, Determination of amino
sugar linkage in glycolipids by methylation: Amino sugar linkage
of cevamide pentasaccharides of rabbit erythmocytes and of Forrsman
antigen, Arch. Biochem. Biophys., 155:464-472.

Taga, E. M., Waheed, A., and Van Etten, R. L., 1984, Structural and chemical characterization of a homogeneous peptide N-glycosidase from almond, Biochemistry, 23:815-822.

Takasaki, S., Yamashita, K., Suzuki, K., Iwanaga, S., and Kobata, A., 1979, The sugar chains of cold-insoluble globulin: A protein related to fibronectin, J. Biol. Chem., 254:8548-8553.

Takasaki, T., Muzuochi, T., and Kobata, A., 1982, Hydrazinolysis of asparagine-linked sugar chains to produce free oligosaccharides, in: "Methods in Enzymology," V. Ginsburg, ed., Academic Press, New York, vol. 83D, pp. 263-268.

Tarentino, A. L., and Maley, F., 1976, Purification and properties of an endo-β-N-acetylglucosaminidase from hen oviduct, J. Biol. Chem., 252:6337-6543.

Tarentino, A. L., Gomez, C. M., and Plummer, T. H., Jr., 1985, Deglyco-sylation of asparagine-linked glycans by peptide: N-glycosidase F., Biochemistry, 24:4665-4671.

Trimble, R. E., Tarentino, A. L., Aumick, G. E., and Maley F., 1982, Endo-β-N-acetylglucosaminidase L from Streptomyces plicatus, in: "Methods in Enzymology," V. Ginsburg, ed., Academic Press, New York, vol. 83, pp. 603-610.

Turco, S. J., 1981, Rapid separation of high mannose-type oligosaccharides by high-pressure liquid chromatography, Analyt. Biochem., 118: 278-283.

Umemoto, J., Bhavanandan, V. P., and Davidson, E. A., 1977, Purification and properties of an endo-α-N-acetyl-D-galactosaminidase from Diploccus pneumoniae, J. Biol. Chem., 252:8609-8614.

Wagh, P. V., and Bahl, O. P., 1981, Sugar residues on proteins, Crit. Rev. Biochem., 10:307-377.

Wong, K.-L., Charlwood, P. A., Hatton, M. W. C., and Regoeczi, E., 1974, Studies of the metabolism of asialotransferrins: Evidence that transferrin does not undergo desialylation in vivo, Clin. Sic. Mol. Med., 46:763-774.

Yamashita, K., Mizuocho, T., and Kobata, A., 1982, Analysis of oligosac-charides by gel filtration, in: "Methods in Enzymology," V. Ginsburg, Academic Press, New York, vol. 83D, pp. 105-126.

BIOSYNTHETIC CONTROLS THAT DETERMINE THE BRANCHING AND
MICROHETEROGENEITY OF PROTEIN-BOUND OLIGOSACCHARIDES

Harry Schachter

Research Institute, Hospital for
Sick Children, and Department of Biochemistry
University of Toronto
Toronto, Ontario M5G 1X8

INTRODUCTION

The past ten years have witnessed an unprecedented explosion in our knowledge of complex carbohydrate fine structure. The techniques which have been responsible for these advances include high resolution gel filtration chromatography, high performance liquid chromatography, the availability of a battery of exo- and endo-glycosidases with well-characterized specificities, new chemical methods for cleaving carbohydrate-amino acid linkages, high field nuclear magnetic resonance spectrometry, and high resolution mass spectrometry. It is clear that two properties distinguish complex carbohydrates from proteins and nucleic acids, the other two classes of biological macromolecules, namely: 1) branching and 2) microheterogeneity. These properties pose both challenges and problems for researchers interested in the function and biosynthesis of complex carbohydrates.

The functions of complex carbohydrates do not fall within the scope of the present review. Suffice it to say that we are only beginning to understand some of these functions in higher organisms, e.g., the role of cell surface carbohydrates in the interactions of cells with their cellular and fluid environments. There is preliminary evidence that the branching patterns of cell surface carbohydrate may be important in oncogenic transformation (Yamashita et al., 1983b, 1985b; Kobata and Yamashita, 1984; Hitoi et al., 1984). However, the oligosaccharides on some glycoproteins show extreme microheterogeneity, e.g., highly purified hen ovomucoid has more than 20 different oligosaccharides (Yamashita et al., 1982, 1983d). This phenomenon is common and is difficult to reconcile with the hypothesis that complex carbohydrates mediate an information-transfer role.

The highly branched nature of complex carbohydrates has important implications concerning their synthesis. Proteins and nucleic acids are linear molecules and are synthesized by a highly accurate template mechanism; DNA acts as a template for RNA, and RNA as a template for protein. Because they are branched, complex carbohydrates cannot be made in this way. Genetic information is transferred via an indirect non-template pathway. Genes code for the protein backbone of the glycoprotein, for the glycosyltransferases and glycosidases that form the oligosaccharides, for substrate and co-factor availability, and for the construction of the endomembrane assembly lines within which all complex carbohydrates are synthesized. The end-result is that nucleic acid and protein assembly tends to be very accurate, while the assembly of complex carbohydrates is prone to error and leads to the phenomenon of microheterogeneity. There is a more optimistic view-point which states that there are no errors involved in microheterogeneity and that the large diversity of structures has a purpose not yet understood.

BIOSYNTHETIC FACTORS INVOLVED IN COMPLEX CARBOHYDRATE MICROHETEROGENEITY

Figure 1 attempts to outline the various factors which contribute to the microheterogeneity of complex carbohydrates.

Mendelian Populations - Species

Sufficient data is now available to indicate that species may differ in the nature of their complex carbohydrates. For example, Kobata and his colleagues detected species-specific differences when they compared the carbohydrate structures present on gamma-glutamyltranspeptidase purified from the kidney and liver of rats, mice, and cows (Yamashita et al., 1983a, b, c, e, f, 1985a; Kobata and Yamashita, 1984), on fibronectin from bovine and human plasma, and on α_1-acid glycoprotein from rat and human plasma (Kobata, 1982, 1984). We have observed similar species-specific differences in the activities of glycosyltransferases involved in the assembly of mucin cores (Table I; Brockhausen et al., 1985). The functional significance of these findings is not clear.

54

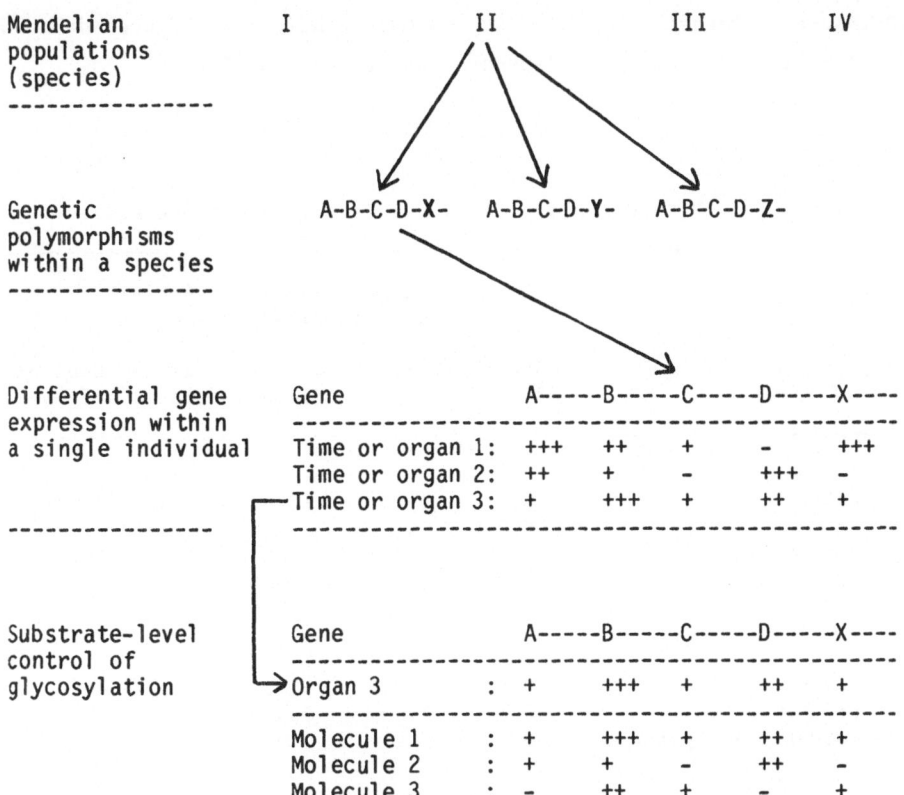

```
Mendelian                  I             II            III            IV
populations
(species)
----------------

Genetic                 A-B-C-D-X-   A-B-C-D-Y-   A-B-C-D-Z-
polymorphisms
within a species
----------------


Differential gene     Gene              A-----B-----C-----D-----X----
expression within     ------------------------------------------------
a single individual   Time or organ 1:  +++    ++    +     -      +++
                      Time or organ 2:  ++     +     -     +++    -
----------------      Time or organ 3:  +      +++   +     ++     +
                      ------------------------------------------------


Substrate-level       Gene              A-----B-----C-----D-----X----
control of            ------------------------------------------------
glycosylation         Organ 3        :  +      +++   +     ++     +
                      ------------------------------------------------
                      Molecule 1     :  +      +++   +     ++     +
                      Molecule 2     :  +      +     -     ++     -
                      Molecule 3     :  -      ++    +     -      +
```

Fig. 1. Hierarchies of control involved in complex
carbohydrate assembly. Species differ in the genes
controlling glycosylation. Within a single
species, there is genetic polymorphism (genes X, Y,
and Z in the above scheme may, for example,
respectively code for the blood group A-dependent
α3-GalNAc-transferase, the blood group B-dependent
α3-Gal-transferase or the absence of either
transferase). Differential gene expression between
organs or at different times in the same organ
occurs during differentiation. Even in a single
organ at a particular time, substrate-level factors
may modify the activity of a glycosyltransferase.

Genetic Polymorphism Within A Species

    Individuals within a Mendelian population, and even within a
subspecies, can differ in their genetic make-up. Only true clones
obtained by asexual reproduction are genetically identical. The best
example of genetic polymorphism in man is due to complex carbohydrates,
i.e., the ABO and Lewis human blood group antigens. The genes responsible
for these antigens code for a series of glycosyltransferases (A-dependent

```
```

α3-GalNAc-transferase, B-dependent α3-Gal-transferase, H-dependent α2-fucosyltransferase and Lewis-dependent α4-fucosyltransferase).

Table I.   Tissue and species survey for GlcNAc-transferases[a]

| Enzyme source | GlcNAc-Transferase Activity (nmoles/hour/mg) | | | | |
	Core 3 β3-Gn-T	Core 4 β6-Gn-T	Core 2 β6-Gn-T	2/4 ratio	Elongation β3-Gn-T
Rat:					
Colon	19.7	108	135	1.3	17
Stomach	0.4	25.2	29.6	1.2	1
Submaxillary gland	0	4.1	4.4	1.1	0
Small intestine	3.0	38.6	48.6	1.3	1
Liver microsomes	1.0	<0.6	<0.6	--	<0.4
Pig:					
Colon	20.5	13.4	51.2	3.8	14
Stomach	0.8	167.0	334.0	2.0	9
Submaxillary gland	0	0	0	--	0
Dog:					
Submaxillary gland	0	17.8	114.0	6.4	0
Monkey:					
Colon	2.4	4.7	11.9	2.5	--
Stomach	0.4	9.5	25.9	2.7	--
Human:					
Colon	5.5	10.3	24.0	2.3	0-9.6
Serum	--	--	--	--	0
Sheep:					
Stomach	0.6	16.9	21.3	1.3	--

[a]Enzyme reactions:

Core 3 β3-Gn-T (Brockhausen et al., 1983a, 1985):
UDP-GlcNAc + GalNAc-R → GlcNAcβ1-3GalNAc-R + UDP

Core 4 β6-Gn-T (Brockhausen et al., 1983a, 1985):
UDP-GlcNAc + GlcNAcβ1-3GalNAc-R →
                        GlcNAcβ1-3(GlcNAcβ1-6)GalNAc-R + UDP

Core 2 β6-Gn-T (Williams and Schachter, 1980; Williams et al., 1980):
UDP-GlcNAc + Galβ1-3GalNAc-R →
                        Galβ1-3(GlcNAcβ1-6)GalNAc-R + UDP

Elongation β3-Gn-T (Brockhausen et al., 1983b, 1984):
UDP-GlcNAc + Galβ1-3(Rβ1-6)GalNAc-R →
                        GlcNAcβ1-3Galβ1-3(Rβ1-6)GalNAc-R + UDP

There are two major forces for the above type of genetic variability: 1) mutation and recombination due to sexual matings followed by natural selection, and 2) genetic drift due to random or adaptively neutral loss of genes in sexual matings. It is interesting that not only are many apparently adaptively neutral genetic polymorphisms preserved but, in fact, some decidedly harmful polymorphisms are also preserved, e.g., sickle cell anemia. One possible reason is the concept of balanced polymorphism due to hybrid vigor, i.e., that the heterozygote has a higher survival potential than either homozygote. In the case of sickle cell anemia, the heterozygote survives malaria better than normal homozygotes. Natural selection ensures optimum survival of the species, not of the individual.

The lesson to be learned is that the survival of certain molecules through evolution may be due to a relatively minor and undetectable selective advantage. A large amount of ATP is required to synthesize a complex carbohydrate. This implies that the organism probably derives some advantage from evolutionary preservation of at least some of these large and complex structures.

Differential Gene Expression Within An Individual

The development of the fertilized egg into a complex organism (embryogenesis) requires organ development and differentiation. This process involves quantitative and qualitative differences in gene expression rates between different organs at any one time and in a particular organ at different times. Although the controls involved in eukaryotes are not yet established, it is clear that the environment of a cell must play a role during differentiation by influencing which genes are turned off and which are turned on.

There are several lines of evidence to suggest that complex carbohydrates on the cell surface are involved in the embryogenesis and differentiation processes. Hematopoietic tissue involves the differentiation of pluripotent stem cells into T and B lymphocytes, erythrocytes, granulocytes, platelets, etc. Marked changes have been observed in glycoprotein and glycolipid patterns during both erythropoiesis and granulocyte (myeloid cell) differentiation (Fukuda, 1985; Fukuda and Fukuda, 1984; Feizi, 1984; Hakomori, 1985; Hakomori et al., 1982; Kannagi et al., 1983; Testa et al., 1982). For example, human fetal erythrocytes carry the blood group i antigen, which is a linear oligosaccharide consisting of (Galβl-4GlcNAcβl-3) repeating units, whereas most adults carry on their red cells the I antigen containing the branched Galβl-4GlcNAcβl-3(Galβl-4GlcNAcβl-6)Gal- structure. Narasimhan et al. (1983 and unpublished data) have observed that human B cell lineage lymphoma lines involved in immunoglobulin synthesis contain GlcNAc-transferase III (Narasimhan, 1982), the enzyme which inserts a bisecting GlcNAc in βl-4 linkage to the β-linked Man of the core of N-glycosyl oligosaccharides; resting T and B lymphocytes and T-lineage acute lymphatic leukemia cells were essentially devoid of GlcNAc-transferase III activity. Changes in complex carbohydrate structure have also been observed as embryogenesis progresses (Feizi, 1984; Fenderson et al., 1984; Gooi et al., 1981; Hakomori, 1985; Ivatt, 1984).

Changes in the expression of complex carbohydrate antigens have been detected during oncogenesis. These tumor-associated antigens often represent retrogenetic expression of carbohydrate synthesis to a particular stage of fetal development and are called oncofetal antigens. It is therefore not surprising that these antigens are not tumor-specific since they are synthesized in normal fetal tissues (Feizi, 1984; Hakomori, 1985).

Several studies have shown differential gene expression between different organs for complex carbohydrate synthesis. Kobata and colleagues observed consistent differences between kidney and liver in their studies on gamma-glutamyltranspeptidase (Yamashita et al., 1983a, b, c, e, f, 1985a; Kobata and Yamashita, 1984). The most dramatic observation was the presence of bisecting GlcNAc residues (linked βl-4 to the β-linked Man of N-glycosyl oligosaccharide cores) in the enzyme from rat, cow, and mouse kidney, but the complete absence of this residue from rat and mouse liver enzyme. Rat hepatoma enzyme, however, showed the presence of bisecting GlcNAc residues. We have also observed organ-specific differences in our studies on the synthesis of mucin cores (Table I; Brockhausen et al., 1985). For example, the activity of β3-GlcNAc-transferase responsible for synthesis of core type 3 (GlcNAcβl-3GalNAc) is high in rat, pig, monkey and human colon but is very low in stomach and submaxillary gland.

Endomembrane Factors

Even in a particular organ at a particular time, a polypeptide core may acquire different types of oligosaccharide, i.e., the individual molecules of a so-called pure glycoprotein preparation will have identical amino acid sequences but yet will differ from one another in oligosaccharide side chains (e.g., α_1-acid glycoprotein, hen ovalbumin, hen ovomucoid, transferrin, etc.). Or a single organ may make two glycoproteins with totally different oligosaccharides, e.g., the hen oviduct (Yamashita et al., 1982, 1983d, 1984) makes ovalbumin with high mannose and bisected hybrid N-glycosyl oligosaccharides, but ovomucoid, in the same organ, ends up with truncated (primarily GlcNAc-terminal antennae) bisected highly branched complex N-glycosyl oligosaccharides (Fig. 2).

A relatively simple explanation for microheterogeneity is that polypeptides travel through different endomembrane assembly lines, either in the same cell or in different cells. This explanation applies both to the situation in which different oligosaccharides appear at a particular amino acid position of a particular polypeptide chain or to the situation in which a single organ makes different polypeptides with different oligosaccharides. This hypothesis requires a mechanism for sorting peptides to the different assembly lines. If all molecules made by a particular organ move through identical assembly lines, then factors operative at the endomembrane level must be responsible for microheterogeneity. The remainder of this review will deal with these endomembrane or "substrate-level" factors.

THE ENDOMEMBRANE ASSEMBLY LINE

A great deal of information is now available on the structures of N- and O-glycosyl oligosaccharides (Kobata, 1984; Carver and Brisson, 1984; Kornfeld and Kornfeld, 1985) and on the detailed enzymatic steps involved in their synthesis (Snider, 1984; Sadler, 1984; Kornfeld and Kornfeld, 1985; Schachter et al., 1983, 1985; Schachter and Williams, 1982). Oligosaccharide synthesis occurs on a membranous assembly line. The polypeptide backbones of all glycoproteins are synthesized on membrane-bound polyribosomes, and all carbohydrate addition occurs on the luminal side of the endomembrane system.

Figure 3, taken from the excellent review by Kornfeld and Kornfeld (1985), outlines the basic steps in the assembly of biantennary N-glycosyl oligosaccharides. Synthesis starts in the rough endoplasmic reticulum with the transfer of a large oligosaccharide from dolichol pyrophosphate oligosaccharide to an asparagine residue in the polypeptide backbone (step 1, Fig. 3). This is followed by oligosaccharide processing within both the rough endoplasmic reticulum (steps 2 to 4, Fig. 3) and

$$M \xrightarrow{\alpha 2} M_{\alpha 6}$$
$$M \xrightarrow{\alpha 2} M_{\alpha 3} \searrow M \xrightarrow{\alpha 6} M\text{-}R$$
$$M \xrightarrow{\alpha 2} M \xrightarrow{\alpha 2} M_{\alpha 3}$$

HIGH MANNOSE (M9)

$$M_{\alpha 6}$$
$$M_{\alpha 3} \searrow M_{\alpha 6}$$
$$M_{\alpha 3} \searrow M\text{-}R$$
$$M_{\alpha 3}$$

HIGH MANNOSE (M5)

$$X\!-\!Gn \xrightarrow{\beta 2} M_{\alpha 6}$$
$$\qquad\qquad\qquad M\text{-}R$$
$$X\!-\!Gn \xrightarrow{\beta 2} M_{\alpha 3}$$

BIANTENNARY COMPLEX (GnGn if X = H)

$$X\!-\!Gn \xrightarrow{\beta 2} M_{\alpha 6}$$
$$\qquad\qquad\qquad\qquad M\text{-}R$$
$$X\!-\!Gn \xrightarrow{\beta 2} M_{\alpha 3}$$
$$X\!-\!Gn \diagup^{\beta 4}$$

TRIANTENNARY COMPLEX (GnGnGn if X = H)

$$X\!-\!Gn \xrightarrow{\beta 2} M_{\alpha 6}$$
$$X\!-\!Gn \diagup^{\beta 6} M\text{-}R$$
$$X\!-\!Gn \xrightarrow{\beta 2} M_{\alpha 3}$$
$$X\!-\!Gn \diagup^{\beta 4}$$

TETRAANTENNARY COMPLEX (GnGnGnGn if X = H)

$$X\!-\!Gn \xrightarrow{\beta 2} M_{\alpha 6}$$
$$\boxed{Gn} \xrightarrow{\beta 4} M\text{-}R$$
$$X\!-\!Gn \xrightarrow{\beta 2} M_{\alpha 3}$$

BISECTED BIANTENNARY COMPLEX (GnGn(Gn) if X = H)

$$X\!-\!Gn \xrightarrow[\beta 4]{\beta 2} M_{\alpha 6}$$
$$\boxed{Gn} \xrightarrow{\beta 2} M\text{-}R$$
$$X\!-\!Gn \diagup M_{\alpha 3}$$
$$X\!-\!Gn \diagup^{\beta 4}$$

BISECTED TRIANTENNARY COMPLEX (GnGnGn(Gn) if X = H)

$$X\!-\!Gn \xrightarrow{\beta 2}$$
$$X\!-\!Gn \diagup^{\beta 6} M_{\alpha 6}$$
$$\boxed{Gn} \xrightarrow{\beta 4} M\text{-}R$$
$$X\!-\!Gn \diagup M_{\alpha 3}$$
$$X\!-\!Gn \diagup^{\beta 4}$$

BISECTED TETRAANTENNARY COMPLEX (GnGnGnGn(Gn) if X = H)

$$M_{\alpha 6}$$
$$M_{\alpha 3} \searrow^{\alpha 6}$$
$$\boxed{Gn} \xrightarrow{\beta 4} M\text{-}R$$
$$X\!-\!Gn \diagup M_{\alpha 3}$$
$$X\!-\!Gn \diagup^{\beta 4}$$

BISECTED TRIANTENNARY HYBRID (Gn(I,III,IV)M5 if X = H)

$$M_{\alpha 6}$$
$$M_{\alpha 3} \searrow M\text{-}R$$
$$X\!-\!Gn \xrightarrow{\beta 2} M_{\alpha 3}$$

NON-BISECTED BIANTENNARY HYBRID (Gn(I)M5 if X = H)

$$M_{\alpha 6}$$
$$M_{\alpha 3} \searrow^{\alpha 6} M\text{-}R$$
$$X\!-\!Gn \xrightarrow{\beta 2} M_{\alpha 3}$$
$$X\!-\!Gn \diagup^{\beta 4}$$

NON-BISECTED TRIANTENNARY HYBRID (Gn(I,IV)M5 if X = H)

$$M_{\alpha 6}$$
$$M_{\alpha 3} \searrow^{\alpha 6}$$
$$\boxed{Gn} \xrightarrow{\beta 4} M\text{-}R$$
$$X\!-\!Gn \xrightarrow{\beta 2} M_{\alpha 3}$$

BISECTED BIANTENNARY HYBRID (Gn(I,III)M5 if X = H)

$$(G \xrightarrow{\beta 4} Gn \xrightarrow{\beta 3})_n \!-\! G \xrightarrow{\beta 4} Gn \xrightarrow{\beta 2} M_{\alpha 6}$$
$$\qquad\qquad\qquad\qquad\qquad\qquad M\text{-}R$$
$$(G \xrightarrow{\beta 4} Gn \xrightarrow{\beta 3})_m \!-\! G \xrightarrow{\beta 4} Gn \xrightarrow{\beta 2} M_{\alpha 3}$$

POLY-N-ACETYLLACTOSAMINOGLYCAN (i)

$$-\!G \xrightarrow{\beta 4} Gn_{\beta 6}$$
$$-\!G \xrightarrow{\beta 4} Gn \xrightarrow{\beta 3} G \xrightarrow{\beta 4} Gn \xrightarrow{\beta 3}$$

Fig. 2. Examples of N-glycosyl oligosaccharides (OS). High mannose OS are named according to the number of Man (M) residues, e.g., M_5. Complex OS are named according to the sugars at the non-reducing termini, with the Manα1-6 arm being named first [M, Man; Gn, GlcNAc; G, Gal; S, sialic acid; (Gn), bisecting GlcNAc]. Hybrid OS are named as shown, using the numbering system in Fig. 6 to designate the arms. X can be H, Galβ1-4(3), or sialylα2-3(6)Galβ1-4(3).

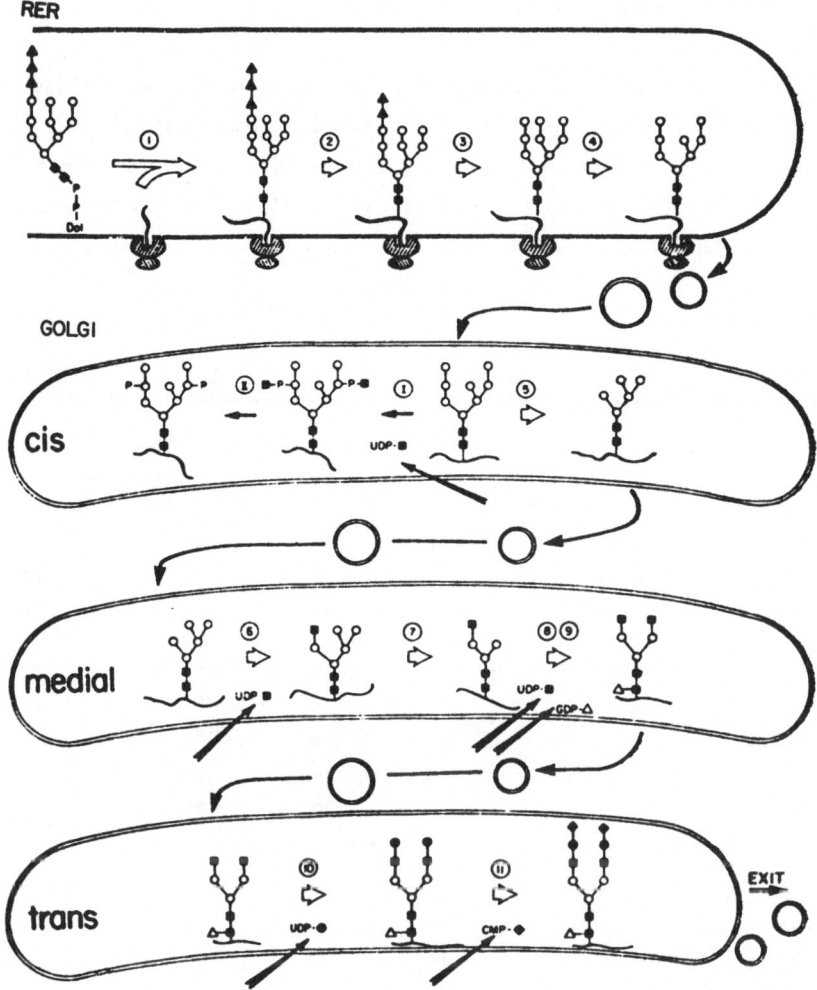

Fig. 3. Biosynthesis of N-glycosyl non-bisected biantennary
complex oligosaccharide, from the review by Kornfeld
and Kornfeld (1985), with kind permission. Processing enzymes: (1) transfer of $Glc_3Man_9GlcNAc_2$ from
dolichol pyrophosphate oligosaccharide to peptide
nascently bound to ribosome, (2) α-glucosidase
I, (3) α-glucosidase II, (4) rough endoplasmic
reticulum (RER) α2-mannosidase, (5) α-mannosidase I,
(6) GlcNAc-transferase I, (7) α-mannosidase II,
(8) GlcNAc-transferase II, (9) α6-fucosyltransferase,
(10) Gal-transferase, (11) sialyltransferase,
(I) N-acetylglucosaminylphosphotransferase, which
acts on hydrolases destined for the lysosomes,
(II) N-acetylglucosamine-1-phosphodiester α-N-acetyl-
glucosaminidase, which exposes the Man-6-phosphate
signal required for movement of hydrolases to the
lysosome. Transport of nucleotide-sugars into the
lumen is indicated. Symbols: ■ , GlcNAc; ○ , Man; .
▲ Glc; △ , Fuc; ● , Gal; ◆ , sialic acid.

Golgi apparatus (steps 5 and 7, Fig. 3) as the glycoprotein moves through the endomembrane assembly line.

Our laboratory has been concerned primarily with the mechanisms which determine the branching patterns of complex and hybrid N-glycosyl oligosaccharides (Fig. 2) and with the synthesis of the four major core types of O-glycosyl oligosaccharides (Fig. 4). In particular, as mentioned in the previous section, we have been interested in the

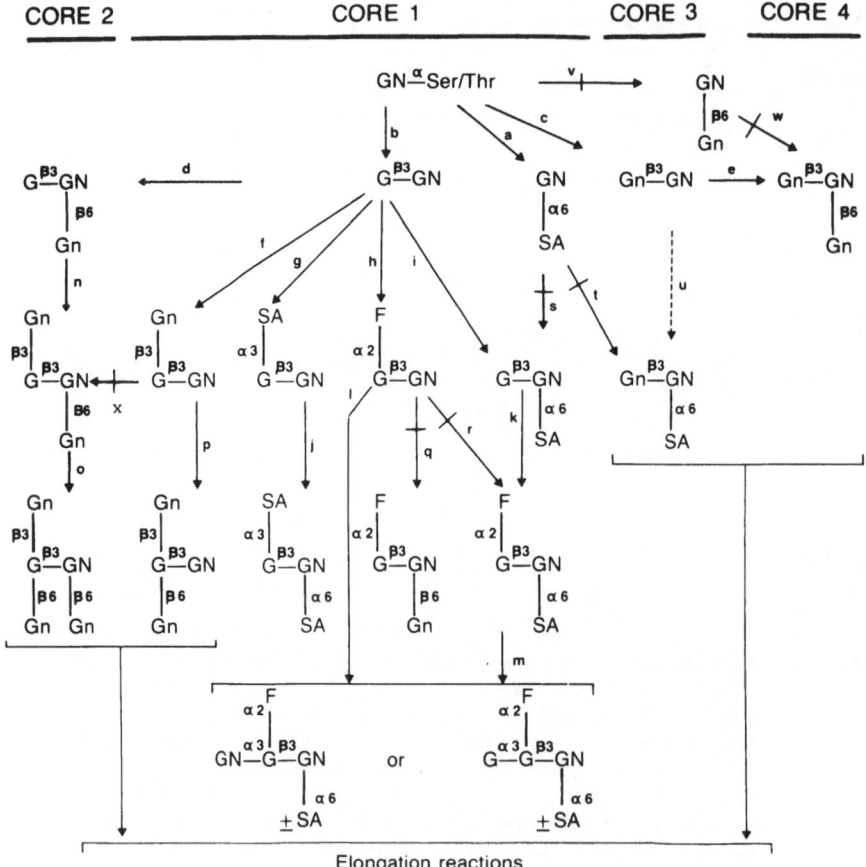

Fig. 4. Biosynthesis of O-glycosyl oligosaccharides. Four major core classes are shown: (1) Galβ1-3GalNAc, (2) Galβ1-3(GlcNAcβ1-6)GalNAc, (3) GlcNAcβ1-3GalNAc, and (4) GlcNAcβ1-3(GlcNAcβ1-6)GalNAc. Abbreviations: GN, GalNAc; Gn, GlcNAc; G, Gal; SA, sialic acid; F, Fuc. All steps except u have been proven by in vitro experiments (Brockhausen et al., 1983a, 1983b, 1984, 1985, and in preparation; Sadler, 1984; Schachter and Williams, 1982).

Table II. Control of glycoprotein synthesis at the substrate level.

---

1. Competition for a common substrate.

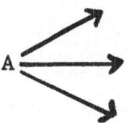

2. Substrate specificity of glycosyltransferases.

    (i)   GO-NO GO.

    (ii) NO GO-GO.

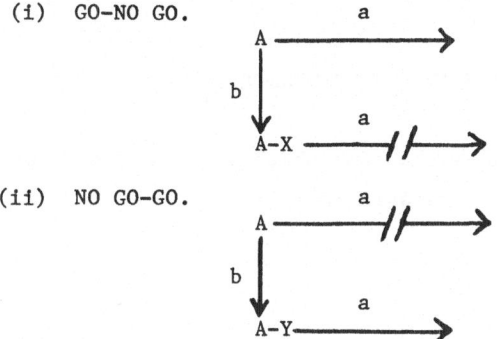

    (iii) Recognition site distinct from catalytic site.

    (iv) Branch specificity.

    (v)   Role of polypeptide.

3. Substrate Availability.

    (i)   Subcellular compartments as assembly lines.

    (ii) Transport of nucleotide-sugars across membranes.

4. Other factors: pH, cations, phosphatases, glycosidases. Phospholipid environment.

"substrate-level" controls which operate within the endomembrane system to produce the large variety of oligosaccharides that occur on mammalian and avian glycoproteins. These steps occur within the Golgi apparatus.

The Golgi apparatus is now believed to be subdivided into sub-compartments with different functions (Fig. 3; Kornfeld and Kornfeld, 1985; Rothman, 1985; Dunphy and Rothman, 1985). The steps involved in assembly of the two antennae of a typical biantennary complex N-glycosyl oligosaccharide probably occur in the medial and trans-Golgi compartments (steps 6, 8-11, Fig. 3). Although Figure 3 transmits a great deal of information in a clear and concise manner, it is quite incomplete, since it lacks the pathways for the synthesis of N-glycosyl oligosaccharides of the hybrid type and of complex types larger than biantennary. An expansion of the synthetic scheme which includes the latter structures is shown in Figure 5. It is necessary to use a short-hand notation in order to fit all the pathways into a single page, but the visual impact is not as clear as the diagrams used in Figure 3. Even Figure 5 is not complete, since it omits GlcNAc-transferases V, VI, and VII (Fig. 6).

CONTROL OF GLYCOPROTEIN SYNTHESIS AT THE SUBSTRATE LEVEL

Table II lists the various mechanisms which control the assembly of protein-bound oligosaccharide at the substrate level.

Competition for a Common Substrate

At many points in the synthetic schemes for both N- and O-glycosyl oligosaccharides (Figs. 4 and 5), a particular substrate can be acted on by more than one enzyme. Although the schemes shown in Figures 4 and 5 are composite schemes based on in vitro assays carried out in different tissues, it is assumed that they approximate the in vivo situation and that competition for a common substrate will also occur in vivo.

Several important "cross-roads" can be pointed out (Fig. 5). $Gn(I)M_5$ can be acted on by (1) GlcNAc-transferase III to form bisected biantennary 5-Man hybrids (Allen et al., 1984), or by (2) GlcNAc-transferase IV to form non-bisected triantennary 5-Man hybrids (Allen et al., 1984), or by

Fig. 5. Biosynthesis of N-glycosyl oligosaccharides (OS) showing the conversion of M_5 into: (A) 5-Man non-bisected biantennary hybrid OS; (B) 3-Man non-bisected biantennary hybrid OS (or incomplete biantennary complex OS); (C) non-bisected biantennary complex OS; (D) 5-Man bisected triantennary hybrid OS; (E) 3-Man bisected triantennary hybrid OS (or incomplete bisected triantennary complex OS); (F) bisected triantennary complex OS; (G) 5-Man non-bisected triantennary hybrid OS; (H) 3-Man non-bisected triantennary hybrid OS (or incomplete triantennary complex OS); (I) non-bisected triantennary complex; (J) 5-Man bisected biantennary hybrid OS; (K) 3-Man bisected biantennary hybrid OS (or incomplete bisected biantennary complex OS); (L) bisected biantennary complex. Abbreviations are explained in the legend to Figure 2.

(3) α6-fucosyltransferase (Longmore and Schachter, 1982), or by
(4) α-mannosidase II to form MGn, or (5) can move from the medial to the
trans-Golgi to form non-bisected biantennary 5-Man hybrids. Once the
bisecting GlcNAc has been inserted by GlcNAc-transferase III,
α-mannosidase II can no longer act (Harpaz and Schachter, 1980) and the
pathway is fixed into the production of hybrid structures (Fig. 5). The

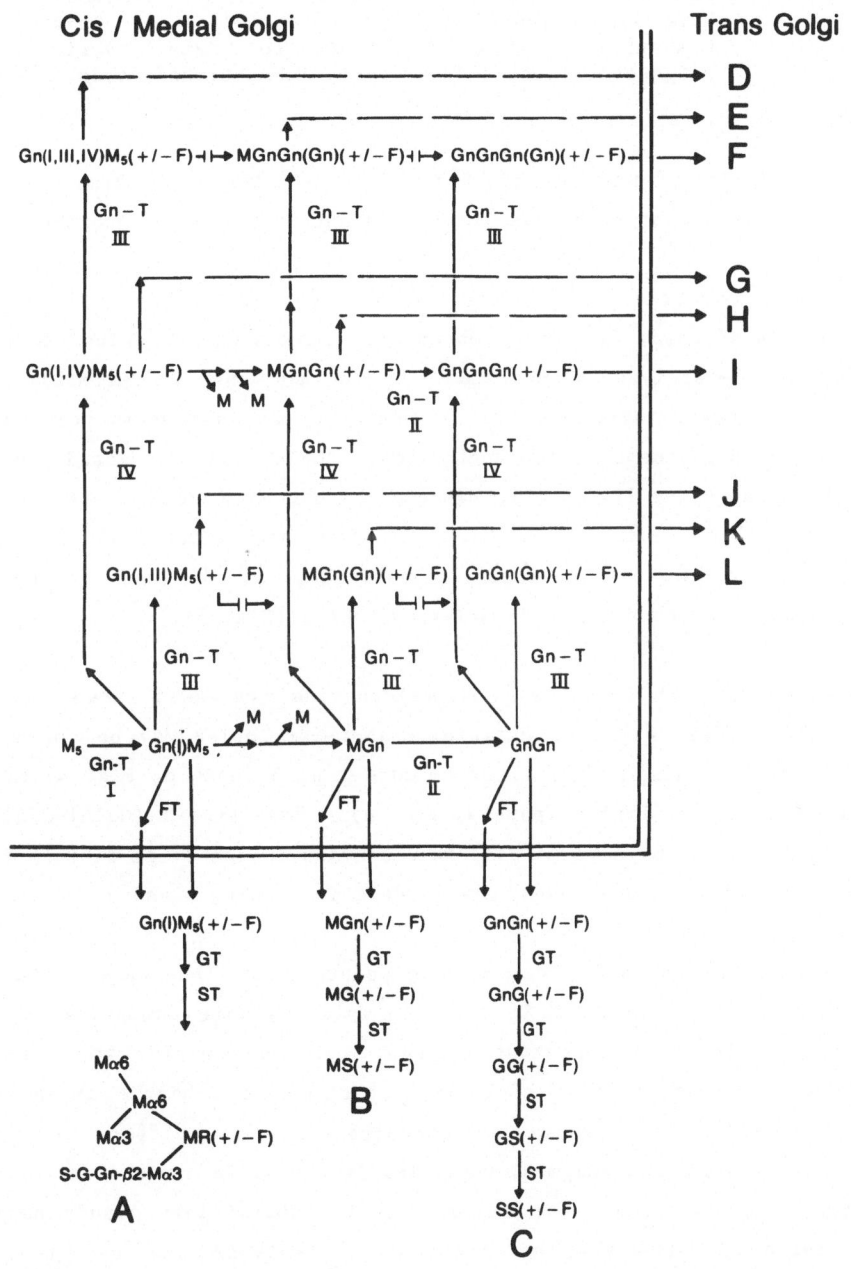

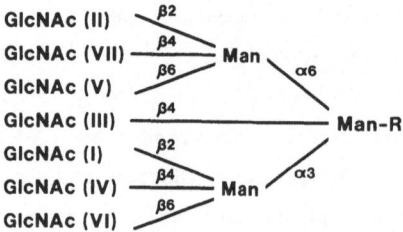

Fig. 6. A hypothetical structure showing all the antennary GlcNAc residues described to date. GlcNAc linked 1-3 to either arm of the core, 1-2 to the β-linked Man, and in α linkages are theoretically possible but have not yet been described. GlcNAc residues have been numbered arbitrarily as indicated. This numbering system is used to name hybrid oligosaccharides (Fig. 2) and GlcNAc-transferases.

rules permit the formation of 5-Man hybrid structures with a core α6-fucose residue, but such structures have not as yet been reported.

MGn can be acted on by GlcNAc-transferases II, III, or IV, or by α6-fucosyltransferase, or can enter the trans-Golgi. The GlcNAc-transferase II-catalyzed conversion of MGn to GnGn is probably the "main-line" pathway (step 8, Fig. 3). GnGn is the main entry point into all complex N-glycosyl oligosaccharides. GnGn can be acted on by 1) GlcNAc-transferase III to form bisected biantennary complex structures, 2) either GlcNAc-transferase IV or V to form non-bisected triantennary complex structures, 3) α6-fucosyltransferase, or 4) can enter the trans-Golgi to be acted on by β4-Gal-transferase (Fig. 5).

The pathway for O-glycosyl oligosaccharide synthesis shows similar "cross-roads" (Fig. 4). For example, GalNAc-Ser(Thr)-R may be converted to core type 1 (path b, Fig. 4) or to core type 3 (path c, Fig. 4) or to sialylα2-6GalNAc-Ser(Thr)-R (path a, Fig. 4). Core type 1 (Galβ1-3GalNAc-R) may be acted on by several different enzymes (paths f, g, h, i in Fig. 4). Various other "cross-roads" are evident in Figures 4 and 5.

The existence of these competition points means that many different oligosaccharides can be made in a single assembly line depending on the relative activities of competing enzymes, and may explain, at least in part, the microheterogeneity of oligosaccharides at a single amino acid site. In addition, tissues and species probably differ in their complement of competing enzymes, resulting in the differences in structure that have been observed. For example, ovine submaxillary glands have a higher ratio of CMP-sialic acid:GalNAc-R α6-sialyltransferase (path a,

Table III. Substrate specificity: GO—NO GO*

X	a	References
-Mα6 **Gn**β4M- Gnβ2Mα3	α-mannosidase II Gn-T II, IV α6-Fuc-T	Harpaz and Schachter, 1980 Gleeson and Schachter, 1983 Longmore and Schachter, 1982
-Mα6 →M- **G**β4Gnβ2Mα3	α-mannosidase II Gn-T II, III, IV Gn-T V[δ] α6-Fuc-T[δ]	Harpaz and Schachter, 1980 Narasimhan, 1982; Gleeson and Schachter, 1983 Cummings et al., 1982 Longmore and Schachter, 1982
Gβ4(3)Gn- α3(4) **Fuc**	Blood Group Enzymes: A-dependent α3-GalNAc-T to Gal B-dependent α3-Gal-T to Gal H-dependent α2-Fuc-T to Gal	Beyer et al., 1981
GalNAc-Ser(Thr) α2,6 **SA**	β3-Gal-T to GalNAc	Schachter et al., 1971
Gβ4Gn- **G**β4Gnβ3	Blood Group I-dependent β6-Gn-T to Gal	Piller et al, 1984

*M, mannose; Gn, N-acetylglucosamine; G, galactose; SA, sialic acid; Fuc, fucose; Gn-T, N-acetylglucosaminyl-transferase.
The residue in bold print is X, the "NO GO" signal.
The arrows indicate the site of action of the enzymes listed under "a".

δThese enzymes have only been tested with substrates in which there is a Gal residue on both antennae.

Fig. 4) to UDP-Gal:GalNAc-R β3-Gal-transferase (path b, Fig. 4) than do porcine submaxillary glands; the α6-sialyl residue prevents β3-Gal-transferase action (path s, Fig. 4) and, therefore, ovine submaxillary mucin contains primarily sialylα2-6GalNAc chains, whereas porcine submaxillary mucin contains mainly larger chains. The interesting point is that these two mucins differ quantitatively rather than qualitatively in their oligosaccharide compositions. Table I lists other tissue and species differences for several of the mucin glycosyltransferases.

Substrate Specificity: GO-NO GO

One of the most important directing forces in oligosaccharide biosynthesis are the constraints imposed by the specific substrate requirements of many of the glycosyltransferases. Some of these properties will now be reviewed. The first example has been named the GO-NO GO effect. This simply means that the insertion of a single critical sugar residue (X) into an oligosaccharide will convert it from a substrate to a non-substrate. Table III lists some of the most important NO GO residues.

The first two examples refer to a group of enzymes acting on N-glycosyl oligosaccharides (illustrated in Fig. 7) which have in common the following properties (Schachter et al., 1983; Vella et al., 1984): 1) they require the presence of the GlcNAcβ1-2Manα1-3Manβ1-4 grouping,

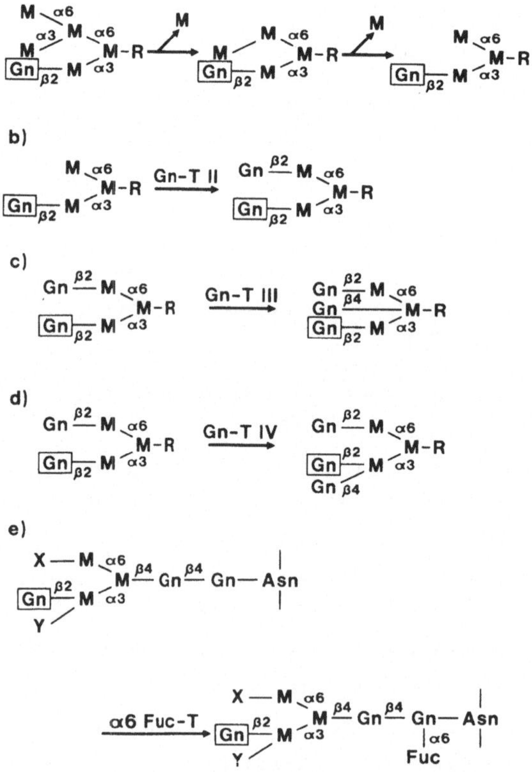

Fig. 7. Reactions catalyzed by five enzymes requiring the prior action of GlcNAc-transferase I which inserts a GlcNAc in β1-2 linkage into the Manα1-3 arm. (a) Mannosidase II; (b) GlcNAc-transferase II; (c) GlcNAc-transferase III; (d) GlcNAc-transferase IV; (e) α6-fucosyltransferase. Abbreviations as in Figure 2.

2) activity is blocked if the substrate contains a bisecting GlcNAc (Fig. 7), and 3) activity is blocked if the GlcNAcβ1-2Manα1-3Manβ1-4 antenna is covered at its non-reducing end by a Galβ1-4- residue. We explain these data by assuming that these enzymes all require the GlcNAcβ1-2Manα1-3Manβ1-4 grouping as a recognition site distinct from the catalytic site and that this recognition site is blocked either by a bisecting GlcNAc or by a Gal residue. The structures shown in Figures 8 and 9 are based on three-dimensional studies by nuclear magnetic resonance spectrometry (Carver and Brisson, 1984) and illustrate the steric hindrance of the recognition site by either the bisecting GlcNAc or Gal residues. Figure 8 shows that the sites of action of GlcNAc-transferases II and IV and of α6-fucosyltransferase are accessible and well-removed from the sterically-blocked GlcNAcβ1-2Manα1-3Manβ1-4 site; Figure 9 shows a similar situation for α-mannosidase II.

Other NO GO residues are: 1) a fucose linked α1-3 or α1-4 to a penultimate GlcNAc residue will prevent the action of blood group A-dependent α3-GalNAc-transferase, blood group B-dependent α3-Gal-transferase, and blood group H-dependent α2-fucosyltransferase. 2) As mentioned in the previous section, a sialyl residue linked α2-6 to GalNAc prevents the action of the mucin core 1 β3-Gal-transferase. 3) A Galβ1-4 residue on the GlcNAcβ1-3Galβ1-4- terminus prevents the action of blood

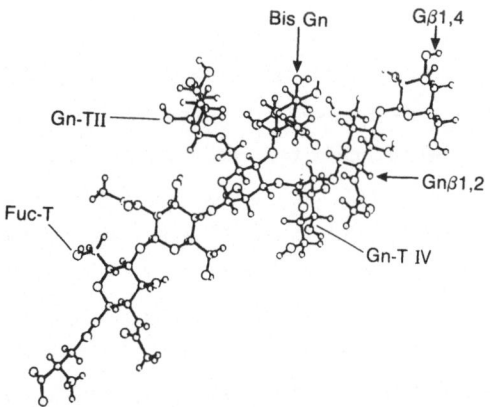

Fig. 8. Computer-drawn structure for MG(Gn), ω = +180°, based on NMR data of Carver and Brisson (1984). The sites of action of GlcNAc-transferases II and IV (Gn-T II and IV) and of α6-fucosyltransferase (Fuc-T) are indicated. Abbreviations: M, Man; Gn, GlcNAc; Bis Gn, bisecting GlcNAc; G, Gal. The diagram was prepared by D.Cummings.

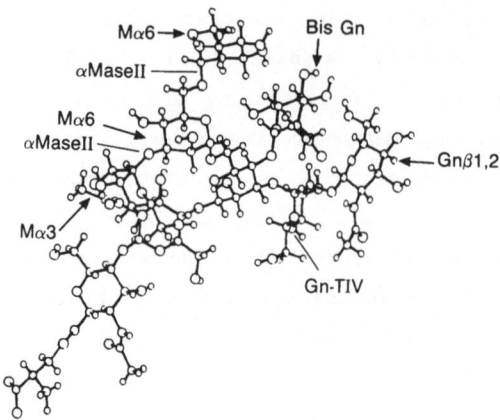

Fig. 9.   Computer-drawn structure for $Gn(I,III)M_5$, based on
          NMR data of Carver and Brisson (1984). The sites
          of action of α-mannosidase II are indicated (αMase
          II).   Abbreviations as for Figures 2 and 8.   The
          diagram was prepared by D. Cummings.

group I β6-GlcNAc-transferase.   These examples lend support to the concept
of carbohydrate recognition sites remote from the catalytic sites.

Substrate Specificity: NO GO-GO

    The NO GO-GO effect is the reverse phenomenon, i.e. the insertion of
a single critical sugar residue (Y) into an oligosaccharide will convert
it from a non-substrate to a substrate.   Two typical examples are shown in
Table IV.   The first example is at an important decision point in the
synthesis of N-glycosyl oligosaccharides.   If GlcNAc-transferase I does
not insert a GlcNAc residue on the Manα1-3 arm (step 6, Fig. 3), synthesis
of hybrid and complex N-glycosyl oligosaccharides cannot occur.   This
GlcNAc residue is a GO signal for all enzymes shown in Figure 7 and
provides a recognition site for them, as explained in the previous
section.   The second example in Table IV has been called the "3 before 6"
rule and applies to three enzymatic activities which are probably due to
the action of a single β6-GlcNAc-transferase (Brockhausen et al., 1985,
and unpublished data).   The rule states that the β6-GlcNAc-transferase(s)
responsible for synthesis of mucin core 2 (path d, Fig. 4) and core 4
(path e, Fig. 4) and for the branch point of the blood group I determinant
(paths o, p, Fig. 4) requires substitution of carbon 3 of the target sugar
residue before carbon 6 can be substituted.   Put another way, core 1
synthesis precedes core 2 synthesis, and core 3 synthesis precedes core 4
synthesis.

70

Table IV. Substrate specificity: NO GO-GO*

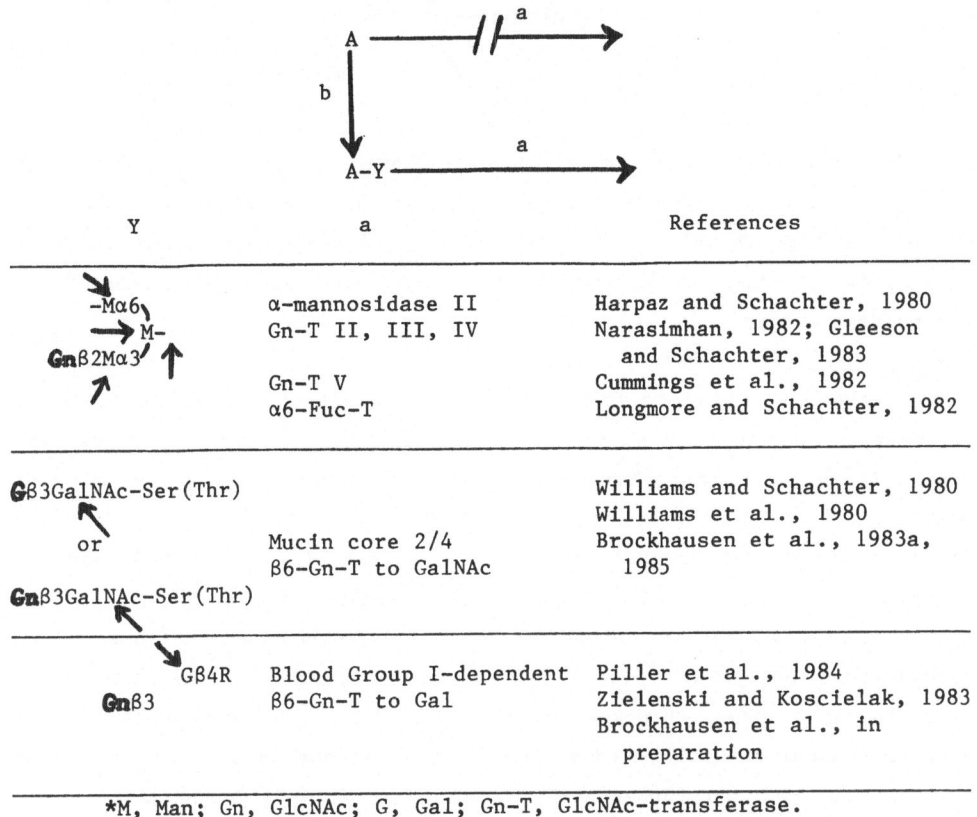

Y	a	References
-Mα6⟍ →M- Gnβ2Mα3✓	α-mannosidase II Gn-T II, III, IV Gn-T V α6-Fuc-T	Harpaz and Schachter, 1980 Narasimhan, 1982; Gleeson and Schachter, 1983 Cummings et al., 1982 Longmore and Schachter, 1982
Gβ3GalNAc-Ser(Thr) or Gnβ3GalNAc-Ser(Thr)	 Mucin core 2/4 β6-Gn-T to GalNAc	Williams and Schachter, 1980 Williams et al., 1980 Brockhausen et al., 1983a, 1985
Gβ4R Gnβ3	Blood Group I-dependent β6-Gn-T to Gal	Piller et al., 1984 Zielenski and Koscielak, 1983 Brockhausen et al., in preparation

*M, Man; Gn, GlcNAc; G, Gal; Gn-T, GlcNAc-transferase.
The residue in bold print is Y, the "CO" signal.
The arrows indicate the site of action of the enzymes
listed under "a".

Since Gal residues appear to be needed for iI-antigenic activities, it is not strictly correct to state that blood group i synthesis precedes blood group I synthesis. However, the β3-GlcNAc-transferase must act before the β6-GlcNAc-transferase. For example, pig stomach mucosal extracts can transfer GlcNAc in β1-3 linkage to the terminal Gal of Galβ1-3(GlcNAcβ1-6)GalNAc-R but not to GlcNAcβ1-6Galβ1-3(GlcNAcβ1-6)GalNAc-R (Brockhausen et al., 1983b). Further, we have detected both the GlcNAcβ1-3Gal- and GlcNAcβ1-3(GlcNAcβ1-6)Gal- structures but not the GlcNAcβ1-6Gal structure (Brockhausen et al., 1983b, 1984; unpublished data).

The above rules, like all the other rules stated in this review, may have exceptions. For example, Yamashita et al. (1983d) reported the presence in hen ovomucoid of the following structure:

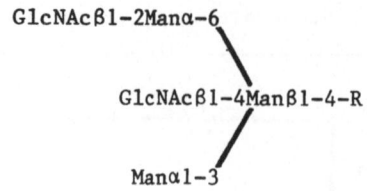

GlcNAcβ1-2Manα-6

GlcNAcβ1-4Manβ1-4-R

Manα1-3

This structure does not fit the synthetic rules because it implies that both GlcNAc-transferases II and III have acted without the presence of the GlcNAcβ1-2Manα1-3- grouping. The above compound represented less than 4.6% of the total carbohydrate and may have been formed by glycosidase action.

Another exception to the rules is the report of GlcNAcβ1-6GalNAc in human kappa-casein (Fiat et al., 1984). We have been unable to detect the synthesis of this disaccharide in vitro in our systems. For example, GalNAc-mucin was incubated with UDP-GlcNAc and enzyme from either pig colon or rat colon; product was released by alkaline borohydride and was purified by gel filtration and high performance liquid chromatography. Analysis by high resolution nuclear magnetic resonance spectrometry and methylation analysis revealed GlcNAcβ1-3GalNAcOH to be the only disaccharide product (Brockhausen et al., 1985). Perhaps human tissues have an unusual β6-GlcNAc-transferase or the residue on carbon 3 of GalNAc was incorporated and subsequently lost. What is reassuring is that the great majority of structures isolated from many different glycoproteins fit the synthetic rules outlined above.

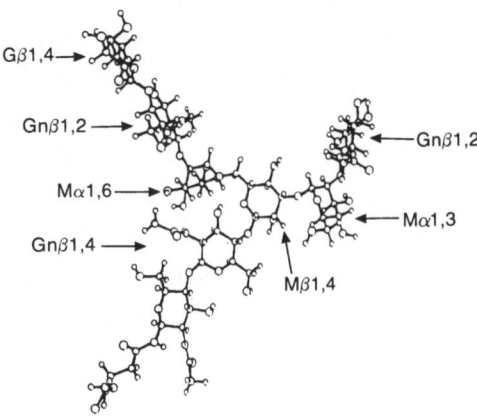

Fig. 10. Computer-drawn structure for GGn, ω = -60° based on NMR data of Carver and Brisson (1984). Abbreviations as for Figures 2 and 8. The diagram was prepared by D. Cummings.

Substrate Specificity: Recognition Sites Distinct from Catalytic Sites

Table V gives three examples which support the concept of recognition sites distinct from catalytic sites. We have already discussed the first example in which GlcNAcβ1-2Manα1-3Manβ1-4- serves as a recognition site for the enzymes listed in Figure 7. The second example is the requirement shown by GlcNAc-transferase I for a Manα1-3Manβ1-4GlcNAc- moiety (Vella et al., 1984). The third example was reported by Joziasse et al. (1984) for bovine colostrum CMP-sialic acid:Galβ1-4GlcNAc-R α2-6-sialyltransferase; they noted a partial requirement for an intact GlcNAc ring in the Manβ1-4GlcNAc sequence of the N-glycosyl oligosaccharide core and postulated that the sialyltransferase required this region for optimum positioning on the substrate.

Substrate Specificity: Branch Specificity

Van den Eijnden and co-workers have coined the term branch specificity to describe the preferential action of a glycosyltransferase on a particular arm of a branched oligosaccharide (Table VI). For example, elongation of biantennary complex N-glycosyl oligosaccharides by both β4-Gal-transferase and α2-6-sialyltransferase occurs preferentially on the Manα1-3 arm. This may be due, in part, to the fact that the Manα1-6 arm can exist in two states in equilibrium with each other; one of these states may not be as readily accessible as the other (Carver and Brisson, 1984; Figs. 10, 11). However, the fact that α3-Gal-transferase from calf thymus prefers the Manα1-6 arm (Blanken et al., 1984) makes this

Table V. Substrate specificity: Recognition sites distinct from catalytic sites.

Recognition Site	Enzymes	References

R'
  α6
    Manβ4GlcNAc-R
    α3
GlcNAcβ2Man
	α-mannosidase II
	GlcNAc-T II, III
	IV
	GlcNAc-T V
	α6-Fuc-T
		Harpaz and Schachter,1980
		Narasimhan, 1982; Gleeson and Schachter, 1983
		Cummings et al., 1982
		Longmore and Schachter, 1982

R2
  α6
    Manβ4GlcNAc-R
    α3
R1-Man
	GlcNAc-T I
	
	α2,6-sialyl-T to Galβ4GlcNAc-
		Vella et al., 1984
		
		Joziasse et al., 1984

73

Table VI. Substrate specificity: branch specificity.

Branched oligosaccharide	Enzyme	Preference	References
R1—Mα6 ⎢ ⎢ Mβ4—R ⎢ R2—Mα3	β4—GalT α6—Sialyl—T α3—GalT	Mα3 arm Mα3 arm Mα6 arm	Paquet et al., 1984; Narasimhan et al., 1985 Van den Eijnden et al., 1980 Blanken et al., 1984
R1—Gnβ6 ⎢ Gal—R ⎢ R2—Gnβ3	α3—GalT β4—GalT	Gnβ6 arm Gnβ6 arm	Blanken et al., 1984 Blanken et al., 1982

argument somewhat suspect. The indications are that branch specificities found in vitro reflect structural variations that occur in vivo.

Substrate Specificity: Role of Polypeptide

The in vitro glycosyltransferase studies described above were carried out either with well characterized low molecular weight acceptors or relatively poorly characterized glycoprotein acceptors. It is as yet not possible to obtain a pure large polypeptide with a homogeneous oligosaccharide at a single amino acid position. Thus, the role of the polypeptide in glycosyltransferase action has not been properly studied. Although in vitro transferase studies are usually supported by the structures isolated from tissues, it is clear that the polypeptide backbone can modify oligosaccharide assembly.

For example, hen ovalbumin and ovomucoid (Kobata, 1984) are made in the same organ and yet ovalbumin contains mainly high mannose and bisected hybrid structures whereas ovomucoid is very rich in bisected highly branched truncated (mainly GlcNAc terminal) complex structures. Both glycoproteins are rich in bisected structures and indeed we have found that hen oviduct is a rich source of GlcNAc-transferase III (Narasimhan, 1982) whereas rat and pig liver lack this enzyme. Why the difference between the two proteins? One explanation is that the two proteins pass through separate and different assembly lines. Figure 12 suggests another possible explanation. The polypeptide backbone may interact with oligosaccharide in such a way as to make ovalbumin a good substrate and ovomucoid a poor substrate for GlcNAc-transferase III at the 5-mannose stage whereas ovomucoid becomes a good substrate for the transferase at the 3-mannose stage.

The reason for the under-galactosylation of ovomucoid and various other glycoproteins is not known. It is interesting that Narasimhan et al. (1985) found that not only does a Gal residue on the Manα1-3 arm inhibit GlcNAc-transferase III but β4-Gal-transferase is inhibited by a bisecting GlcNAc; this is perhaps not surprising on stereochemical grounds (Fig. 8). Thus prior addition of a bisecting GlcNAc may lead to under-galactosylation of the GlcNAcβ1-2Manα1-3- and GlcNAcβ1-2Manα1-6- antennae relative to the other antennae; the structures reported for ovomucoid tend to support this hypothesis (Kobata, 1984).

Savvidou et al. (1984) have provided evidence for the role of polypeptide-oligosaccharide interaction in the synthetic process. They showed that a human IgG$_1$k monoclonal protein (Hom) had oligosaccharide moieties not only at the usual position (Asn-297 of the H chain) but also at Asn-107 on the L chain. All oligosaccharides were of the complex biantennary N-glycosyl type with core fucose residues and, therefore, processing at both sites had entered the Golgi phase (Fig. 3). At this point, the H and L chains are already covalently associated and therefore both glycosylation sites must pass through the same endomembrane assembly line and encounter the same array of processing enzymes. However, the

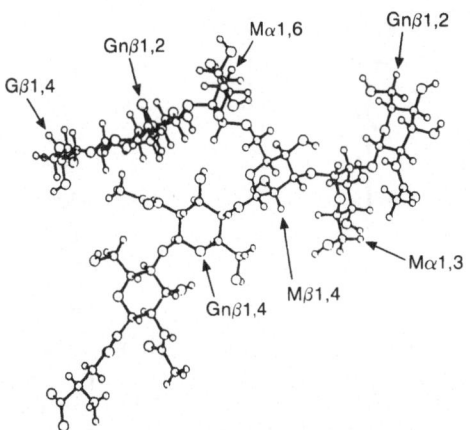

Fig. 11. Computer-drawn structure for GGn, ω = +180°, based on NMR data of Carver and Brisson (1984). Abbreviations as for Figs. 2 and 8. The diagram was prepared by D. Cummings.

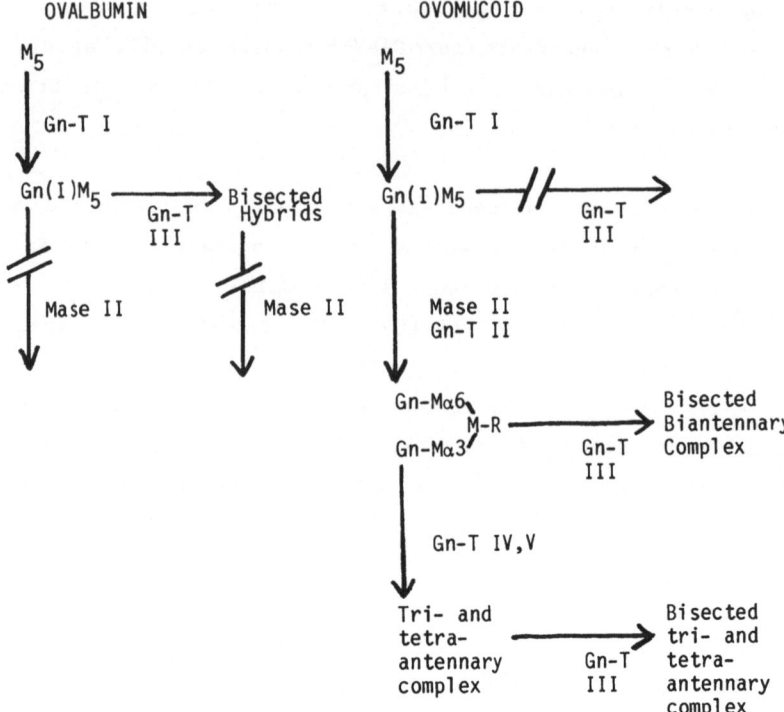

Fig. 12. Suggested mechanism for the difference between ovalbumin and ovomucoid (see text). M, Man; Gn, GlcNAc; Gn-T, GlcNAc-transferase; Mase, mannosidase.

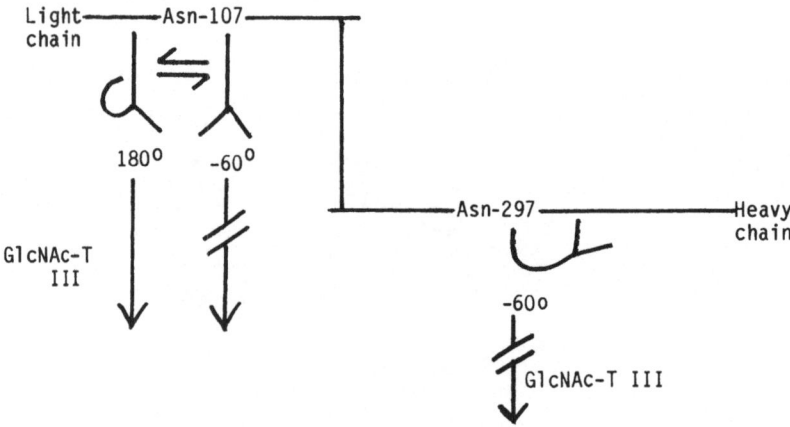

Fig. 13. A monoclonal IgG studied by Savvidou et al., 1984, was found to have bisected oligosaccharides at Asn-107 of the light chains but mainly non-bisected oligosaccharides at Asn-297 of the heavy chains. Interaction of the oligosaccharide at Asn-297 with the protein backbone is believed to be the basis of this effect (see text).

Asn-107 site was occupied entirely by mono- and di-sialyl bisected biantennary complex structures, whereas the Asn-297 site had mainly (73%) non-bisected sialylated biantennary structures and 27% bisected neutral biantennary structures. Since galactosylation and sialylation occur after the action of GlcNAc-transferase III (Fig. 5), it cannot be argued that the oligosaccharide at Asn-297 is buried in the protein and therefore not accessible to GlcNAc-transferase III. The clone of cells making this IgG was obviously rich in GlcNAc-transferase III since the Asn-107 site is fully bisected. Why is the Asn-297 site so poorly bisected?

The explanation hinges on the assumption that GlcNAc-transferase III cannot act on the substrate conformation in which the torsional angle ω about the C5-C6 bond of the Manα1-6 linkage equals $-60°$ (the two antennae form a Y; Fig. 10). The substrate for GlcNAc-transferase III is the conformation in which the angle ω is $+180°$ (the Manα1-6 arm is bent backwards towards the core; Fig. 11). The basis for this assumption (Brisson and Carver, 1983) is that whereas the non-bisected biantennary oligosaccharide exists in an equilibrium between two thermodynamically favorable conformations ($\omega = -60°$ or $+180°$; Figs. 10, 11), the bisected biantennary structure is found exclusively in the orientation corresponding to $\omega = +180°$ (Fig. 8). It was suggested (Brisson and Carver, 1983) that the oligosaccharide at Asn-107 exists in equilibrium between both conformations and is therefore accessible to GlcNAc-transferase III, whereas most of the oligosaccharide at Asn-297 is stabilized by interaction with the polypeptide in the $\overset{\text{v}}{\omega} = -60°$ conformation and cannot serve as a substrate for GlcNAc-transferase III. The reason for the 27% bisected structures at Asn-297 is not clear. Figure 13 illustrates the hypothesis.

Substrate Availability and other Factors

It has already been pointed out that the endomembrane system is highly compartmentalized (Figs. 3 and 5). The analogy to an assembly line is obvious. The transferases and glycosidases which shape the oligosaccharide structure are arranged along the endomembrane in the order in which they are needed. This arrangement minimizes the need for substrate to seek its proper enzyme and also exercises a certain amount of control over the assembly process.

For example, the fact that the GlcNAc-transferases which control branching (Fig. 6) are in the medial-Golgi compartment, whereas the terminal glycosyltransferases (Gal- and sialyl-transferases) are in the trans-Golgi (Fig. 5) means that branching is determined before the antennae are completed. This control must be important since the cell uses yet another mechanism to ensure that it is carried out. Once the glycoprotein enters the trans-Golgi, it is galactosylated (Fig. 5) and this immediately prevents further action of GlcNAc-transferases II, III, IV, and V and of α6-fucosyltransferase. Thus, even if there are small amounts of these medial-Golgi enzymes in the trans-Golgi, they are quickly prevented from acting. The reason for this stringent control is not known.

Several laboratories have reported transport proteins within the membranes of the endomembrane system which allow the movement of nucleotide-sugars from the cytoplasm into the luminal spaces. These transport mechanisms are indicated in Figure 3 and obviously play a major role in controlling glycoprotein synthesis. Defects in these transport proteins have been shown to lead to disruption of glycoprotein and glycolipid synthesis. However, a discussion of this important topic is beyond the scope of this review.

Finally, a variety of other substrate-level factors may exert control on the synthetic process: 1) the pH of the lumenal compartment may affect transferase activity, 2) the availability of cations is usually needed for transferase action, 3) the presence in the Golgi of pyrophosphatases allows conversion of the nucleotide diphosphate formed after glycosyl-transferase action into nucleotide monophosphate; the nucleotide monophosphate is exported out of the Golgi apparatus in a counter-transport system with nucleotide-sugars, 4) several laboratories have reported on the importance of phospholipids in glycosyltransferase activity, and 5) post-translational modifications of glycosyl-transferases, e.g., phosphorylation, may play a role in modifying their activity. These factors will not be discussed in this review.

CONCLUSIONS

Many different factors enter into the construction of a complex carbohydrate. It is therefore not surprising that a large variety of structures is synthesized by the cell. We are beginning to understand the biosynthetic machinery involved. However, it has not been ruled out that much of the microheterogeneity of these structures is indeed a random process with little functional significance. The arguments presented in this review indicate that a great deal of control is exerted and raises the hope that as we understand the process better, the functional aspects will become more evident.

What is needed are new approaches to the problem. Figure 1 emphasizes the genetic aspects and suggests that one approach must be to obtain genetic probes for the enzyme machinery involved in biosynthesis. Within the next few years, a battery of genetic probes will undoubtedly become available for the detailed analysis of the biosynthetic pathways outlined in Figures 4 and 5.

Another area of future research is the three-dimensional aspect of protein-oligosaccharide interaction, particularly the role of branching. Such studies are important not only for model systems such as lectin-oligosaccharide interactions (Carver et al., 1985), but also for proper understanding of how the biosynthetic machinery operates, how complex carbohydrates at the cell surface control the interactions of a cell with its environment, and, in fact, for all functional aspects of complex carbohydrates.

In conclusion, although tremendous advances have been made in recent years in the structure and biosynthesis of complex carbohydrates, we are only beginning to attack the problem of function.

REFERENCES

Allen, S. D., Tsai, D.. and Schachter, H., 1984, Control of glycoprotein synthesis. X. The in vitro synthesis by hen oviduct membrane preparations of hybrid asparagine-linked oligosaccharides containing 5 mannose residues. J. Biol. Chem., 259:6984-6990.

Beyer, T. A., Sadler, J. E., Rearick, J. I., Paulson, J. C., and Hill R. L., 1981, Glycosyltransferases and their use in assessing oligosaccharide structure and structure-function relationships, Adv. in Enzymol., 52:23-175.

Blanken, W. M., Hooghwinkel, G. J. M., and Van den Eijnden, D. H., 1982, Biosynthesis of blood-group I and i substances. Specificity of bovine colostrum β-N-acetyl-D-glucosaminide β1-4galactosyl-transferase, Eur. J. Biochem., 127:547-552.

Blanken, W. M., Van Vliet, A., and Van den Eijnden, D. H., 1984, Acceptor specificity of calf thymus N-acetyllactosaminide α1-3-galactosyltransferase, in: "Carbohydrates, 1984", Abstracts of the XIIth International Carbohydrate Symposium, J. F. G. Vliegenthart, J. P. Kamerling, and G. A. Veldink, eds., Vonk Publishers, Utrecht, p. 229.

Brisson, J-R., and Carver, J. P., 1983, The relation of three-dimensional structure to biosynthesis in the N-linked oligosaccharides, Can. J. Biochem. Cell Biol., 61:1067-1078.

Brockhausen, I., Rachaman, E. S., Matta, K. L., and Schachter, H., 1983a, Mucin Synthesis. IV. The separation by high performance liquid chromatography of phenyl, benzyl and ortho-nitrophenyl oligosaccharide glycosides. Analysis of substrates and products for four N-acetyl-D-glucosaminyltransferases involved in mucin synthesis, Carbohyd. Res., 120:3-16.

Brockhausen, I., Williams, D., Matta, K. L., Orr, J., and Schachter, H., 1983b, Mucin Synthesis. III. UDP-GlcNAc:Galβ1-3(GlcNAcβ1-6)GalNAc-R (GlcNAc to Gal) β3-N-acetylglucosaminyltransferase, an enzyme in porcine gastric mucosa involved in the elongation of mucin-type oligosaccharides, Can. J. Biochem. Cell Biol., 61:1322-1333.

Brockhausen, I., Orr, J., and Schachter, H., 1984, Mucin synthesis. V., The action of pig gastric mucosal UDP-GlcNAc:Galβ1-3(R_1)GalNAc-R_2 (GlcNAc to Gal) β3-N-acetylglucosaminyltransferase on high molecular weight substrates, Can. J. Biochem. Cell Biol., 62:1081-1090.

Brockhausen, I., Matta, K. L., Orr, J., and Schachter, H., 1985, Mucin Synthesis. VI. UDP-GlcNAc:GalNAc-R β3-N-acetylglucosaminyltransferase and UDP-GlcNAc:GlcNAcβ1-3GalNAc-R (GlcNAc to GalNAc) β6-N-acetylglucosaminyltransferase from pig and rat colon mucosa, Biochemistry, 24:1866-1874.

Carver, J. P., and Brisson, J-R., 1984, The three-dimensional structure of N-linked oligosaccharides, in: "Biology of Carbohydrates", Vol.2, V. Ginsburg and P. W.Robbins, eds., John Wiley & Sons, New York, pp.289-331.

Carver, J. P., Mackenzie, A. E., and Hardman, K. D., 1985, Molecular model for the complex between concanavalin A and a biantennary-complex class glycopeptide, Biopolymers, 24:49-63.

Cummings, R. D., Trowbridge, I. S., and Kornfeld, S., 1982, A mouse lymphoma cell line resistant to the leukoagglutinating lectin from Phaseolus vulgaris is deficient in UDP-GlcNAc:α-D-mannoside β1,6 N-acetylglucosaminyltransferase, J. Biol. Chem., 257:13421-13427.

Dunphy, W. G., and Rothman, J. E., 1985, Compartmental organization of the Golgi stack, Cell, 42:13-21.

Feizi, T., 1984, Monoclonal antibodies reveal saccharide structures of glycoproteins and glycolipids as differentiation and tumor-associated antigens, Biochem. Soc. Trans., 12:545-549.

Fenderson, B. A., Zehavi, U., and Hakomori, S-I., 1984, A multivalent lacto-N-fucopentaose III-lysyllysine conjugate decompacts preimplantation-stage mouse embryos, while the free oligosaccharide is ineffective, J. Exp. Med., 160:1591-1596.

Fiat, A. M., Jolles, P., Vliegenthart, J. F. G., and Van Halbeek, H., 1984, Structural aspects concerning the prosthetic sugar groups of bovine and human kappa-caseinoglycopeptides, in: "Carbohydrates, 1984", Abstracts of the XIIth International Carbohydrate Symposium, J. F. G. Vliegenthart, J. P. Kamerling, and G. A. Veldink, eds., Vonk Publishers, Utrecht, p.426.

Fukuda M., 1985, Cell surface glycoconjugates as onco-differentiation markers in hematopoietic cells, Biochem. Biophys. Acta, 780:119-150.

Fukuda, M., and Fukuda, M. N., 1984, Cell surface glycoproteins and carbohydrate antigens in development and differentiation of human erythroid cells, in: "The Biology of Glycoproteins", R. J. Ivatt, ed., Plenum Press, New York, pp.183-234.

Gleeson, P. A., and Schachter, H., 1983, Control of Glycoprotein Synthesis. VIII. UDP-GlcNAc:GnGn (GlcNAc to Manα1-3) β4-N-acetylglucosaminyltransferase IV, an enzyme in hen oviduct which adds GlcNAc in β1-4 linkage to the α1-3-linked Man residue of the trimannosyl core of N-glycosyl oligosaccharides to form a tri-antennary structure, J. Biol. Chem., 258:6162-6173.

Gooi, H. C., Feizi, T., Kapadia, A., Knowles, B. B., Solter, D., and Evans, J. M., 1981, Stage-specific embryonic antigen involves α1-3fucosylated type 2 blood group chains, Nature (London), 292:156-158.

Hakomori, S-I., 1985, Aberrant glycosylation in cancer cell membranes as focused on glycolipids: Overview and perspectives, Cancer Res., 45:2405-2414.

Hakomori, S-I., Fukuda, M., and Nudelman, E., 1982, Role of cell surface carbohydrates in differentiation: Behavior of lactosaminoglycan in glycolipids and glycoproteins, in: "Teratocarcinoma and embryonic cell interactions", T. Muramatsu, G. Gachelin, A. A. Moscona, and Y. Ikawa, eds., Japan Scientific Soc. Press, Tokyo, pp.179-200.

Harpaz, N., and Schachter, H., 1980, Control of glycoprotein synthesis. V. Processing of asparagine-linked oligosaccharides by one or more rat liver Golgi α-D-mannosidases dependent on the prior action of

UDP-N-acetylglucosamine:α-D-mannoside   β-2-N-acetylglucosaminyl-transferase I. J. Biol. Chem., 255:4894-4902.

Hitoi, A., Yamashita, K., Ohkawa, J., and Kobata, A., 1984, Application of a Phaseolus vulgaris erythroagglutinating lectin agarose column for the specific detection of human hepatoma gamma-glutamyl transpeptidase in serum. Gann, 75:301-304.

Ivatt, R. J., 1984, Role of glycoproteins during early mammalian embryogenesis, in: "The Biology of Glycoproteins", R. J. Ivatt, ed., Plenum Press, New York, pp.95-181.

Joziasse, D. H., Schiphorst, W. E. C. M., Van den Eijnden, D. H., Van Kuik, J. A., Van Halbeek, H., and Vliegenthart, J. F. G. , 1984, Branch specificity of bovine colostrum α2-6-sialyltransferase: Interaction with biantennary oligosaccharides and glycopeptides of N-glycosylproteins, in: "Carbohydrates, 1984", Abstracts of the XIIth International Carbohydrate Symposium, J. F. G. Vliegenthart, J. P. Kamerling and G. A. Veldink, eds., Vonk Publishers, Utrecht, p. 228.

Kannagi, R., Levery S. B., and Hakomori, S-I., 1983, Sequential change of carbohydrate antigen associated with differentiation of murine leukemia cells: Ii antigenic conversion and shifting of glycolipid synthesis, Proc. Nat. Acad. Sci. USA, 80:2844-2848.

Kobata, A., 1982, "Structures of the sugar chains of cell surface glycoproteins, in: "Structure, Dynamics and Biogenesis of Biomembranes", R. Sato and S. Ohnishi, eds., Japan Scientific Society Press, Tokyo, pp. 97-112.

Kobata, A., 1984, The carbohydrates of glycoproteins, in: "Biology of Carbohydrates", Vol. 2, V. Ginsburg and P. W. Robbins, eds., John Wiley & Sons, New York, pp. 87-161.

Kobata, A., and Yamashita, K., 1984, The sugar chains of gamma-glutamyl transpeptidase, Pure & Appl. Chem., 56:821-832.

Kornfeld, R., and Kornfeld S., 1985, Assembly of asparagine-linked oligosaccharides, Ann. Rev. Biochem., 54:631-664.

Longmore, G. D., and Schachter, H., 1982, Control of Glycoprotein Synthesis. VI. Product identification and substrate specificity studies of the GDP-L-Fucose:2-acetamido-2-deoxy-β-D-glucoside (Fuc to Asn-linked GlcNAc) 6-α-L-fucosyltransferase in a Golgi-rich fraction from porcine liver, Carbohydrate Research, 100:365-392.

Narasimhan, S., 1982, Control of glycoprotein synthesis. VII. UDP-GlcNAc:glycopeptide β4-N-acetylglucosaminyltransferase III, an enzyme in hen oviduct which adds GlcNAc in β1-4 linkage to the β-linked mannose of the trimannosyl core of N-glycosyl oligosaccharides, J. Biol. Chem., 257:10235-10242.

Narasimhan, S., Shirley, M., Hewitt, J. M., Freedman, M. H., Gelfand, E. W., and Schachter, H., 1983, UDP-GlcNAc:GnGn (GlcNAc to Manβ1-4) β4-GlcNAc-transferase III (Gn-T III) in human tissues, Fed.Procs., 42:2199.

Narasimhan, S., Freed, J. C., and Schachter, H., 1985, Control of glycoprotein synthesis. XI. Bovine milk UDP-galactose: N-acetylglucosamine β4-galactosyltransferase catalyzes the preferential transfer of galactose to the GlcNAcβ1,2Manα1,3- branch of both bisected and non-bisected complex biantennary asparagine-linked oligosaccharides, Biochemistry, 24:1694-1700.

Paquet, M. R., Narasimhan, S., Schachter, H., and Moscarello, M. A., 1984, Branch specificity of purified rat liver Golgi UDP-galactose: N-acetylglucosamine β-1,4-galactosyltransferase. Preferential transfer of galactose on the GlcNAcβ1,2-Manα1,3- branch of a complex biantennary Asn-linked oligosaccharide. J. Biol. Chem., 259:4716-4721.

Piller, F., Cartron, J-P., Maranduba, A., Veyrieres, A., Leroy. Y., and Fournet, B., 1984, Biosynthesis of blood group I antigens. Identification of a UDP-GlcNAc:GlcNAc-β1-3Gal (GlcNAc to Gal) β1-6-N-acetylglucosaminyltransferase in hog gastric mucosa, J. Biol. Chem., 259:13385-13390.

Rothman, J. E., 1985, The compartmental organization of the Golgi apparatus, Scientific American, 253:74-89.

Sadler, J. E., 1984, Biosynthesis of glycoproteins: Formation of O-linked oligosaccharides, in: "Biology of Carbohydrates", Vol.2, V. Ginsburg and P. W. Robbins, eds., John Wiley & Sons, New York, pp. 199-288.

Savvidou, G., Klein, M., Grey., A. A., Dorrington, K. J., and Carver, J. P., 1984, Possible role for peptide-oligosaccharide interactions in differential oligosaccharide processing at asparagine-107 of the light chain and asparagine-297 of the heavy chain in a monoclonal IgG_{1k}, Biochemistry, 23:3736-3740.

Schachter, H., and Williams, D., 1982, Biosynthesis of mucus glycoproteins, in: "Mucus in Health and Disease", Vol.II, E. N. Chantler, J. B. Elder, and M. Elstein, eds., Adv. Exp. Medicine and Biology, Vol. 144, Plenum Press, New York and London, pp. 3-28.

Schachter H., McGuire, E. J., and Roseman, S., 1971, Sialic Acids. XIII. A uridine diphosphate D-galactose: mucin galactosyltransferase from porcine submaxillary gland, J. Biol. Chem., 246:5321-5328.

Schachter, H., Narasimhan, S., Gleeson, P., and Vella, G. J., 1983, Control of branching during the biosynthesis of asparagine-linked oligosaccharides, Can. J. Biochem. Cell Biol., 61:1049-1066.

Schachter, H., Narasimhan, S., Gleeson, P., Vella, G., and Brockhausen, I., 1985, Glycosyltransferases involved in the biosynthesis of protein-bound oligosaccharides of the asparagine-N-acetyl-D-glucosamine and serine(threonine)-N-acetyl-D-galactos-amine types, in: "The Enzymes of Biological Membranes," Second Edition, Volume 2, Biosynthesis and Metabolism, A. N. Martonosi, ed., Plenum Press, New York, pp. 227-277.

Snider, M. D., 1984, Biosynthesis of glycoproteins: Formation of N-linked oligosaccharides, in: "Biology of Carbohydrates", Vol. 2, V. Ginsburg and P. W.Robbins, eds., John Wiley & Sons, New York, pp. 163-198.

Testa, U., Henri, A., Bettaieb, A., Titeux, M., Vainchenker, W., Tonthat, H., Docklear, M. C., and Rochant, H., 1982, Regulation of i- and I-antigenic expression in the K562 cell line, Cancer Res., 42:4694-4700.

Van den Eijnden, D. H., Joziasse, D. H., Dorland, L., Van Halbeek, H., Vliegenthart, J. F. G., and Schmid, K., 1980, Specificity in the enzymic transfer of sialic acid to the oligosaccharide branches of bi- and triantennary glycopeptides of α_1-acid glycoprotein, Biochem. & Biophys. Res. Commun., 92:839-845.

Vella, G. J., Paulsen, H., and Schachter, H., 1984, Control of glycoprotein synthesis. IX. A terminal Manα1-3Manβ1- sequence in the substrate is the minimum requirement for UDP-N-acetylglucos-amine:α-D-mannoside (GlcNAc to Manα1-3-) β2-N-acetylglucosaminyl-transferase I, Can. J. Biochem. Cell5 Biol., 62:409-417.

Williams, D., and Schachter, H., 1980, Mucin synthesis. I. Detection in canine submaxillary glands of an N-acetylglucosaminyltransferase which acts on mucin substrates, J. Biol. Chem., 255:11247-11252.

Williams, D., Longmore, G., Matta, K. L., and Schachter, H., 1980, Mucin synthesis. II. Substrate specificity and product identification studies on canine submaxillary gland UDP-N-acetylglucosamine: Galβ1-3GalNAc (GlcNAc to GalNAc) β6-N-acetylglucosaminyl-transferase, J. Biol. Chem., 255:11253-11261.

Yamashita, K., Kamerling, J. P., and Kobata, A., 1982, Structural study of the carbohydrate moiety of hen ovomucoid. Occurrence of a series of pentaantennary complex-type asparagine-linked sugar chains, J. Biol. Chem., 257:12809-12814.

Yamashita, K., Hitoi, A., Matsuda, Y., Tsuji, A., Katunuma, N., and Kobata, A., 1983a, Structural studies of the carbohydrate moieties of rat kidney gamma-glutamyltranspeptidase. An extremely heterogeneous pattern enriched with nonreducing terminal N-acetylglucosamine residues, J. Biol. Chem., 258:1098-1107.

Yamashita, K., Hitoi, A., Taniguchi, N., Yokosawa, N., Tsukada, Y., and Kobata, A., 1983b, Comparative study of the sugar chains of gamma-glutamyltranspeptidases purified from rat liver and rat AH-66 hepatoma cells, Cancer Res., 43:5059-5063.

Yamashita, K., Hitoi, A., Tateishi, N., Higashi, T., Sakamoto, Y., and Kobata, A., 1983c, Organ-specific difference in the sugar chains of gamma-glutamyltranspeptidase, Arch. Biochem. Biophys., 225:993-996.

Yamashita, K., Kamerling, J. P., and Kobata, A., 1983d, Structural studies of the sugar chains of hen ovomucoid. Evidence indicating that they are formed mainly by the alternate biosynthetic pathway of asparagine-linked sugar chains, J. Biol. Chem., 258:3099-3106.

Yamashita, K., Tachibana, Y., Hitoi, A., Matsuda, Y., Tsuji, A., Katunuma, N., and Kobata, A., 1983e, Difference in the sugar chains of two subunits and of isozymic forms of rat kidney gamma-glutamyltranspeptidase, Arch. Biochem. Biophys., 227:225-232.

Yamashita, K., Tachibana, Y., Shichi, H., and Kobata, A., 1983f, Carbohydrate structures of bovine kidney gamma-glutamyltranspeptidase, J. Biochem., 93:135-147.

Yamashita, K., Tachibana, Y., Hitoi, A., and Kobata, A., 1984, Sialic acid-containing sugar chains of hen ovalbumin and ovomucoid, Carbohydrate Res., 130:271-288.

Yamashita, K., Hitoi, A., Tateishi, N., Higashi, T., Sakamoto, Y., and Kobata, A., 1985a, The structures of the carbohydrate moieties of mouse kidney gamma-glutamyltranspeptidase: Occurrence of X-antigenic determinants and bisecting N-acetylglucosamine residues, Arch. Biochem. Biophys., 240:573-582.

Yamashita, K., Tachibana, Y., Ohkura, T., and Kobata, A., 1985b, Enzymatic basis for the structural changes of asparagine-linked sugar chains of membrane glycoproteins of baby hamster kidney cells induced by polyoma transformation, J. Biol. Chem., 260:3963-3969.

Zielenski, J., and Koscielak, J., 1983, Sera of I subjects have the capacity to synthesize the branched GlcNAc-β(1-6)[GlcNAc-β(1-3)]Gal ...structure, FEBS Letters, 163:114-118.

MECHANISMS AND REGULATION OF TSH GLYCOSYLATION

Neil Gesundheit and Bruce D. Weintraub

Molecular, Cellular and Nutritional Endocrinology Branch
National Institute of Arthritis, Diabetes, and
  Digestive and Kidney Diseases
National Institutes of Health
Bethesda, Maryland 20892

INTRODUCTION

Thyroid-stimulating hormone (TSH) is a glycoprotein composed of two noncovalently linked subunits, α and β. The structure of TSH from a variety of species has been well characterized, including the amino acid sequence and carbohydrate composition (Pierce and Parsons, 1981). The α-subunit of bovine TSH has a molecular weight of 13,600, of which 10,800 is composed of a protein core of 96 amino acids and 2,800 (21%) represents two oligosaccharide units linked to asparagine residues. Bovine TSH-β has a molecular weight of 14,700, of which 13,100 is comprised of a protein core of 113 amino acids and 1,600 (12%) represents one asparagine-linked oligosaccharide unit.

TSH is structurally related to the pituitary gonadotropins, luteinizing hormone (LH), and follicle-stimulating hormone (FSH), as well as to the placental hormone chorionic gonadotropin (CG). Within a single species, the α-subunits from each of these glycoprotein hormones are virtually identical, while the β-subunits are unique and confer immunologic and biologic specificity. Full hormonal activity is dependent on proper assembly and carbohydrate processing of the TSH heterodimer, while the free subunits are essentially devoid of receptor-binding and biologic activity.

The α- and β-subunits of TSH are synthesized from separate mRNA's coded by DNA contained on separate chromosomes (Chin et al., 1981a;

Kourides et al., 1984). The nucleotide sequence for TSH-α and TSH-β has confirmed the presence of an amino terminal "signal" peptide (but no other precursor forms) for each subunit, previously identified and partially characterized by cell-free translation, peptide analysis, and micro-sequencing studies (Giudice et al., 1979; Giudice and Weintraub, 1979). The initial steps in TSH biosynthesis are shown schematically in Figure 1. TSH subunit messenger RNA levels appear to be regulated to a major extent by thyroid hormone (Gurr et al., 1983; Shupnik et al., 1983) but to little or no extent by thyrotropin-releasing hormone (TRH; Kourides et al., 1984; Lippman et al., 1986).

INITIAL GLYCOSYLATION OF TSH

We and others have used various strains of mouse thyrotropic tumors, first developed by Jacob Furth (Furth et al., 1973), for detailed studies of TSH biosynthesis and processing. These tumors synthesize and secrete large amounts of TSH, and most lines do not synthesize other pituitary hormones. Initial glycosylation of TSH subunits occurs in the endoplasmic reticulum by the co-translational transfer en bloc of high mannose oligosaccharide units from the dolichol-phosphate carrier, as has been demonstrated for other glycoproteins (Behrens and Leloir, 1970). However, studies from our (Weintraub and Stannard, 1978; Weintraub et al., 1980) and other laboratories (Chin and Habener, 1981; Chin, et al., 1981b) have suggested a two-step process for α-subunit glycosylation. TSH biosynthesis was studied in thyrotropic tumor cells using selective immunoprecipitation combined with sodium dodecyl sulfate gradient gel electrophoresis under reducing conditions (Weintraub et al, 1980). After a 10-minute pulse with [^{35}S]methionine, the predominant [^{35}S]- labeled α form was of apparent molecular weight 18,000 with a second component of 21,000. When the pulse was followed by variable chase periods with excess unlabeled methionine, the 18,000 α form was converted progressively to the 21,000 form, implying a precursor-product relationship between the two. [^{35}S]-labeled TSH-β subunit began to combine selectively with the larger (21,000) α form within 10 to 30 minutes of the chase period, and secretion of the TSH dimer was observed between 60 and 240 minutes of chase. At early labeling times both α-subunits, of molecular weights 18,000 and 21,000, were converted to apparent molecular weight 11,000 by endoglycosidase H (endo H) treatment, consistent with the weight of non-glycosylated α-subunit cleaved of its signal peptide (Chin and Habener, 1981; Magner and Weintraub, 1982; Weintraub et al., 1983). These data suggest that for α-subunit glycosylation, one asparagine-linked high

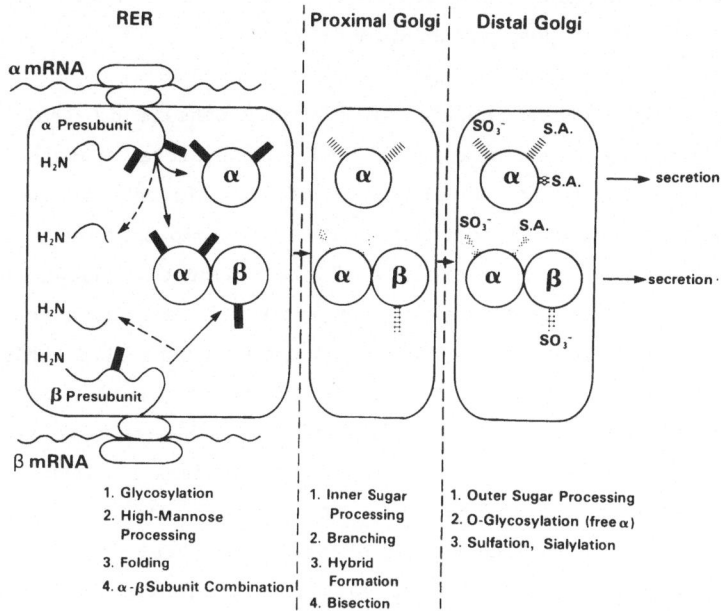

RER	Proximal Golgi	Distal Golgi
1. Glycosylation	1. Inner Sugar Processing	1. Outer Sugar Processing
2. High-Mannose Processing	2. Branching	2. O-Glycosylation (free α)
3. Folding	3. Hybrid Formation	3. Sulfation, Sialylation
4. α-β Subunit Combination	4. Bisection	

Fig. 1. Synthesis and carbohydrate processing of thyroid-stimulating hormone (TSH; reviewed in Weintraub et al., 1985). The α- and β-subunits are synthesized from separate mRNA's coded on separate chromosomes. Each presubunit contains a signal peptide that is cleaved co-translationally upon insertion of the protein into the rough endoplasmic reticulum (RER). Glycosylation of the two subunits occurs either co-translationally or immediately post-translationally within the RER. The biochemical signals that direct combination are poorly understood; however, part of the carbohydrate moiety has been shown to be needed for subunit combination, either by directing subunit folding or by providing specific structural determinants for combination. α-β combination begins in the RER and continues throughout the secretory pathway. The oligosaccharide moieties on TSH-α, TSH-β, and free α-subunits incorporate radiolabeled carbohydrate precursors differentially, suggesting different sugar composition of each secreted subunit. While there has not been shown to be clear compartmentalization of carbohydrate processing in the Golgi, it is generally observed that initial sugar trimming occurs in the RER and that further processing, including addition of GlcNAc residues that determine the number of branches, requires translocation of this latter structure into the Golgi. The addition of the terminal "capping" residues, sulfate and sialic acid, confers additional negative charge to TSH and occurs late in the secretory pathway. Neuroendocrine tissue may contain several distal secretory pathways, some dependent on and some independent of the formation of secretory granules.

mannose carbohydrate unit is added co-translationally, while the second is added soon thereafter post-translationally.

Another distinctive feature of TSH biosynthesis is that the high-mannose processing of asparagine-linked carbohydrate appears to differ substantially from that of total glycoproteins. In contrast to the rapid processing of membrane glycoproteins to final complex forms, we found a slower processing rate of secretory TSH carbohydrate. Using pulse-chase techniques in a mouse thyrotropic tumor system, it was shown that after an 11-minute pulse, most α- and β-subunit precursors were of the high mannose type, as shown by sensitivity to endo H (Weintraub et al., 1983). At 30 minutes of chase, only 19% of intracellular α- subunits were resistant to endo H; and while the percentage of resistant forms increased progressively with increasing chase time, they did not comprise the majority (76%) until 18 hours of chase. Total cellular glycoproteins showed more advanced processing at each time point studied. Media forms of TSH were predominantly resistant to endo H at all time points, suggesting that under physiologic conditions TSH carbohydrate is processed to complex forms prior to secretion.

In contrast to the slower rate of processing of TSH carbohydrate to endo H resistant forms compared to total cellular glycoproteins, the initial post-translational processing of TSH carbohydrate in the endoplasmic reticulum appears to occur rapidly. In order to study the very earliest processing steps, TSH subunit glycosylation was compared to that of total cell glycoproteins in mouse thyrotropic tumors utilizing metabolic labeling with [^{3}H]mannose, [^{3}H]galactose, or [^{3}H]glucose in pulse and pulse-chase experiments (Ronin et al., 1984). Oligosaccharides were isolated from dolichol-phosphate intermediates by lipid extraction and mild acid hydrolysis; and from TSH subunits and total proteins by selective immunoprecipitation and acid precipitation, respectively. Proteins thus isolated were digested with trypsin and treated with endo H. Oligosaccharide structure was assessed by migration in paper chromatography compared to known standards, relative incorporation of different precursors, as well as susceptibility to α-mannosidase. At 60 minutes, lipid-linked oligosaccharides were comprised of $Glc_{3-2}Man_9GlcNAc_2$, $Man_{9-8}GlcNAc_2$, and $Man_5GlcNAc_2$. At 10 or 60 minutes of labeling, total cellular glycoproteins contained $Glc_3Man_9GlcNAc_2$, $Glc_1Man_9GlcNAc_2$, $Man_9GlcNAc_2$, $Glc_1Man_8GlcNAc_2$, $Man_8GlcNAc_2$, $Man_7GlcNAc_2$. In contrast, no $Glc_3Man_9GlcNAc_2$ was found on either TSH + α-, or β-subunits even during very brief (10 minute) pulses. Instead, primarily

$Man_9GlcNAc_2$ was found after a 10-minute pulse on both TSH + α-, and β-subunits. When the pulse was followed by a chase up to 2 hours, there was progressive increase in $Man_8GlcNAc_2$ in higher amount on TSH + α carbohydrate chains than on β. These data suggest a differential carbohydrate processing rate for secretory TSH subunits compared to certain nonsecretory cellular glycoproteins and also demonstrate a more rapid processing rate or a different transfer mechanism for TSH + α compared to free β subunits. Taken together with the endo H experiments summarized above, these data suggest that the earliest processing of TSH in the endoplasmic reticulum occurs rapidly, but that processing to endo H-resistant forms in the Golgi apparatus occurs slowly when compared to processing rates for nonsecretory glycoproteins (see Figure 1 for subcellular compartments that are believed to have specialized roles in carbohydrate processing).

In nearly all TSH-synthesizing tissue that has been studied, the α-subunit is synthesized in excess of β; part of the α-subunit appears to be stored uncombined in the cell while the rest is secreted as the free α-subunit (Weintraub and Stannard, 1978; Weintraub et al, 1980; Chin et al., 1981b). Various lines of evidence had suggested that secreted free α-subunit had a slightly higher molecular weight than those α forms combined with β (Weintraub et al, 1980). Recently it was reported (Parsons et al., 1983; Corless and Boime, 1985) that the free α-subunit derived from bovine pituitaries is glycosylated at an additional site, the threonine residue at position 43. This is compatible with studies in mouse thyrotropic tumor cells in which endo H treatment of TSH-α precursors resulted in two components, one of M_r = 13,000 in addition to the form of M_r = 11,000 that is usually detected (Weintraub et al, 1983). The form with M_r = 13,000 appeared to increase progressively with chase periods of 30 minutes to 18 hours after an 11-minute pulse and might represent an α-subunit that had undergone a further post-translational modification, such as glycosylation at threonine 43. Compositional data based on incorporation of labeled precursors (see next section) suggest still other differences in processing of free and combined α-subunits.

PROCESSING OF TSH CARBOHYDRATE AND α-β SUBUNIT COMBINATION

Recent studies have shown that incubation of mouse thyrotropic tumor cells with tunicamycin resulted in the synthesis of TSH subunits that were unable to combine (Weintraub et al., 1983). Since tunicamycin prevents en bloc glycosylation of proteins at asparagine sites, it was clear that at least some portion of the asparagine-linked carbohydrate was necessary for α-β combination. The use of subcellular fractionation techniques has provided new insight into the intracellular location of subunit combination and the distinct steps in carbohydrate processing (Magner and Weintraub, 1982; Magner et al., 1984). Subcellular separation of mouse thyrotropic tumor cells was achieved by sucrose-gradient centrifugation techniques and three microsomal fractions were compared: "heavy" microsomes, enriched in rough endoplasmic reticulum (RER); "intermediate" microsomes, enriched in proximal Golgi elements; and "light" microsomes, enriched in distal Golgi. Mouse thyrotropic tumor cells were pulsed with [^{35}S]methionine and chased for various periods with excess unlabeled methionine. Combination of high mannose forms of α of M_r = 21,000 and β of M_r = 18,000 began in the RER and continued in the smooth endoplasmic reticulum (SER), Golgi, and post-Golgi compartments. Attainment of endoglycosidase H resistance, signifying processing of the carbohydrate to complex type, was not a prerequisite for subunit combination. The β-subunit appears to be limiting in TSH biosynthesis and its combination with α began soon after subunits were synthesized within the RER. After only 20 minutes of pulse-labeling, 19% of TSH-β subunits were combined with TSH-α. Combination of the α- and β-subunits continued throughout the secretory pathway such that after 20 minutes of pulse-labeling and 60 minutes of chase, 61% of TSH-β subunits isolated from the distal Golgi fraction were combined.

The intracellular transport rates for the free and combined subunits of TSH differed when studied by these subcellular fractionation techniques. Free and combined β precursors moved from the RER to a light-density SER/Golgi fraction faster than free α-subunit precursors (Magner and Weintraub, 1982). This finding is consistent with detailed carbohydrate analysis showing more rapid carbohydrate processing for combined TSH subunits than for free α in unfractionated cellular homogenate (Ronin et al., 1984).

DIFFERENCES IN CARBOHYDRATE COMPOSITION OF SECRETED TSH-α,
TSH-β, AND FREE α-SUBUNITS

Previous studies of the carbohydrate composition of TSH have analyzed purified intrapituitary forms of the molecule (Condliffe et al., 1969; Liu and Ward, 1975; Sairam and Li, 1977). Little is known, however, about the carbohydrate composition of the more biologically relevant, secreted form of the hormone because of the difficulty in purifying it from serum. In order to explore whether each of the secreted TSH subunits has a distinctive carbohydrate composition, we incubated hypothyroid mouse hemi-pituitaries (Gesundheit and Weintraub, 1985) with a variety of tritiated sugar precursors: [^{3}H]glucosamine; [^{3}H]mannose; [^{3}H]L-fucose; and ^{3}H]N-acetylmannosamine, a specific precursor of the sialic acid [^{3}H]N-acetylneuraminic acid. In each case dual labeling was performed by co-incubation with [^{35}S]methionine, to permit normalization to the apoprotein moiety. Similarly, tissue was incubated with [^{35}S]sulfate and [^{3}H]methionine to examine possible differential labeling of subunits with [^{35}S]sulfate. Mouse hemi-pituitaries removed after decapitation of hypothyroid LAF mice were incubated for 18 hours in either methionine- or sulfate-free Dulbecco's Modified Eagles Medium in an environment of 95% O_2/5% CO_2. Secreted TSH present in medium was immunoprecipitated by anti-TSH-β antisera and dissociated into TSH-α and TSH-β subunits prior to electrophoresis in sodium dodecyl sulfate polyacrylamide gels (SDS-PAGE; Fig. 2). As is shown in Figure 3, the TSH-α to TSH-β ratio of [^{3}H]glucosamine and [^{3}H]mannose was 2.16 ± 0.62 and 2.48 ± 0.49, respectively, consistent with the expected value of 2.0, since there are two asparagine-linked carbohydrate chains on TSH-α and one on TSH-β. For [^{3}H]fucose, [^{3}H]N-acetylneuraminic acid, and [^{35}S]sulfate, however, the TSH-α to TSH-β ratio was significantly below 2.0 (0.54 ± 0.03, 0.75 ± 0.01, 1.54 ± 0.08, respectively), suggesting that the TSH-β chain contains more of these substituents per carbohydrate chain than does TSH-α. It should be pointed out that these calculations pertain to a hypothyroid model and that TSH carbohydrate structure may vary with the physiological state of the animal (Taylor and Weintraub, 1985a). Figure 4 illustrates the electrophoresis of TSH-α (22,000 peak) and TSH-β (18,000 peak) initially precipitated as the dimer and dissociated under the conditions of SDS-PAGE. Note that despite the presence of two asparagine- linked oligosaccharide chains on TSH-α and only one on TSH-β, there is nearly equal labeling of the two secreted subunits with [^{35}S]sulfate (triangles and solid line, lower panel) and more labeling of TSH-β than TSH-α with [^{3}H]N-acetylmannnosamine (circles and broken line, top panel). Since both sulfate and sialic acid contain negative charge, these data suggest that the TSH-β subunit contains more net negative charge on its oligosaccharide moiety than the TSH-α subunit. In summary, although the two TSH subunits, α and β, combine soon after synthesis (Magner and Weintraub, 1982) and are secreted as a dimeric protein, there are notable differences in the carbohydrate processing of the two subunits.

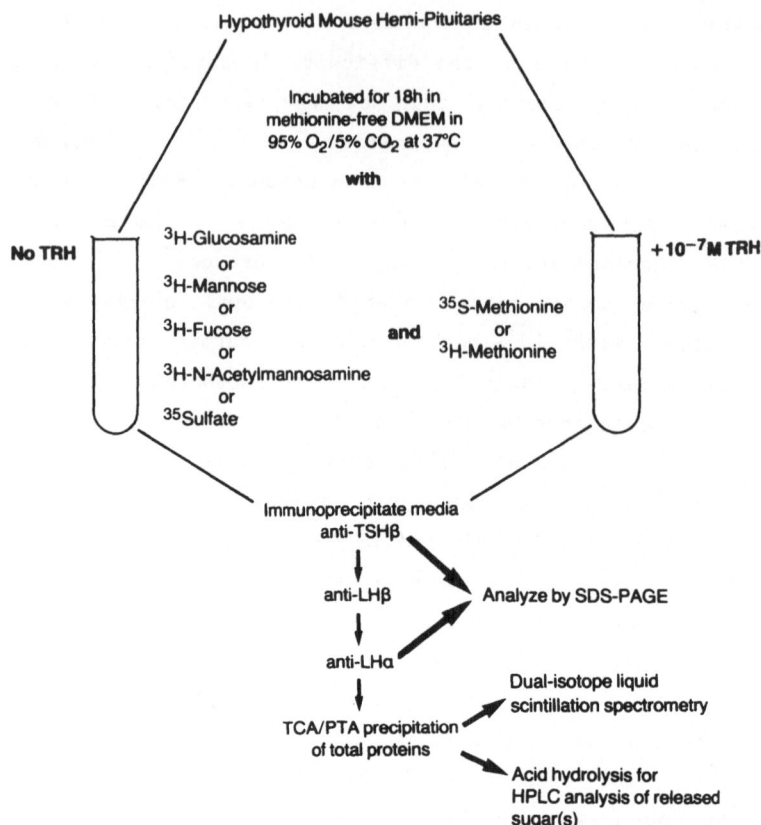

Fig. 2. Experimental method to analyze the incorporation of radiolabeled precursors into secreted TSH-α, TSH-β, and free α-subunits. LAF mice were decapitated and their pituitaries placed into methionine-free Dulbecco's Modified Eagles Medium (DMEM) in an environment of 95% O_2/5% CO_2. In separate tubes the tritium-labeled sugar precursors [^{3}H]glucosamine, [^{3}H]mannose, [^{3}H]fucose, [^{3}H]N-acetylmannosamine were co-incubated with [^{35}S]methionine to study incorporation of the sugar moiety relative to the apoprotein. [^{3}H]N-acetylmannosamine is a specific precursor, in this system, of [^{3}H]N-acetylneuraminic acid. To study [^{35}S]sulfate incorporation, $MgCl_2$ was substituted for $MgSO_4$ in the DMEM and samples were co-incubated with [^{35}S]sulfate and [^{3}H]methionine. Incubations were terminated after 18 hours, and the following immunoprecipitation sequence was followed: anti-TSH-β to precipitate the TSH dimer; anti-LH-β to precipitate LH; anti-LH-α to precipitate remaining free α-subunits. FSH is present in such small amount compared to the other glycoprotein hormones in this system that its removal is not necessary. Precipitated subunits were analyzed by electrophoresis in SDS-polyacrylamide gels. Total non-TSH secreted proteins were acid precipitated and hydrolyzed to release the incorporated sugars, which were analyzed by HPLC to check for possible interconversion of labeled sugar precursors.

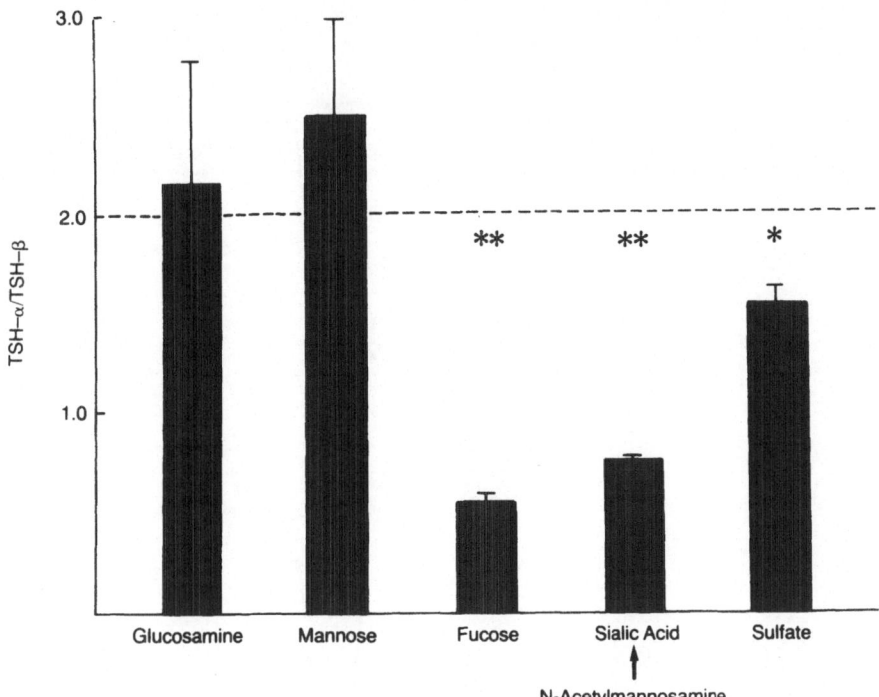

Fig. 3. Incorporation of sugar and sulfate precursors into secreted TSH. Biosynthetic labeling of secreted TSH was performed, as shown in the previous figure, to determine the relative incorporation of the sugar moieties into the TSH-α and TSH-β subunits. This method was chosen because more rigorous chemical methods require amounts of secreted TSH that are not currently available. Since TSH-α contains two N-linked oligosaccharide chains while TSH-β contains only one, the expected ratio α to β is 2.0, as shown by the discontinuous horizontal line. For incorporation of tritium-labeled glucosamine and mannose, the expected ratio was observed. However, for tritium-labeled fucose and N-acetyl-mannosamine, a specific precursor of N-acetyl-neuraminic acid, and for [^{35}S]sulfate, this ratio was significantly lower than 2.0, suggesting greater amounts of fucose, sialic acid, and sulfate per carbohydrate chain on TSH-β than TSH-α. * signifies $p < 0.05$, ** signifies $p < 0.02$. N=3 in these experiments.

Secreted TSH-α and free α-subunits also incorporate radiolabeled precursors differentially. In Figure 5 the incorporation of various labeled precursors into free α and TSH-α is compared; the ratio on the ordinate is derived by normalizing for methionine incorporation into the common α-subunit apoprotein. As can be seen, there was no statistical difference in [^{3}H]glucosamine incorporation between secreted free

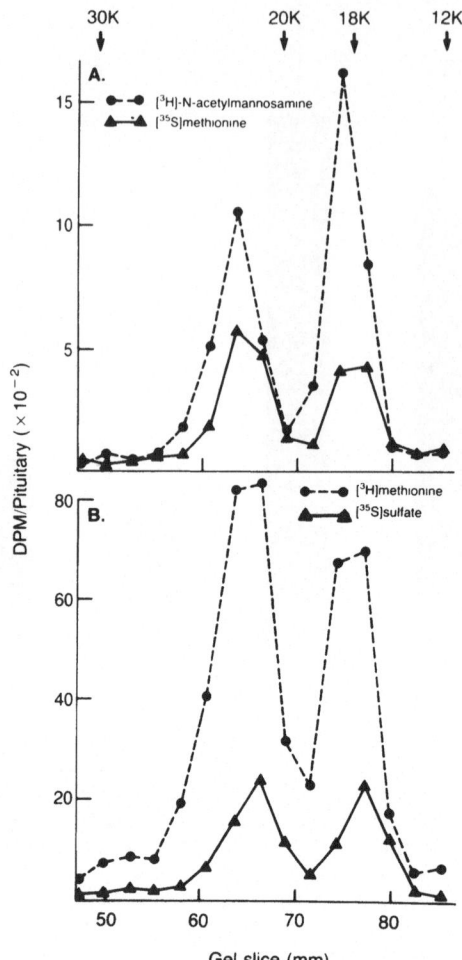

Fig. 4. SDS-PAGE of secreted mouse TSH subunits. Both in
the top panel (A), where sialic acid incorporation
was studied, and in the lower one (B), where
sulfate incorporation was examined, secreted
thyroid-stimulating hormone (TSH) is immunoprecipitated by TSH-β antisera and then dissociated
into α- and β-subunits prior to electrophoresis.
TSH-α migrates with an apparent molecular weight of
21,000-22,000 while TSH-β migrates with an apparent
molecular weight of 18,000, even though the
apoprotein moiety on TSH-β is slightly larger.
This discrepancy is due to the presence of two
asparagine-linked carbohydrate residues on TSH-α
and only one on TSH-β. Methionine labeling in the
top panel (closed triangles) and in the lower panel
(closed circles) shows a ratio of TSH-α to TSH-β of
3:2, consistent with 3 methionine residues in the
α- and 2 in the β-subunit in the mouse. [^{3}H]N-
acetylmannosamine is a specific precursor of
[^{3}H]N-acetylneuraminic acid. Note that despite the
presence of two oligosaccharide chains on α and
only one on β, there is equal or greater incorporation of [^{35}S]sulfate and [^{3}H]N-acetylmannosamine
into the β-subunit.

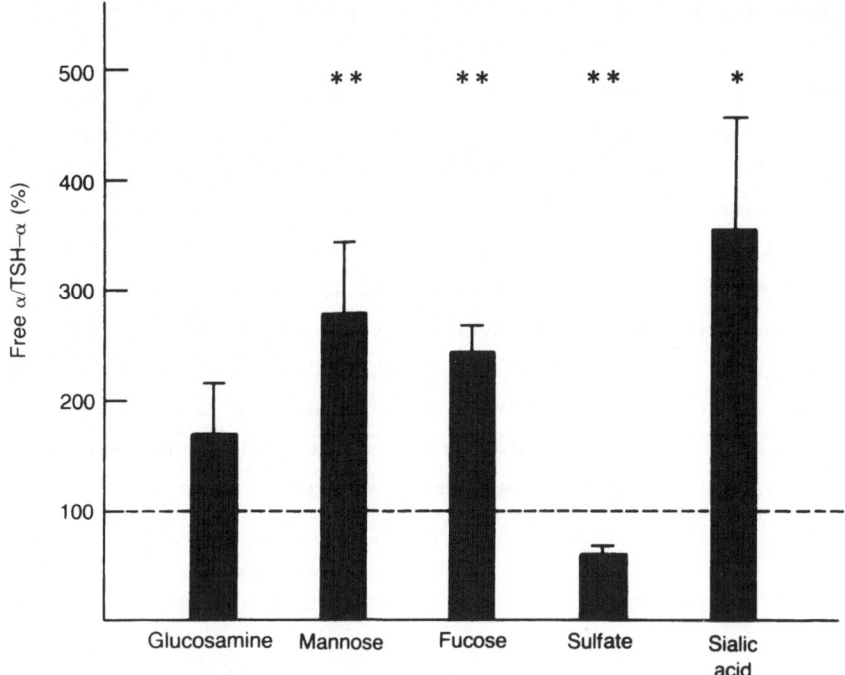

Fig. 5. Incorporation of sugar precursors and sulfate into
TSH-α and free α-subunits in hypothyroid mouse
pituitaries. TSH dimer and free α-subunits were
immunoprecipitated by the method outlined in Figure
2 and analyzed by electrophoresis in SDS-PAGE.
Incorporation of radiolabeled precursors was
compared in the combined and free α-subunits by
normalization to methionine incorporation into the
common apoprotein moiety. Free α-subunit appeared
to be enriched in mannose, fucose, and sialic acid,
but contained less sulfate than the combined
α-subunit. Since the free α-subunit contains an
extra O-glycosylation site, this may represent an
additional source of fucose and sialic acid. These
differences were observed in a hypothyroid model
and may vary with the physiologic state of the
animal. * $p < 0.05$; ** $p < 0.02$. (N = 3).

α-subunit and TSH-α. However, free α-subunit incorporated more of the
neutral sugars [^{3}H]mannose and [^{3}H]fucose. With regard to the negatively
charged moieties, free α incorporated more [^{3}H]N-acetylneuraminic acid but
less [^{35}S]sulfate than did combined α, suggesting a difference in the
predominant "capping group" present in the free versus combined
α-subunits. Some of the increase in neuraminic acid and fucose on free
α-subunit may be attributable to the additional O-linked chain attached to
threonine 43, while the increased mannose content on free α-subunit might
be explained by a greater proportion of hybrid structures; these
possibilities are currently under investigation.

Figure 6 illustrates schematically a proposed "average" structure for secreted mouse TSH. The α-subunit is N-glycosylated on asparagine residues 56 and 82, while the β-subunit is N-glycosylated at asparagine 23. The average TSH-β subunit oligosaccharide moiety contains more fucose, sulfate, and sialic acid than does its counterpart on TSH-α. The oligosaccharide moiety on free α contains, on average, more fucose and sialic acid but less sulfate compared to TSH-α. In addition, there is a carbohydrate residue O-linked to threonine 43 on free α but not on TSH-α. These proposed structures are derived from experiments in a hypothyroid model and may differ in the euthyroid and other physiologic states. Similarly, these structures for secreted mouse TSH may differ from the average structures of intrapituitary mouse TSH; both secreted and intrapituitary structures may vary among species.

EFFECT OF TSH CARBOHYDRATE ON HORMONAL BIOACTIVITY AND METABOLIC CLEARANCE

The high mannose carbohydrate moiety of TSH has been shown to perform multiple functions, including permitting α-β subunit combination and minimizing intracellular proteolysis and subunit aggregation (Weintraub et al., 1980, 1983). This was shown by experiments in which in vitro tunicamycin treatment of mouse thyrotropic tumor cells resulted in the appearance of α-subunits which incorporated [^{35}S]methionine but not [^{3}H]glucosamine and had M_r = 11,000 and M_r = 12,000, corresponding to the weights of the α and β apoproteins, respectively. The nonglycosylated subunits produced by tunicamycin treatment showed a high degree of aggregation, especially after heating at 37° C under nonreducing conditions. Nonglycosylated subunits were 50-65% degraded intracellularly before secretion, quite distinct from normally glycosylated subunits which show negligible degradation. Incubation of various [^{35}S]-labeled α forms with excess unlabeled TSH-β showed high combining activity for intracellular α-subunit with two high mannose units; intermediate activity for media α-subunit with two complex units; and low activity for intracellular α-subunit with one high mannose unit or nonglycosylated media α-subunit. These data suggest that the initial glycosylation with high mannose carbohydrate units prevents intracellular aggregation, decreases degradation, and enhances attainment of specific conformations necessary for TSH-α and TSH-β combination.

The final processed or complex structure of secreted TSH carbohydrate appears to be important in determining the intrinsic biological activity

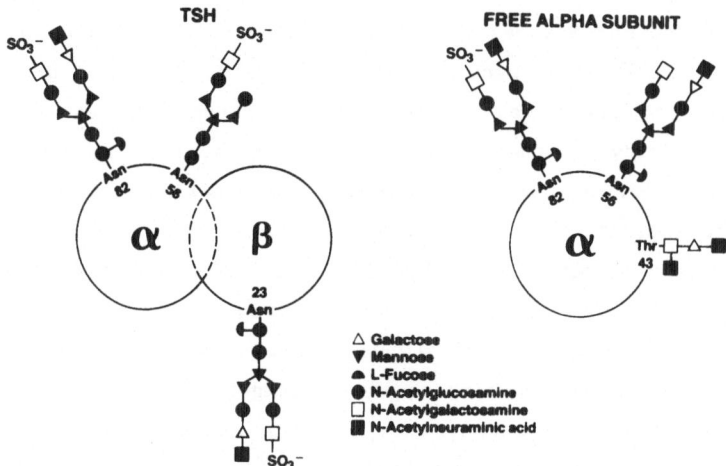

Fig. 6. Proposed average structure for secreted mouse TSH.
While there are heterogeneous carbohydrate moieties
in secreted TSH, the structures shown above
represent our current working model of the
"average" TSH structure.   In the mouse, there are
two asparagine residues that are glycosylated in
the α-subunit (Asn 82 and 56) and one in the
β-subunit (Asn 23).   The average β-linked
carbohydrate chain contains proportionally more
fucose, sulfate, and sialic acid than does its
counterpart on α. The carbohydrate composition of
the free α-subunit seems to resemble combined
β-subunit more than it does combined α-subunit,
except that free α-subunit contains less sulfate.
The free α-subunit is secreted into the circulation, but no clear function has yet been assigned
to this protein; it contains an additional 0-linked
chain attached to threonine residue 43.   Not
depicted   above   are   bisecting,   triantennary,
multiantennary, hybrid, and high mannose forms of
TSH carbohydrate, which have been observed in
lectin analysis of glycopeptides (see text).   The
carbohydrate moieties may vary with the physiological state of the animal and may be under the
influence   of   neuroendocrine   factors.   These
proposed structures have not yet been rigorously
demonstrated and differ somewhat from previously
suggested   structures   for   pituitary   glycoprotein
hormones (Parsons and Pierce, 1980; Anumula and
Bahl, 1983).   It is assumed that the N-acetylneuraminic acid moieties (Fig. 4) are added to
conventional galactose residues, while the acceptor
of the sulfate moieties (Fig. 4) is N-acetylgalactosamine (Green et al., 1985).

and the metabolic clearance rate of the secreted, circulating hormone. Mouse thyrotropic tumor extract, serum from tumor bearing mice, culture medium from dispersed cell incubations, and two preparations of purified bovine TSH (Sigma and Pierce) were fractionated on Sephadex G-100 (Joshi and Weintraub, 1983). For each fraction, TSH bioactivity was measured by the stimulation of adenylate cyclase activity in human thyroid membranes, while immunoactivity of these fractions was determined by radioimmuno-assay. Pierce bovine TSH had several immunoactive components with partition coefficients (K_{av}) of 0.28-0.32 and ratios of biological to immunological activities (B/I) of 0.59-1.42. Sigma bovine TSH, mouse tumor, serum, and medium were even more heterogeneous (K_{av} = 0.23-0.32) and had a lower range of B/I (0.04-1.01). Double reciprocal plots comparing TSH from different sources showed competitive inhibition for the low B/I forms from all sources, except for a medium form which showed mixed inhibition. To determine whether these forms of TSH with competing agonistic/antagonistic properties differed in their carbohydrate structure, affinity chromatography on concanavalin A, wheat germ agglutinin, and soybean agglutinin was employed. The apparent lower molecular weight form of TSH with lower B/I contained decreased amounts or availability of N-acetylgalactosamine and/or β-linked galactose compared to the higher molecular weights form(s) with higher bioactivity. These data suggest that the complex carbohydrate moiety of TSH may modulate its bioactivity and that natural antagonists exist within the heterogeneous TSH molecules that are secreted by pituitary or thyrotropic tumor tissue. Recently we have demonstrated that after deglycosylation with anhydrous hydrogen fluoride bovine TSH demonstrates markedly reduced bioactivity and also acts as a competitive antagonist to normally glycosylated TSH (Amr et al., 1986); moreover, after intravenous injection into normal rats, the deglycosylated hormone had a more rapid metabolic clearance rate than the native hormone (Constant and Weintraub, unpublished data).

ENDOCRINE REGULATION OF TSH GLYCOSYLATION

Early biosynthetic studies suggested that thyrotropin-releasing hormone (TRH), in addition to promoting secretion of TSH, increased relative [^{3}H]glucosamine incorporation into TSH in whole (Wilber, 1971) and dispersed (Ponsin and Mornex, 1983) rat pituitaries. Using subunit-specific analytic methods and electrophoresis of TSH on SDS-PAGE, Taylor and Weintraub (1985b) showed that normal rat pituitaries stimulated with TRH for 24 hours in vitro showed a 3-fold stimulation of labeled glucosamine incorporation into secreted TSH but no change in labeled alanine incorporation. The increased relative [3H]glucosamine incorporation in the presence of TRH was seen equally in TSH-α and TSH-β but was not seen in secreted free α-subunit or in total secreted proteins.

In vivo administration of TRH into newly thyroidectomized rats resulted in an increase in the high mannose species $Glc_1Man_9GlcNAc_1$ and $Glc_1Man_8GlcNAc_1$ released from intracellular TSH by treatment with endo H (Ronin et al, 1985). This was noted in TSH dimer and free β-subunits but not in free α-subunits. This effect may be due to a TRH effect on the kinetics of carbohydrate processing, as discussed earlier, or to addition of glucose residues post-translationally during high-mannose processing (Ronin and Caseti, 1981).

To explore the structural basis for the apparent increase in TSH incorporation of sugar precursors in the presence of TRH, hypothyroid mouse pituitaries were incubated with [3H]mannose, a marker of the inner core sugars, with or without 10^{-7} M TRH for 18 hours (Gesundheit et al., 1985). Secreted TSH dimer was precipitated by anti-TSH-β antisera, digested with Pronase, and the TSH glycopeptides thus generated were chromatographed on concanavalin A-agarose (con A) by the method of Cummings and Kornfeld (1982). Glycopeptides eluted in three general classes on con A, depending on specific minimal structural requirements of the carbohydrate moiety: unbound glycopeptides consist of bisecting, triantennary, and multiantennary complex structures; bound glycopeptides that elute with 10 mM α-methylglucoside correspond to biantennary complex or truncated hybrid structures; strongly bound glycopeptides that elute with 500 mM α-methylmannoside correspond to high mannose or hybrid carbohydrate structures. Analysis of both intracellular and secreted TSH glycopeptides revealed a 2.5-fold increase by TRH ($p < 0.05$) in one specific class--secreted glycopeptides that elute with 10 mM α-methylglucoside. No change was noted in any intracellular glycopeptide class or in the other two secreted glycopeptide classes. These data suggest that TRH affects the final structure of secreted TSH carbohydrate, although it is not known if this is due to activation or inhibition of specific glycosyltransferases or to stimulation of specialized routes of secretion that result in altered glycosylation.

SUMMARY

Thyroid-stimulating hormone provides an interesting model to study the glycosylation and carbohydrate processing of a heterodimeric glycoprotein with a clear physiological function. The carbohydrate moiety on TSH is required for subunit combination, protection from intracellular proteolysis and aggregation, and for attainment of full biological activity. Recent work, summarized herein, has studied mechanisms and kinetics of TSH carbohydrate maturation and has contrasted processing rates and composition of free and combined subunits. Neuroendocrine factors, such as thyrotropin-releasing hormone, appear to modulate the carbohydrate structure of secreted TSH, which results in a change in the relative bioactivity of the circulating hormone. The biochemical mechanisms by which these carbohydrate alterations occur and how they affect hormone-receptor interaction are currently under investigation.

ACKNOWLEDGMENTS

The authors thank Mrs. Julia Miller and Ms. V. L. Griffin for their secretarial assistance.

REFERENCES

Amr, S., Menezes-Ferreira, M., Shimohigashi, Y., Chen, H. C., Nisula, B., and Weintraub, B. D., 1986, Activities of deglycosylated thyrotropin at the thyroid membrane receptor-adenylate cyclase system, J. Endocrinol. Invest., in press.

Anumula, K. R., and Bahl, O. P., 1983, Biosynthesis of lutropin in ovine pituitary slices: Incorporation of [^{35}S]sulfate in carbohydrate units, Arch. Biochem. Biophys., 220:645-651.

Behrens, N. H., and Leloir, L. F., 1970, Dolichol monophosphate glucose: An intermediate in glucose transfer in liver, Proc. Natl. Acad. Sci., USA, 66:153-159.

Chin, W. W., and Habener, J. F., 1981, Thyroid-stimulating hormone subunits: Evidence from endoglycosidase H cleavage for late presecretory glycosylation, Endocrinology, 108:1628-1633.

Chin, W. W., Kronenberg, H. M., Dee, P. C., Maloof, F., and Habener, J. F., 1981a, Nucleotide sequence of the mRNA encoding the pre-α subunit of mouse thyrotropin, Proc. Natl. Acad. Sci., USA, 78:5329-5333.

Chin, W. W., Maloof, F., and Habener, J. F., 1981b, Thyroid-stimulating hormone biosynthesis, J. Biol. Chem., 256:3059-3066.

Condliffe, P. G., Mochizuki, M., Fontaine, Y. A., and Bates, R. W., 1969,

Purification and properties of thyrotropin from functional pituitary tumors in mice, Endocrinology, 85:453-464.

Corless, C. L., and Boime I., 1985, Differential secretion of O-glyco-sylated gonadotropin α-subunit and luteinizing hormone (LH) in the presence of LH-releasing hormone, Endocrinology, 117:1699-1706.

Cummings, R. D., and Kornfeld, S., 1982, Fractionation of asparagine-linked oligosaccharides by serial lectin-agarose affinity chromatography, J. Biol. Chem., 257:11235-11240.

Furth, J., Moy, P., Hershman, J. M., and Uedo, G., 1973, Thyrotropic tumor syndrome, Arch. Pathol. 96:217-226.

Gesundheit, N., Fink, D. L., Taylor, T., and Weintraub, B. D., 1985, TRH modifies the carbohydrate structure of secreted TSH: Analysis by concanavalin A chromatography, Program of the 67th Annual Meeting of the American Endocrine Society, p. 110, (Abstract).

Gesundheit, N., and Weintraub, B. D., 1985, Effect of thyrotropin-releasing hormone (TRH) on the carbohydrate composition and sulfation of secreted thyrotropin (TSH), Clin. Res., 33, 2:534A (Abstract).

Giudice, L. C., Waxdal, M. J., and Weintraub, B. D., 1979, Comparison of bovine and mouse pituitary glycoprotein pre-α subunits synthesized in vitro, Proc. Natl. Acad. Sci., USA, 76:4798-4802.

Giudice, L. C., and Weintraub, B. D., 1979, Evidence for conformational differences between precursor and processed forms of thyroid-stimulating hormone β subunit, J. Biol. Chem. 254:12679-12683.

Green, E. D., Morishima, C., Boime, I., and Baenziger, J. U., 1985, Structural requirements for sulfation of asparagine-linked oligo-saccharides of lutropin, Proc. Natl. Acad. Sci., U.S.A., 82:7850-7854.

Gurr, J. A., Catterall, J. F., and Kourides, I. A., 1983, Cloning of cDNA encoding the pre-β subunit of mouse thyrotropin, Proc. Natl. Acad. Sci., USA, 80:2122-2126.

Joshi, L. R., and Weintraub, B. D., 1983, Naturally occurring forms of thyrotropin with low bioactivity and altered carbohydrate content act as competitive antagonists to more bioactive forms, Endo-crinology, 113:2145-2154.

Kourides, I. A., Gurr, J. A., and Wolf, O., 1984, The regulation and organization of thyroid stimulating hormone genes, Rec. Prog. Horm. Res., 40:79-120.

Lippman, S. S., Amr, S., and Weintraub, B. D., 1986, Discordant effects of TRH on pre-and post-translational regulation of TSH biosynthesis in rat pituitary, The 18th Miami Winter Symposium (Abstract).

Liu, W-K. and Ward, D. N., 1975, The purification and chemistry of pituitary glycoprotein hormones, Pharmac. Therap., 1:545-570.

Magner, J. A., and Weintraub, B. D., 1982, Thyroid-stimulating hormone subunit processing and combination in microsomal subfractions of mouse pituitary tumor, J. Biol. Chem., 257:6709-6715.

Magner, J. A., Ronin, C., and Weintraub, B. D., 1984, Carbohydrate processing of thyrotropin differs from that of free α-subunit and total glycoproteins in microsomal subfractions of mouse pituitary tumor, Endocrinology, 115:1019-1030.

Parsons, T. F., Bloomfield, G. A., and Pierce, J. G., 1983, Purification of an alternate form of the α subunit of the glycoprotein hormones from bovine pituitaries and identification of its O-linked oligosaccharide, J. Biol. Chem., 258:240-244.

Parsons, T. F., and Pierce, J. G., 1980, Oligosaccharide moieties of glycoprotein hormones: bovine lutropin resists enzymatic deglycosylation because of terminal O-sulfated N-acetylhexosamines, Proc. Natl. Acad. Sci., USA, 77:7089-7093

Pierce, J. G., and Parsons, T. F., 1981, Glycoprotein hormones: Structure and function, Annu. Rev. Biochem., 50:465-495.

Ponsin, G., and Mornex, R., 1983, Control of thyrotropin glycosylation in normal rat pituitary cells in culture: Effect of thyrotropin-releasing hormone, Endocrinology, 113:549-555.

Ronin, C., and Caseti, C., 1981, Transfer of glucose in the biosynthesis of thyroid glycoproteins, Biochim. Biophys. Acta, 674:58-64.

Ronin, C., Stannard, B. S., Rosenbloom, I. L., Magner, J. A., and Weintraub, B. D., 1984, Glycosylation and processing of high-mannose oligosaccharides of thyroid-stimulating hormone subunits: Comparison to nonsecretory cell glycoproteins, Biochemistry, 23:4503-4510.

Ronin, C., Stannard, B. S., and Weintraub, B. D., 1985, Differential processing and regulation of thyroid-stimulating hormone subunit carbohydrate chains in thyrotropic tumors and in normal and hypothyroid pituitaries, Biochemistry, 24:5626-5631.

Sairam, M. R., and Li, C. H., 1977, Human pituitary thyrotropin. Isolation and characterization of the hormone and its subunits, Can. J. Biochem., 55:747-754.

Shupnik, M. A., Chin, W. W., Ross, D. S., Downing, M. F., Habener, J. F., and Ridgway, E. C., 1983, Regulation by thyroxine of the mRNA encoding the α subunit of mouse thyrotropin, J. Biol. Chem., 258:15120-15124.

Taylor, T., and Weintraub, B. D., 1985a, Differential regulation of thyrotropin subunit apoprotein and carbohydrate biosynthesis by thyroid hormone, Endocrinology, 116:1535-1542.

Taylor, T., and Weintraub, B. D., 1985b, Thyrotropin-releasing hormone regulation of TSH subunit biosynthesis and glycosylation in normal and hypothyroid rat pituitaries, Endocrinology, 116:1968-1976.

Weintraub, B. D. and Stannard, B. S., 1978, Precursor-product relationships in the biosynthesis and secretion of thyrotropin and its subunits by mouse thyrotropic tumor cells, FEBS Lett., 92:303-307.

Weintraub, B. D., Stannard, B. S., Linnekin, D., and Marshall, M., 1980, Relationship of glycosylation to de novo thyroid- stimulating hormone biosynthesis and secretion by mouse pituitary tumor cells, J. Biol. Chem., 255:5715-5723.

Weintraub, B. D., Stannard, B. S., and Meyers, L. 1983, Glycosylation of thyroid-stimulating hormones in pituitary cells: influence of high mannose oligosaccharide units on subunit aggregation, combination, and intracellular degradation, Endocrinology, 112:1331-1345.

Weintraub, B. D., Stannard, B. S., Magner J. A., Ronin C., Taylor, T., Joshi, L., Constant R. B., Menezes-Ferreira, M. M., Petrick, P., and Gesundheit, N., 1985, Glycosylation and posttranslational processing of thyroid-stimulating hormone: Clinical implications, Rec. Prog. Horm. Res., 41:577-606.

Wilber, J. F., 1971, Stimulation of ^{14}C-glucosamine and ^{14}C-alanine incorporation into thyrotropin by synthetic thyrotropin-releasing hormone, Endocrinology, 89:873-878.

FUNCTIONAL ADAPTATIONS OF TRANSBILAYER PROTEINS

Vincent T. Marchesi

Department of Pathology
Yale University School of Medicine
New Haven, Connecticut 06510

INTRODUCTION

Many important functions commonly attributed to surface membranes are carried out by membrane-associated proteins that are tightly bound to the lipid bilayer. This class of proteins, which are engineered to extend completely across the lipid bilayer, sometimes in the form of multiple membrane-spanning loops, are generally referred to as transbilayer proteins. Most of the molecules that function as receptors for different proteins also modify cell functions by regulating the rates of transport of different substances across the lipid bilayer. Both types of proteins have unique structural features which allow them to interact with the lipid bilayer. One set of proteins, the receptor type, serve as conduits of information across the bilayer, while the other set, the channel proteins, selectively modify this barrier. This essay will present some generalizations about both types of transbilayer molecules that may help us to understand how they function in the environment of the intact cell.

RECEPTORS

Until recently, our knowledge of receptor biology was based primarily on the characteristics of binding that different ligands displayed when incubated in the presence of membrane fractions prepared from appropriate cell samples. Systems that displayed a "high affinity" association between ligand and membrane fractions were considered most "physiologic" and the factors that regulated this association were assumed to play comparable roles in vivo. The biochemical analysis of the molecules involved in these associations relied primarily on SDS gel electrophoresis and the use of radiolabeled ligands to identify the components assumed to be involved. SDS electrophoresis proved to be the analytic method of choice, since membrane bound receptor molecules were invariably insoluble in the usual solvents used to extract proteins from cells, but they could be solubilized by either non-ionic or anionic detergents such as SDS. Receptors isolated in non-ionic detergents often retained their capacity to bind the appropriate ligand. In contrast, extraction with SDS often caused their inactivation, but since this detergent was able to dissociate most receptor molecules into their constituent subunits, these could be analyzed readily by standard gel electrophoresis. This approach enabled investigators to identify the number and types of polypeptide chains involved in receptor function.

This line of investigation generated many useful suggestions as to the probable identities of many of the molecules involved, but a complete molecular analysis of these receptors was only achieved when modern cloning techniques were applied to the study of each receptor system. At this writing, a number of different receptors have been analyzed, a partial list of which includes receptors for interleukin-2 (IL-2; Leonard et al., 1984; Nikaido et al., 1984), low density lipoproteins (LDL; Yamamoto et al., 1984), transferrin (TF; McClelland et al., 1984), polyimmunoglobulin (PolyIg; Mostov et al., 1984), epidermal growth factor (EGF; Ullrich et al., 1984), and insulin (Ebina et al., 1985; Ullrich et al., 1985).

The primary structures of each of the receptors listed above has been elucidated by DNA sequencing techniques; the details of each can be obtained by consulting the original publications. From the perspective of this discussion a single generalization emerges: Each of these receptors seems to be arranged into three topologically distinct domains, each

INTRA-
MEMBRA-
EXTERNAL   NOUS   CYTOPLASMIC

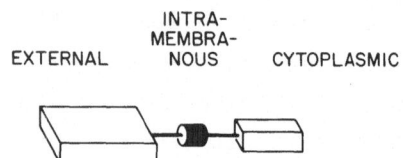

Fig. 1. A schematic representation of the tripartite form
of a membrane-bound receptor.

located in a different compartment of the cell. An extremely simplistic model depicting this relationship is shown in Figure 1.

Two of these domains are shown as units of unspecified dimensions, which are known to vary within wide limits as described below, while the connecting segment is depicted as a relatively short cylindrical form which exists within the interior of the lipid membrane and connects the two extramembranous pieces. The three-dimensional structures of the extramembranous segments are purposely presented as rectangular shapes, emphasizing our almost complete lack of information of their three-dimensional structures. In contrast, the connecting piece is shown as a regular cylindrical unit which is a remarkably uniform component of each of the different receptors as described below.

The amino acid sequence of each of the receptors described above has been determined, and it is instructive to see how their domains correspond to the hypothetical tripartite form. This is presented in Figure 2.

In these figures some attempt has been made to draw the three domain models to an approximate scale, taking into account the amount of peptide which is thought to reside on either side of the lipid bilayer. Of the six receptors shown, which come from a wide variety of cell types, a great deal of variation seems evident when one attempts to estimate the relative

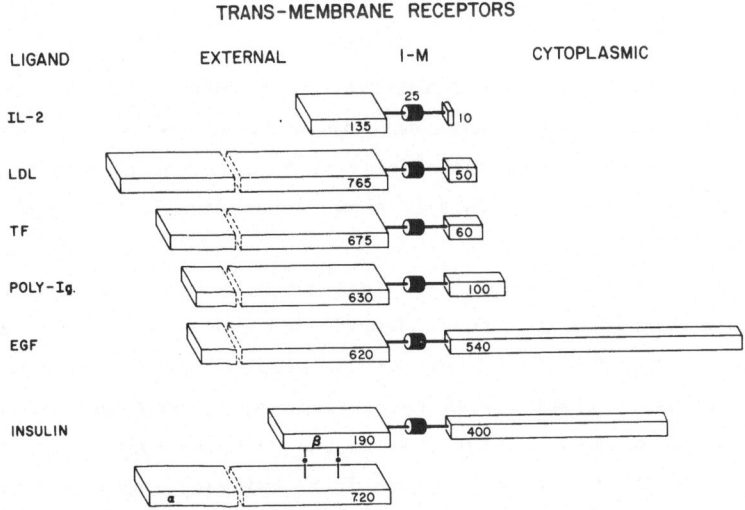

Fig. 2. Hypothetical in situ arrangement of different receptor molecules.

sizes of each of the extramembranous segments. Most of these receptors have sizable extracellular pieces, but the cytoplasmic segments vary from ten amino acids to over five hundred.

The assignment of each extramembranous domain to its respective compartment is based on experimental data, usually radiolabeling studies, which allow one to identify the extracellular domain directly, and in some cases, permit identification of the cytoplasmic domain as well. Cytoplasmic domains are usually identified by their lack of reactivity to exogenous probes, but some have special features, such as the phosphokinase activities associated with the EGF and insulin receptors, that are clearly confined to the cytoplasmic compartment of the cell. Receptor molecules that have enzymatic activity associated with their cytoplasmic domains usually have large cytoplasmic segments, some equal to or larger than their extracellular domains.

The sizes of the extracellular domains depicted in these figures cannot as yet be correlated with the types of ligands that are known to bind to the individual receptors. Most of the extracellular domains of these receptors are composed of far more polypeptide chain than would be needed to create a pocket for a ligand. In some cases, multiple repeating units exist, such as on the LDL receptor, that seem to be correlated with the size and character of the lipoprotein complex that is bound. Multiple disulfide bonded peptide loops are a characteristic feature of many extracellular domains; perhaps this structural motif has evolved to serve as a supporting framework for ligand pockets in general.

The universality of this tripartite arrangement is not confined to transbilayer molecules that serve as complex receptors; a wide variety of transbilayer proteins that serve as cell surface recognition molecules are also organized in this way. This is shown schematically in Figure 3.

Although these molecules listed here and many more like them have the same tripartite form as that described previously, one important difference is evident: Unlike most of the complex receptor molecules, the proteins that serve primarily as recognition sites have, on the average, much smaller cytoplasmic domains than do the receptor types. Two of these molecules, the Th-1 and the plasmodium antigen, may not have any cytoplasmic domain at all, although each seems to have an intramembranous piece similar to that of the others.

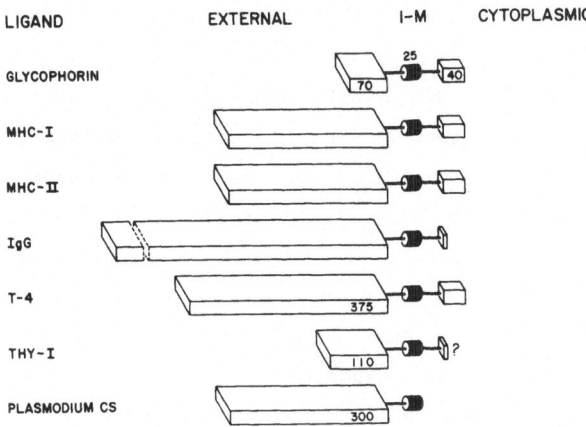

Fig. 3. Transbilayer proteins which serve as recognition molecules can also be arranged in a tripartite form. Based on structural data of glycophorin A (Tomita et al., 1978) histocompatibility molecules and membrane-bound antibodies (Kaufman et al., 1984) to T-4 lymphocyte antigen (Maddon et al., 1985), Thy-I (Seki et al., 1985a; Seki et al., 1985b; Seki et al., 1985c), and the circumsporozoite antigen of plasmodium (Dame et al., 1984; Ozaki et al., 1983).

It is reasonable to assume that these small segments would also have limited functions. Amino acid segments that are 30-40 residues long are not likely to fold into complex catalytic sites such as those that might be required for kinase activity, but polypeptide chains of this length could easily form specific binding sites for other proteins, particularly if these peptides are able to associate with neighboring phospholipids, as has been postulated for the cytoplasmic segment of the human erythrocyte glycophorins (Anderson and Marchesi, 1985).

In contrast to the great variation in the extramembranous domains of both the receptors and the recognition molecules, the segment of polypeptide that is believed to span the lipid bilayer appears to be remarkably uniform. In each case, the intramembranous segment seems to be 22 or 23 amino acids long.

The actual amino acid sequences of the intramembranous segments or what are assumed to be intramembranous segments of each of the molecules are depicted schematically in Figure 4. These sequences have been deduced primarily from DNA sequences of cDNA clones of the receptor molecules in most cases, although in a few instances the information has been derived from direct sequence analysis of the individual peptide segments.

The presentation of the data in this way is to some extent arbitrary, since we really do not know which parts of these molecules are within the lipid bilayer. We do know that at least one segment of each of the molecules must be within the lipid, and in all cases the experimental evidence that is available does support the interpretation shown here. This is not to say that some of these molecules may not violate the generalizations proposed, but if this possibility is ignored for the moment and the sequences are conceived of as a set of common transmembrane segments, some interesting relationships become evident. A second arbitrary feature of this presentation is the decision to place the first positively charged amino acid distal to the putative intramembranous segment at the interface between the cytoplasmic compartment of the membrane and the intramembranous compartment. If the first arginine or lysine is always at the membrane interface as presented in Figure 4, then there would be in every case at least 22 uncharged amino acids available to span the length of an average lipid bilayer.

The composition of these intramembranous segments is exactly what one would predict: There is a complete absence of any charged amino acids, and a great concentration of nonpolar and hydrophobic amino acids such as leucine, isoleucine, valine, etc. A scanning of these sequences shows, however, that not all amino acids within these putative intramembranous domains can be considered hydrophobic. Indeed a significant number of serines and threonines are scattered throughout almost all of the segments

```
EXTERNAL              INTRA-MEMBRANOUS            CYTOPLASM

- - - - - - E - E I T L I I F G V M A G V I G T I L L I S Y G I R R - - K K - - -     Glycophorin A
E - - - - - - E - Q V A V A G C V F L L I S V L L L S G L T W Q R R - R K - R R -     IL-2
- - E K K - - - - R A L S I V L P I V L L V F L C L G V F L L W K - - R - K - - -     LDL
- - - - - - - - K V L I S T L V P L G L V L A A G A M A V A I A R - R - R R - - -     Poly Ig
- - - - - K - - - I A T G M V G A L L L L L V V A L G I G L F M R R R - - - R K R     EGF
- - - - K - D - - V V A V C M V S L V L L L L L G M W G T Y Y Y R - R K - - - - -     EGF-Precursor
E - - K - - - - - G I M F G I L F F V I V A I T G Y C I S G S C R K - K - - - - -     Transferrin
- R - D E - - - - V Q S T I V S V V V L L L I S L G L S L L F L R - - - - - R - -     Asialo-Glycoprotein
E - - - - - - - - A T V A V L V V L G A A I V T G A V V A F V M K - R R R - - - -     H-2
- - - - - E - - E N V V C A L G L T V G L V G I I I G T I F I L K - - R K - - - -     DRα
K - - - - E - K E N V V C A L G L F V G L V G I V V G I I L I M K - - K K R - - -     Eα(K)
- D - E - D - - - T T I T I F I S L F L L S V C Y S A S V T L F K - K - - - - - -     γ2α
E E E - - E - - - T T A S I F I V L F L L S L F Y S T T V T L F K - K                 μ
```

Fig. 4. Amino acid sequences of the intramembranous segments of different receptors and recognition molecules.

of these receptors. Glutamines and asparagines are encountered infrequently, as are prolines, although a few proline residues are seen in some of these segments. To some extent, the absence of prolines is to be expected if these intramembranous segments are to fold in α-helical conformations, as has been suggested for some of them.

The peptide believed to represent the intramembranous segment of glycophorin A has been purified and its conformation analyzed by circular dichroism. This segment appears to be almost entirely α-helical (Schulte and Marchesi, 1979). Unfortunately, the interpretation of these experiments is complicated by the fact that the isolated intramembranous peptides are only soluble in detergents, and either triton or SDS must be present during the analysis. The need for detergent to be present during the analysis has caused some to question whether the conformation of a peptide in detergent necessarily reflects the conformation it might assume while in an intact membrane. Arguments favoring a positive correlation between the two rely on the claim that transmembrane segments of membrane proteins probably assume the same conformation in detergents that they assume in a membrane, since the internal environment of a membrane is similar to the internal environment of a detergent micelle. The only direct experimental evidence that transmembrane proteins may indeed exist as α-helical loops comes from x-ray diffraction and electron diffraction studies of a complex channel-forming protein called bacteriorhodopsin found in unusual microorganisms. This interesting molecule aggregates in membrane patches to form paracrystalline arrays that can be analyzed by diffraction techniques. This factor has been exploited by Henderson et al. (1975) who have been able to obtain x-ray and electron diffraction signals by examining unfixed membrane fragments. The results of these studies indicate that almost all of the intramembranous segments of the bacteriorhodopsin molecule exist as α-helical loops. These intramembranous helical loops are approximately 22-24 amino acids in length, and are thus similar in length and presumed conformation to the intramembranous segments described here.

Another argument supporting the α-helical conformation of these intramembranous segments comes from simple considerations of geometry and chain length. Twenty-two amino acids packed in an α-helical conformation would fold into 6-7 loops and extend approximately 35-40 A, the length of the nonpolar region of an average lipid bilayer. Thus on the basis of circular dichroism measurements of isolated transmembrane proteins in detergents, the results of x-ray diffraction studies of special transmembrane proteins in isolated membrane fragments, and considerations of geometry and length, we provisionally assume that the most likely arrangement of the intramembranous segments described above is that of an α-helix, long enough to cross the hydrophobic domain of the membrane and yet thermodynamically stable under physiological conditions.

The impressive concentration of positively charged amino acids which border the cytoplasmic ends of the intramembranous segment of these molecules bears special comment. This feature had been noted when these molecules were first studied, and many suggestions have been offered to explain their presence. Most observers favor the idea that these charged groups serve to anchor the molecule within the bilayer, but no acceptable mechanism has been proposed to explain how they might do this. The intramembranous segments of some transbilayer glycoproteins bind phospholipids more tightly than other parts of these molecules, and some recent studies suggest that at least one group of molecules, the glycophorins, bind triphosphoinositide with reasonably high affinity (Armitage et al., 1977). It is reasonable to speculate that the cluster of arginine and lysine amino acids at the membrane interface region may be part of the binding site for specific phospholipids.

It is perhaps disappointing to see that there is no comparable concentration of charged amino acids at the external side of the intramembranous segment. Indeed it is not possible to identify with any certainty where the intramembranous piece ends and the externally disposed portion of the molecule begins. Perhaps there is no one interface segment that is characteristic of each molecule.

CHANNEL-FORMING PROTEINS

The recent application of recombinant DNA techniques to the study of proteins involved in transport substances across membrane barriers has also made an enormous impact on our understanding of how this class of molecules may reside within the membrane. A few proteins whose primary structures have been solved by this approach include rhodopsin (Nathans and Hogness, 1984), the glucose carrier (Mueckler et al., 1985) channels for sodium (Noda et al., 1984) and chloride (Kopito and Lodish, 1985) and the NaK activated ATPase (Shull et al., 1985). These proteins as a group have two features which distinguish them from the transbilayer proteins already described. The most striking difference is the fact that these

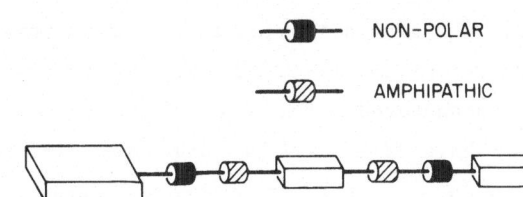

NON-POLAR

AMPHIPATHIC

Fig. 5. Transbilayer proteins that serve a channels or pumps have multiple intramembranous segments that may be composed of nonpolar and amphipathic helical structures with intervening segments of undefined structure.

molecules do not have a simple tripartite domain structure. They do not span the lipid bilayer with a single polypeptide chain composed of nonpolar amino acids. Instead, these proteins have multiple membrane-spanning domains. The membrane-spanning segments of these channels differ from those of the simple receptor type in two ways: The intramembranous segments are not composed solely of nonpolar amino acids, and they are not sandwiched between recognizable external and cytoplasmic domains in a simple regular pattern. Instead of the regular tripartite arrangement depicted schematically in Figure 1, the channel-forming proteins seem to be arranged as shown, again highly schematically, in Figure 5.

These molecules are distinguished by the large number of intramembranous segments that dominate their structures. These intramembranous segments are of two types, one made up of nonpolar amino acid loops similar to those found in the receptors described above, and the second, peptide segments of equivalent length but containing a number of charged amino acids. These mixed amino acid segments have been noted as a characteristic feature of serum lipoproteins and are often referred to as amphipathic helices (Anantharamiah et al., 1985).

Transbilayer proteins of this type are highly variable in terms of the number and types of different segments they contain. The five channel-pump proteins listed earlier reflect this heterogeneity. Although the precise three-dimensional arrangements of these proteins are as yet undetermined, their amino acid sequences predicted from the DNA sequences allow one to conceive of provisional models for their arrangement in the membrane, and these are provided in Figure 6.

These models are even more tentative than those predicted for the receptor class, but they do provide a feeling for the potential complexity that may be involved. The generalization that emerges from this schematic presentation is that there is no simple generalization: All possible arrangements may be encountered. The best characterized member of this class, rhodopsin, may have five nonpolar segments, similar in design to the single nonpolar loops of the receptor group. These are connected by short undefined structures of polypeptide, depicted as boxes in the diagram, which reside on different sides of the lipid bilayer. According to the amino acid sequence predicted from the DNA sequence, and confirmed by peptide amino acid sequence, two intramembranous segments of the rhodopsin molecule have charged amino acids within their sequences and are depicted in the diagram in the form of amphipathic helices. Although we do not know for certain that these seven intramembranous domains are necessarily α-helical in conformation, arguments by analogy with the x-ray diffraction findings of bacteriorhodopsin support this assumption. Again the polypeptide segments within these intramembranous segments are long enough to span the nonpolar regions of the lipid bilayer as helically coiled peptides, and the hydrogen-binding properties of the helix contribute to their thermodynamic stability.

The postulated arrangement of the intramembranous segments of the glucose carrier protein seems similar in overall design to the rhodopsin arrangement, but all the membrane-spanning segments are of the nonpolar type, and long stretches of polypeptide, predicted to be α-helical loops, are linked together without recognizable intervening extramembranous segments. The protein believed to be responsible for anion channel functions in red blood cell membranes has a similar arrangement, but most of the membrane-spanning segments are amphipathic rather than nonpolar loops. This protein also has a relatively large (over 400 amino acid)

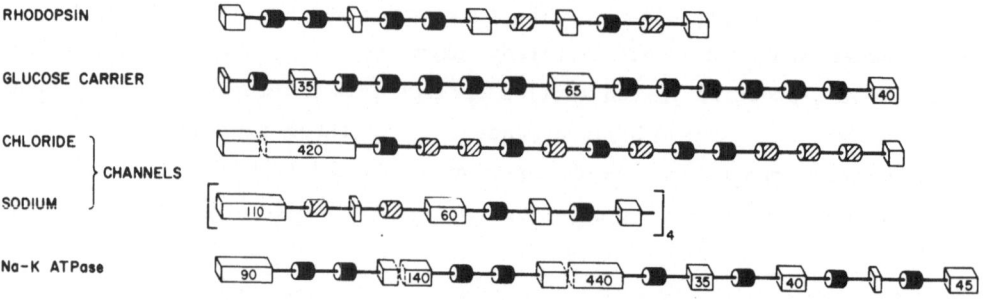

Fig. 6. Hypothetical arrangement of the different segments of transport proteins.

segment that extends into the cytoplasm of the cell, similar to the large cytoplasmic domains of some of the more complex receptor molecules. The postulated arrangements of polypeptide loops of the sodium channel and the sodium-potassium ATPase molecule have their own special variations. The point to be emphasized from this type of modelling is not that we have a clear understanding of the molecular anatomy of these molecules, but rather that their predicted amino acid sequences suggest an extremely heterogeneous arrangement of polypeptide chain within and around the lipid bilayer.

CONCLUSIONS

We now have an infinitely rich supply of structural information about most of the physiologically important transbilayer proteins that are of interest to students of the life sciences. This newly acquired structural information has done much to dispel the mysteries that have surrounded the study of these complex molecules. Based on this information we can explain our previous inability to isolate them using standard techniques, and we can understand now why they proved so resistant to analysis using conventional approaches. Using the principles of membrane biology that have emerged during the past four decades, we have formulated tentative and extremely simplistic models of their arrangements in the membrane, but the structural facts about these molecules still outstrip our capacity to formulate more than primitive guesses as to how they function. However, the newly acquired structural data will help to focus structure-function studies to the most appropriate sites, using direct and targeted modifications rather than the indirect and hit-or-miss approaches attempted in the past.

REFERENCES

Anantharamaiah, G. M., Jones, J. L., Brouillette, C. G., Schmidt, C. F., Chung, B. H., Hughes, T. A., Bhown, A. S., and Segrest, J. P., 1985, Studies of synthetic peptide analogs of the amphipathic helix: Structure of complexes with dimyristoyl phosphatidyl- choline, J. Biol. Chem., 260:10248-10255.

Anderson, R. A., and Marchesi, V. T., 1985, Associations between glycophorin and protein 4.1 are modulated by polyphosphoinositides: A mechanism for membrane skeletal regulation, Nature, 318:295-298.

Armitage, I., Shapiro, D. L., Furthmayr, H., and Marchesi, V. T., 1977, P^{31} nuclear magnetic resonance evidence for polyphosphoinositide associated with the hydrophobic segment of glycophorin A., Biochem., 16:1317-1320.

Dame, J. B., Williams, J. L., McCutchan, T. F., Weber, J. L., Wirtz, R. A., Hockmeyer, W. T., Maloy, W. L., Haynes, J. D., Schneider, I., Roberts, D., Sanders, G. S., Reddy, E. P., Diggs, C. L., and Miller, L. H., 1984, Structure of the gene encoding the immunodominant surface antigen on the sporozoite of the human malaria parasite Plasmodium falciparum, Science, 225:593-599.

Ebina, Y., Ellis, L., Jarnagin, K., Edery, M., Graf, L., Clauser, E., Ou, J.-H., Masiarz, F., Kan, Y. W., Goldfine, I. D., Roth, R. A., and Rutter, W. J., 1985, The human insulin receptor cDNA: The structural basis for hormone-activated transmembrane signalling, Cell, 40:747-758.

Henderson, R., and Unwin, P. N. T., 1975, Three-dimensional model of purple membrane obtained by electron microscopy, Nature, 257:28-32.

Kaufman, J. F., Auffray, C., Korman, A. J., Shackelford, D. A., and Strominger, J., 1984, The class II molecules of the human and murine major histocompatability complex, Cell, 36:1-13.

Kopito, R. R., and Lodish, H. F., 1985, Primary structure and trans-membrane orientation of the murine anion exchange protein, Nature, 316:234-238.

Leonard, W. J., Depper, J. M., Crabtree, G. R., Rudikoff, S., Pumphrey, J., Robb, R. J., Kronke, M., Svetlik, P. B., Peffer, N. J., Waldmann, T. A., and Greene, W. C., 1984, Molecular cloning and expression of cDNAs for the human interleukin-2 receptor, Nature, 311:626-630.

Maddon, P. J., Littman, D. R., Godfrey, M., Maddon, D. E., Chess, L., and Axel, R., 1985, The isolation and nucleotide sequence of a cDNA encoding the T cell surface protein T4: A new member of the immunoglobulin gene family, Cell, 42:93-104.

McClelland, A., Kuhn, L., and Ruddle, F. H., 1984, The human transferrin receptor gene: Genomic organization, and the complete primary structure of the receptor deduced from a cDNA sequence, Cell, 39:267-274.

Mostov, K. E., Friedlander, M., and Blobel, G., 1984, The receptor for transepithelial transport of IgA and IgM contains multiple immunoglobulin-like domains, Nature, 308:37-43.

Mueckler, M., Caruso, C., Baldwin, S. A., Panico, M., Blench, I., Morris, H. R., Allard, W. J., Lienhard, G. E., and Lodish, H. F., 1985,

Sequence of structure of a human glucose transporter, Science, 229:941-945.

Nathans, J., and Hogness, D. S., 1984, Isolation and nucleotide sequence of the gene encoding human rhodopsin, Proc. Natl. Acad. Sci. USA, 81:4851-4855.

Nikaido, T., Shimizu, A., Ishida, N., Sabe, H., Teshigawara, K., Maeda, M., Uchiyama, T., Yodoi, J., and Honjo, T., 1984, Molecular cloning of cDNA encoding human interleukin-2 receptor, Nature, 311:631-635.

Noda, M., Shimizu, S., Tanabe, T., Takai, T., Kayano, T., Ikeda, T., Takahashi, H., Nakayama, H., Kanaoka, Y., Minamino, N., Kangawa, K., Matsuo, H., Raftery, M. A., Hirose, T., Inayama, S., Hayashida, H., Miyata, T., and Numa, S., 1984, Primary structure of Electrophorus electricus sodium channel deduced from cDNA sequence, Nature, 312:121-127.

Ozaki, L. S., Svec, P., Nussenzweig, R. S., Nussenzweig, V., and Godson, G. N., 1983, Structure of the plasmodium knowlesi gene coding for the circumsporozoite protein, Cell, 34:815-822.

Schulte, T. H., and Marchesi, V. T., 1979, Conformation of human erythrocyte glycophorin A and its constituent peptides, Biochem., 18:275-280.

Seki, T., Spurr, N., Obata, F., Goyert, S., Goodfellow, P., and Silver, J., 1985a, The human Thy-1 gene: Structure and chromosomal location, Proc. Natl. Acad. Sci. USA, 82:6657-6661.

Seki, T., Chang, H.-C., Moriuchi, T., Denome, R., Ploegh, H., and Silver, J., 1985b, A hydrophobic transmembrane segment at the carboxyl terminus of Thy-1, Science, 227:649-651.

Seki, T., Moriuchi, T., Chang, H.-C., Denome, R., and Silver, J., 1985c, Structural organization of the rat thy-1 gene, Nature, 313:485-487.

Shull, G. E., Schwartz, A., and Lingrel, J. B., 1985, Amino-acid sequence of the catalytic subunit of the (Na^+-K^+)ATPase deduced from a complementary DNA, Nature, 316:691-700.

Tomita, M., Furthmayr, H., and Marchesi, V. T., 1978, Primary structure of human erythrocyte glycophorin A. Isolation and Characterization of peptides and complete amino acid sequence, Biochem., 17:4756-4770.

Ullrich, A., Bell, J. R., Chen, E. Y., Herrera, R., Petruzzelli, M., Dull, T. J., Gray, A., Coussens, L., Liao, Y.-C., Tsubokawa, M., Mason A., Seeburg, P. H., Grunfeld, C., Rosen, O. M., and Ramachandran, J., 1985, Human insulin receptor and its relationship to the tyrosine kinase family of oncogenes, Nature, 313:756-761.

Ullrich, A., Coussens, L., Hayflick, J. S., Dull, T. J., Gray, A., Tam, A. W., Lee, J., Yarden, Y., Liberman, T. A., Schlessinger, J.,

Downward, J., Mayes, E. L. V., Whittles, N., Waterfield, M. D., and Seeburg, P. H., 1984, Human epidermal growth factor receptor cDNA sequence and aberrant expression of the amplified gene in A431 epidermoid carcinoma cells, Nature, 309:418-425.

Yamamoto, T., Davis, C. G., Brown, M. S., Schneider, W. J., Casey, M. L., Goldstein, J. L., and Russell, D. W., 1984, The human LDL receptor: A cysteine-rich protein with multiple alu sequences in its mRNA, Cell, 39:27-38.

CHARACTERIZATION OF A SPERM MEMBRANE GLYCOPROTEIN

David W. Hamilton, John C. Wenstrom, and Alison Moore

Department of Anatomy
University of Minnesota
Minneapolis, Minnesota 55455

INTRODUCTION

The sperm plasma membrane has been the object of intense research activity in recent years (Moore, 1985). As a consequence, a large amount of data has accumulated, particularly on its structure and biochemistry, that is now beginning to yield information that may be generally useful in cell biology. Of particular interest are the restricted structural domains that overlie specific regions of the cell (such as the acrosome, the mid-piece mitochondria, etc.).

Immunocytochemical studies have shown that membrane antigens can have distributions that either are restricted to specific plasma membrane domains, overlap into other domains, or span a number of domains (Feuchter et al., 1981; Miles et al., 1981; Gaunt, 1982; Olson and Orgebin-Crist, 1982; Schmell et al., 1892; Brooks and Tiver, 1983; Primakoff and Myles, 1983). The distribution of antigens among domains is complicated by the fact that the actual components of the membrane vary during epididymal maturation (e.g., Voglmayr et al., 1980; Brooks and Tiver, 1984; Dacheux and Voglmayr, 1983), and the final product in the mature sperm may depend as much on molecules synthesized and secreted into the lumen by the epididymal epithelium as on molecules present in the membrane when sperm are released from the seminiferous epithelium.

The complexity of the sperm plasma membrane, both in terms of its constituents and in the changes that it goes through during maturation, makes it a difficult subject for study. Structural and biochemical simplicity in a sperm plasma membrane, therefore, would be a useful attribute in studying dynamic aspects of post-testicular sperm maturation.

The rat sperm surface, at least comparatively, appears to meet the criterion of simplicity. Studies on the glycoproteins of the plasma membrane of rat cauda epididymal sperm (Hamilton et al., 1986) have revealed that there is a single major glycoprotein that can be labeled in its carbohydrate moieties. Additionally, the studies by Hamilton et al. (1983, 1986) and others (Jones et al., 1981; Brown et al., 1983; Zeheb and Orr, 1984) confirm the earlier report by Olson and Hamilton (1978) that it is not possible to detect the same molecule on caput sperm using carbohydrate labeling techniques. Thus, at least one maturational event involves glycosylation of the sperm surface.

In this review, the major surface glycoprotein of rat cauda epididymal sperm will be discussed with respect to its characterization and its possible origin.

IDENTIFICATION OF THE MAJOR GLYCOPROTEIN ON RAT SPERM SURFACES

When rat cauda epididymal sperm are treated with either galactose oxidase-NaB[^{3}H]$_4$ or NaIO$_4$-NaB[^{3}H]$_4$ (to label surface galactose or sialic acid moieties, respectively; Gahmberg and Hakomori, 1973; van Lenten and Ashwell, 1971), ^{3}H is incorporated primarily into two macromolecules that can be visualized by SDS-polyacrylamide gel electrophoresis on gels with >10% acrylamide concentration. The first is a protease sensitive peak that runs on the gel at an apparent molecular weight of 24,000 daltons. The second is a chloroform/methanol-extractable peak that runs at the dye-front. Other small peaks are present along the length of the gel, but in total, these represent only a small part of the applied radioactivity.

The major glycoprotein was first described by Olson and Hamilton (1978) and assigned a molecular weight of 37,000 daltons based on SDS-PAGE. Subsequently it has been studied by a number of laboratories, and reported molecular weights have varied from 37,000 (Olson and Danzo, 1981) to 32,000 (Jones et al., 1981; Brown et al., 1983; Zeheb and Orr, 1984). In recent studies it was shown that the actual molecular weight of the molecule is closer to 24,000 (Hamilton et al., 1983, 1986) based on experiments in which labeled extracts were electrophresed in a variety of acrylamide concentrations (Table I). Extracts centrifuged on sucrose density gradients in the presence and absence of detergent also confirmed that the molecule's molecular weight is approximately 24,000 daltons (Hamilton et al., 1986).

CHARACTERIZATION OF THE M_r = $\sim$24,000 GLYCOPROTEIN

In order to investigate the origin of the M_r = $\sim$24,000 membrane polypeptide it was necessary to characterize it as fully as possible, bearing in mind that it was only detectable through labeled carbohydrates in early stages of the study. Initial experiments focused on whether or not it is an integral membrane molecule. Studies by Hamilton et al. (1983, 1986), as well as by Brown et al. (1983), Olson and Orgebin-Crist (1982), and Zeheb and Orr (1984) clearly showed that it is a hydrophobic molecule that is not extracted by high or low salt, high or low pH, EDTA, dithiothreitol or betamercaptoethanol, or washing by centrifugation. It is partially extracted with 6M guanidine hydrochloride and 8M urea, and is completely removed from the sperm with both ionic and non-ionic detergents. Thus, it is an integral membrane molecule, and there is evidence (Hamilton et al. 1983, 1986; Brown et al. 1983; Zeheb and Orr, 1984) that it is a lipoprotein.

Table I. Molecular weight determination of the major glycoproteins.

Percent Gel	Apparent Molecular Weight	Number
5.6	37,900 ± 830	5
11.2	23,000	1
14.0	24,875 ± 1,274	24
16.8	23,300 ± 300	3

Data derived from Hamilton et al. (1986). Galactose oxidase-NaB$[^3H]_4$ or NaIO$_4$-NaB$[^3H]_4$ labeled sperm were extracted with 40 mM octylbeta glucopyranoside. After dialysis the extracts were electrophoresed in 10 cm tube gels in gel percentages as given above. Gels were sliced/pulverized into Aquasol-2 and counted by liquid scintillation spectroscopy. Peak migration was measured and compared to migration of radioactive standards (Phosphorylase B, M_r = 97,400; Bovine serum albumin, M_r = 69,000; Ovalbumin, M_r = 46,000; Carbonic anhydrase, M_r = 30,000; Cytochrome C, M_r = 12,300) for molecular weight determinations. A plot of percent acrylamide against relative mobility (Hamilton et al., 1986) shows that there is a significant deflection from linearity with the samples run in 5.6% gels.

The polypeptide is trypsin sensitive, and when supernatants from trypsin-treated sperm are electrophoresed on SDS-PAGE tube gels, a major peak is found at $M_r = \sim 10,000$. This is interpreted to mean that about half of the molecular weight of the molecule is accessible to trypsin and therefore is exposed on the cell surface.

IMMUNOCYTOCHEMICAL STUDIES ON THE DISTRIBUTION OF THE MOLECULE

As noted in the introduction, there is a large body of data that indicates that proteins and glycoproteins in the luminal fluid interact with sperm in the epididymis, either by inserting into the membrane or by becoming more or less tightly bound to the surface (Kohane, et al., 1980; Vernon et al., 1982).

Data concerning the rat sperm surface are somewhat contradictory. Jones and his colleagues (Op. cit.) have shown that antibodies raised against fluid molecules ($M_r = 23,000$, $M_r = 19,000$, and $M_r = 18,500$) also interact with the sperm surface, and Klinefelter and Hamilton (1984, 1985)

Table II. Immunoreactivity of antigen using monoclonal antibodies.

Location	B109	Serum
TESTIS	+	+
RETE TESTIS		
Sperm	+	+
Fluid	−	−
CAPUT		
Sperm	+	+
Fluid	−	−
CORPUS		
Sperm	+	+
Fluid	−	−
CAUDA		
Sperm	+	+
Fluid	−	−
EPITHELIUM		
Caput	−	+
Corpus	−	+
Cauda	−	+

Localization (+) or lack of localization (−) of antigen using either monoclonal antibody (B109) or rabbit serum antibodies raised against epididymal alpha-lactalbumin-like polypeptides (Hamilton, 1981). B109 recognizes only the $M_r = \sim 24,000$ molecule. Serum antibodies recognize both the $M_r = \sim 24,000$ and $M_r = \sim 18,000$ molecules.

and Ensrud et al. (1985) have shown that both rabbit serum antibodies and a monoclonal antibody raised against the epididymal alpha-lactalbumin-like polypeptides (Hamilton, 1981) recognize sperm (Table II). Zeheb and Orr (1984), on the other hand, were not able to show immunological cross-reactivity between luminal and sperm surface M_r = 32,000 molecules.

Recent immunocytochemical studies (Ensrud et al. 1985; unpublished data) have shown that the ~24,000 dalton antigen is rather widely distributed in the rat male reproductive tract. Using the rabbit serum antibody (Klinefelter and Hamilton, 1984) and a monoclonal antibody (B109) raised against the epididymal alpha-lactalbumin-like polypeptides, it has been possible to show that the antigen is detected in the testis in late spermatids (Fig. 1) and in residual bodies, but not in Sertoli cells (unpublished data). An unexplained localization is that the antigen is found also in Leydig cells.

STUDIES ON THE ORIGIN OF THE M_r = ~24,000 MOLECULE

The major issue in establishing the origin of the M_r = ~24,000 molecule is whether it is present on caput epididymal sperm or not, since if it is then the fact that it cannot be detected using carbohydrate

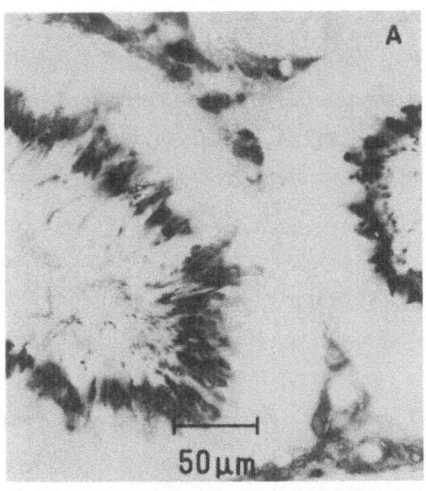 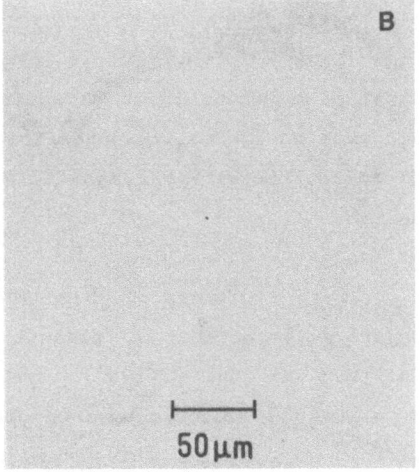

Fig. 1. Immunocytochemical staining of rat testis with monoclonal antibody B109. Note that staining is restricted to periluminal sites. Other sections show clearly that this represents staining in late spermatid cytoplasm. (A) positively stained material. (B) control.

labeling techniques means that it must be glycosylated in the epididymis. If the molecule is not present on caput sperm, then the implication is that it must come from the epididymis. A number of possible mechanisms have been proposed to explain earlier observations that the molecule can be detected only on cauda sperm (Olson and Hamilton, 1978; Brown et al., 1983), such as unmasking of a preexisting molecule by protease or glycosidase activity, or glycosylation of an existing glycoprotein with incomplete oligosaccharide chains. Some data exist (Hamilton and Gould, 1982) which show that caput sperm surfaces can be galactosylated in vitro. Another possibility is that the molecule originates by insertion into the sperm membrane of a molecule synthesized and secreted by the epididymal epithelium.

The epididymis can synthesize and secrete a M_r = ~24,000 protein in vitro (both in explants and in perifusion chambers; Brooks and Tiver, 1983; Klinefelter and Hamilton, 1984, 1985; unpublished data), but it is not yet clear whether this molecule binds to sperm. In experiments using perifusion culture of caput epididymal tubule segments in which either [^{35}S]methionine or [^{14}C]amino acid mixture were present as substrates (Klinefelter and Hamilton, 1984, 1985), a labeled M_r = ~24,000 molecule could be detected in fluid obtained from minced tubule segments, but the labeled molecule could not be immunoprecipitated with a rabbit antibody against the epididymal alpha-lactalbumin-like polypeptides either from fluid or from detergent extracts of sperm (Klinefelter and Hamilton, 1985). An M_r = ~18,000 polypeptide was immunoprecipitated from both fluid and sperm, however. In preliminary experiments using short-term incubation of caput tubule fragments, Moore and Hamilton (unpublished data) also were able to detect secretion of an M_r = ~24,000 polypeptide, but were unable to immunoprecipitate it. Again, however, it was possible to immunoprecipitate a labeled M_r = ~18,000 polypeptide from the culture medium.

There are numerous observations that support the contention that the major glycoprotein is present on sperm in the caput epididymidis. Hamilton and Gould (1982) were able to enzymatically galactosylate an M_r = ~24,000 molecule (among others) on rat caput epididymal sperm in vitro. Also, after three days of perifusion culture, it is possible to label sperm with the galactose oxidase-NaB[^{3}H]$_4$ technique and recover a single major peak that migrates at ~24,000 daltons (as well as a peak at the dye-front, Klinefelter and Hamilton, 1985). The M_r = ~24,000 peak is also able to be immunoprecipitated with the rabbit serum antibody against epididymal alpha-lactalbumin-like polypeptides. Finally, detergent extracts of caput sperm have a band running at M_r = ~24,000 that is able to be immunostained with our rabbit serum antiboby (Klinefelter and Hamilton, 1985). Thus, sperm in the caput possess a plasma membrane molecule of the proper molecular weight and it can be glycosylated in vitro.

These data, then, lead to the conclusion that although an M_r = ~24,000 polypeptide is synthesized and secreted in the epididymis, it is not the major glycoprotein of the sperm plasma membrane discussed here. The immunocytochemical evidence discussed above, suggests that the molecule first appears in late spermatids and therefore is of testicular origin.

ACKNOWLEDGEMENTS

Original research referred to in this chapter was supported by USPHS grant HD11962 and by a grant from the Andrew W. Mellon Foundation to DWH.

REFERENCES

Brooks, D. E., and Tiver, K., 1983, Localization of epididymal secretory proteins on rat spermatozoa, J. Reprod. Fertil., 69:651-657.

Brooks, D. E., and Tiver, K., 1984, Analysis of surface proteins of rat spermatozoa during epididymal transit and identification of antigens common to spermtozoa, rete testis fluid and cauda epididymal plasma, J. Reprod. Fertil., 71:249-257.

Brown, C. R., von Glos, K. I., and Jones, R., 1983, Changes in plasma membrane glycoproteins of rat spermatozoa during maturation in the epididymis, J. Cell Biol., 96:256-264.

Dacheux, J. L., and Voglmayr, J. K., 1983, Sequence of sperm surface differentiation and its relationship to exogenous fluid proteins in the ram epididymis, Biol. Reprod., 27:1033-1046.

Ensrud, K., Baker, J. B., Wenstrom, J. C., and Hamilton, D. W., 1985, A monoclonal antibody against rat epididymal alpha-lactalbumin-like 24Kd polypeptide recognizes rat cauda sperm surface, J. Androl., 6:54P.

Feuchter, F. A., Vernon, R. B., and Eddy, E. M., 1981, Analysis of the sperm surface with monoclonal antibodies: Topographically restricted antigens appearing in the epididymis, Biol. Reprod. 24:1099-1110.

Gahmberg, C. G., and Hakomori, S., 1973, External labeling of cell surface galactose and galactosamine in glycolipid and glycoprotein of human erythrocytes, J. Biol. Chem., 248:4311-4317.

Gaunt, S. J., 1982, A 28K-dalton cell surface autoantigen of spermatogenesis: Characterization using a monoclonal antibody, Dev. Biol., 89:92-100.

Hamilton, D. W., 1981, Evidence for alpha-lactalbumin-like activity in reproductive tract fluids of the male rat, Biol. Reprod., 25:385-392.

Hamilton, D. W., and Gould, R. P., 1982, Preliminary observations on enzymatic galactosylation of glycoproteins on the surface of rat caput epididymal spermatozoa, Int. J. Androl., Suppl., 5:73-80.

Hamilton, D. W., Wenstrom, J. C., and Baker, J. C., 1983, Partial characterization of an integral M_r 24,000 membrane glycoprotein from rat epididymal spermatozoa, J. Cell Biol., 97:13A.

Hamilton, D. W., Wenstrom, J. C., and Baker, J. B., 1986, Membrane glycoproteins from spermatozoa: Partial characterization of an integral M_r = ∿24,000 molecule from rat spermatozoa that is glycosylated during epididymal maturation, Biol. Reprod., in press.

Jones, R., Pholpramool, C., Setchell, B. P., and Brown, C. R., 1981, Labelling of membrane glycoproteins on rat spermatozoa collected from different regions of the epididymis, Biochem. J., 200:457-460.

Klinefelter, G. R., and Hamilton, D. W., 1984, Organ culture of rat caput epididymal tubules in a perifusion chamber, J. Androl., 5:243-258.

Klinefelter, G. R., and Hamilton, D. W., 1985, Synthesis and secretion of proteins by perifused caput epididymal tubules, and association of secreted proteins with spermatozoa, Biol. Reprod., 33:1017-1027.

Kohane, A. C., Gonzalez Echeverria, F. M. C., Pineiro, L., and Blaquier, J. A., 1980, Interaction of proteins of epididymal origin with spermatozoa, Biol. Reprod., 23:737-742.

Miles, D. G., Primakoff, P., and Bellve, A. R., 1981, Surface domains of the guinea pig sperm defined with monoclonal antibodies, Cell, 23:433-439.

Moore, H. D. M., 1985, Bibliography on the mammalian sperm surface, Bibl. Reprod., 46:A1-A10.

Olson, G. E., and Danzo, B. J., 1981, Surface changes in rat spermatozoa during epididymal transit, Biol. Reprod., 24:431-443.

Olson, G. E., and Hamilton, D. W., 1978, Characterization of the surface glycoproteins of rat spermatozoa, Biol. Reprod., 19:26-35.

Olson, G. E., and Orgebin-Crist, M.-C., 1982, Sperm surface changes during epididymal maturation, in: "Cell Biology of the Testis," C. W. Bardin and R. J. Sherins, eds., New York Academy of Sciences, New York, Annals vol. 383:372-391.

Primakoff, P., and Miles, D. G., 1983, A map of the guinea pig sperm surface constructed with monoclonal antibodies, Dev. Biol., 98:417-428.

Schmell, E. D., Gulyas, B. J., Yuan, L. C., and August, J. T., 1982, Identification of mammalian sperm surface antigens: II. Characterization of an acrosomal cap protein and a tail protein using monoclonal anti-mouse sperm antibodies, J. Reprod. Immunol., 4:91-106.

van Lenten, L., and Ashwell, G., 1971, Studies on the chemical and enzymatic modification of glycoproteins, J. Biol. Chem., 246:1889-1894.

Vernon, R. B., Muller, C. H., Herr, J. C., Feuchter, F. A., and Eddy, E. M., 1982, Epididymal secretion of a mouse sperm surface component recognized by a monoclonal antibody, Biol. Reprod., 26:523-535.

Voglmayr, J. K., Fairbanks, G., Jackowitz, M. A., and Colella, J. R., 1980, Post-testicular developmental changes in the ram sperm cell surface and their relationship to luminal fluid proteins of the reproductive tract, Biol. Reprod., 22:655-667.

Zeheb, R., and Orr, G. A., 1984, Characterization of a maturation-associated glycoprotein on the plasma membrane of rat caudal epididymal sperm, J. Biol. Chem., 259:839-848.

SPERM MEMBRANE AND ZONA PELLUCIDA INTERACTIONS

DURING FERTILIZATION

Michael G. O'Rand, Jeffrey E. Welch, and Susan J. Fisher

Laboratories for Cell Biology
Department of Anatomy
University of North Carolina
Chapel Hill, North Carolina 27514

INTRODUCTION

In most species, fertilization is a necessary and essential step for the passage of one generation to the next. The genome is passed via spermatozoon and ovum into the zygote, and development begins anew. The spermatozoon's passage through the female reproductive tract and its capacitation, which is necessary for successful egg penetration, encompass the fertilization process. Successful fertilization and therefore the survival of the species depends upon a series of steps which has been described as a hierarchy of specificities (O'Rand, 1985): from geographic isolation to the egg plasma membrane. The meeting of the sperm's plasma membrane and the egg's zona pellucida is almost the last step in this hierarchy and is the subject of the present discourse.

THE PLASMA MEMBRANE

The cell's plasma membrane has been the object of intense study over the last fifteen years, with a wealth of data supporting the current concept of a fluid-mosaic model (Singer and Nicolson, 1972), in which glycoproteins are embedded in a lipid bilayer. The bilayer has a fluid-like interior where the tails of the acyl chains are located, and a more ordered exterior where the lipid head groups are located. The glycoproteins may be transmembrane or embedded in only one half of the bilayer. Imposed upon this fluid-mosaic structure are the constraints that each cell type must have in order to organize its membrane components into functional domains. These constraints, which affect the lateral mobility of glycoproteins in the bilayer (Jacobson et al., 1984), may be internal and cytoskeletal in nature, external and proteoglycan in nature, or inherent in the lipid composition of the bilayer itself.

The lateral mobility of glycoproteins in the bilayer is critical for some cell functions such as endocytosis (Goldstein et al., 1979), while the lateral movement of lipids may be critical for certain protein functions such as ion transport, and for establishing areas of the membrane specialized for fusion. Secretory cells in particular require relatively fluid domains in both the plasma membrane and secretory vesicle membrane in order to accomplish fusion (Orci et al., 1977). Similarly, the spermatozoon requires lipid domains ready for fusion events in both the acrosome reaction (Bearer and Friend, 1982) and in subsequent sperm-egg plasma membrane interaction.

During epididymal maturation and capacitation, the sperm's plasma membrane is prepared for zona binding, the acrosome reaction, and subsequent zona penetration through changes in the intrinsic plasma membrane composition and fluidity (O'Rand, 1979, 1982; Parks and Hammerstedt, 1985). Although the bulk of the lipid can diffuse laterally in either direction between head and tail, as measured by the lipid probe C16diI (Wolf and Voglmayr, 1984), there are regional differences in lipid composition (Bearer and Friend, 1982; Friend and Bearer, 1981). The cholesterol/phospholipid molar ratio increases during maturation (Parks and Hammerstedt, 1985) and quantitatively more sterols, as measured by filipin, are found over the sperm head (Friend, 1982). Cardiolipin is also found more abundantly in the head region (Bearer and Friend, 1982; Parks and Hammerstedt, 1985). In preparation for fusion, a patchwork quilt-like topography appears over the anterior sperm head in which there are areas of high fluidity in preparation for fusion interspersed with areas of decreased fluidity rich in protein (see Fig. 6 in O'Rand, 1979). Also, in preparation for fusion, cholesterol content in the anterior head plasma membrane is thought to decrease (Davis et al., 1979), but this observation has not been substantiated in other systems, and the removal or rearrangement of cholesterol alone may not be enough to cause the membrane fusion event (Parks and Hammerstedt, 1985). Indeed, the presence of unique phosphorylcholines (Parks and Hammerstedt, 1985) and phospholipase activity (Conway and Metz, 1976) may be involved in the actual membrane vesiculation of the acrosome reaction (Meizel, 1978;

132

Meizel and Turner, 1984; Ono et al., 1982).

THE ZONA PELLUCIDA

The mammalian zona pellucida is an extracellular matrix of glycoproteins unique to the oocyte (Dunbar and Wolgemuth, 1984; Wassarman et al., 1985; Dunbar et al., 1980; Bleil and Wassarman, 1980). Unlike cell surface glycosaminoglycans (Hook et al., 1984), in which the polysaccharide chains predominate on a core protein (e.g. the chondroitin sulfates), the zona pellucida is predominately glycoprotein in nature, containing about 71% protein, 19% neutral hexoses, sulfate, and sialic acid (Dunbar et al., 1980). Mammalian zonae typically consist of three glycoprotein families, each with a polypeptide chain of differing molecular weight, and the oligosaccharides are both N- and O-linked to the chains (Wasserman et al., 1985). Each polypeptide exhibits charge heterogeneity based on the variation in the amounts of galactose, N-acetylglucosamine, and sialic acid of each family member (Hedrick and Wardrip, 1981).

Structurally, the zona pellucida is a rather uniform Swiss cheese-like layer of branching fibers (Yanagimachi and Philips, 1984; Dietl and Czuppon, 1984; Greve and Wassarman, 1985), with all three families of glycoproteins represented on the surface of the intact structure (Hedrick and Wardrip, 1982). During isolation, the zona appears to retain its shape even after the oocyte itself has been removed. Therefore, it would appear to have a considerable degree of rigidity provided by the individual fibrils and filaments, as seen after LIS extraction or chymotrypsin digestion (Dietl and Czuppon, 1984; Greve and Wassarman, 1985). Morphological observations of spermatozoa penetrating the zona indicate that the lateral aspect of the spermatozoon head binds to the fibrilar network and a penetration slit is formed (e.g., Yanagimachi and Philips, 1984; Dudkiewicz et al., 1976). As the spermatozoon proceeds through the slit, the remnants of the membranes from the acrosome reaction are left behind (Yanagimachi and Philips, 1984).

THE INTERACTION

A number of studies have addressed the topic of sperm-zona interaction (Fig. 1), its specificity, and relationship to zona penetration (e.g., Bedford, 1977; Peterson et al., 1980, 1984; Saling, 1982; Sacco et al., 1984; Swenson and Dunbar, 1982; O'Rand, 1985; O'Rand et al., 1985; Yanagimachi, 1981). The zona is thought to contain a sperm

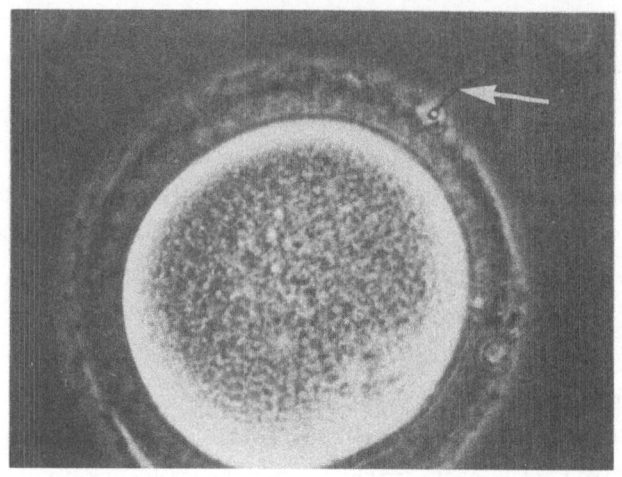

Fig. 1. Sperm-Zona Interaction. Human oocyte showing a
spermatozoon (arrow) bound to the zona pellucida.

receptor: in the mouse it is the glycoprotein ZP3 of 80,000 daltons
(Wassarman et. al., 1985) and in the pig it is the 58,000 dalton
glycoprotein (Sacco et al., 1984). Evidence in the mouse indicates that
small ZP3 glycopeptides which result from protease digestion and which
contain the O-linked oligosaccharides retain sperm receptor activity
(Florman and Wassarman, 1985). The link between the zona's sperm receptor
and the sperm's zona binding proteins (ZBP) is, however, unknown.

Early workers speculated on a possible sperm-lectin binding to zona
oligosaccharides (Yanagimachi, 1977 and earlier references therein), while
the results from studies using protease inhibitors suggested that
proteases might have a function in sperm-zona binding (Saling, 1981).
More recent studies have revealed an additional variety of substances
which inhibit sperm-zona interaction. For example, Huang et al. (1982)
found L-fucose and fucoidin to be strongly inhibitory to guinea pig sperm
binding to zona. Ahuja (1982) reported that acetylated amino sugars, as
well as fucoidin inhibited fertilization in the hamster. He also reported
that a number of plasma glycoproteins inhibited fertilization depending
upon the concentration of substance used. Treatment of capacitated
hamster sperm with fucosidase reduced fertilization to 7.7%, but treatment
with trypsin had no effect (Ahuja, 1982). These results imply that the
important sugar residues are on the spermatozoon. At the moment, these
data are hard to reconcile with those from other laboratories. Shur and
Hall (1982) have suggested that sperm surface N-acetylglucosamine:
galactosyltransferases are responsible for sperm-zona interaction and have
shown that both alpha-lactalbumin and UDP-dialdehyde inhibit sperm binding
to the zona.

134

More recently, Benau and Storey (1985) have clarified the possibility of a trypsin-like enzyme being involved in the binding by showing that unlike trypsin or acrosin, the active site titrant, 4-methylumbelliferyl-p-guanidinobenzoate (MUGB), is turned over by the plasma membrane. This is interpreted to mean that while the binding site may be enzyme-like, it is not enzymatic. In this regard, work from Poirier's laboratory is of interest (Aarons et al., 1984, 1985). They have demonstrated a plasma membrane protein of approximately 15,000 daltons on mouse spermatozoa, which is itself non-enzymatic but which binds an acid-stable protease inhibitor of seminal vesicle origin (SVI). Moreover, both MUGB and soybean trypsin inhibitor will effectively compete with SVI for the plasma membrane binding site. Consequently, the current results suggest a non-enzymatic binding site on the plasma membrane which is sensitive to serine protease inhibitors and which is also capable of binding certain sugars such as N-acetylglucosamine and fucose or fucoidin, which contains alpha-1,3-linked fucose.

Attempts to isolate and study the sperm membrane proteins responsible for the zona binding have begun to meet with some success (Peterson et al., 1985; O'Rand et al., 1985; Sullivan and Bleau, 1985). Using boar sperm, Peterson et al. (1985) have identified several plasma membrane proteins which have a high affinity for zona as well as for dextran sulfate. One of these proteins has a molecular weight of approximately 70,000 daltons and several are below 18,000 daltons. The binding of these proteins to the zona is sensitive to conformational changes and to trypsin treatment. The proteins tend to aggregate upon standing and are typical membrane proteins. These authors suggest a sperm lectin-like protein binding to the zona oligosaccharide chains.

O'Rand et al. (1985) have identified both homologous and heterologous zona binding proteins (ZBP) in the rabbit, pig, mouse, and human using ^{125}I labeled heat solubilized zonae (HSZ) incubated with nitrocellulose blots of sperm proteins following SDS-PAGE. It was found that spermatozoa from each species have their own characteristic set of ZBP, although they may also bind heterologous zona to some extent. The rabbit's ZBP have relative molecular weights of 32K, 18K, 16K, and 14K. The 14,000 molecular weight molecule has been identified as the rabbit sperm autoantigen, RSA-1 (O'Rand, 1985; O'Rand et al., 1985). The pig's major ZBP are 16K and 18K with minor bands between 63K-88K and 45K-53K; human spermatozoa bind human zona at 17K and 18K and mouse epididymal spermatozoa bind mouse zona at 19K, 18K, and 16K. The most obvious

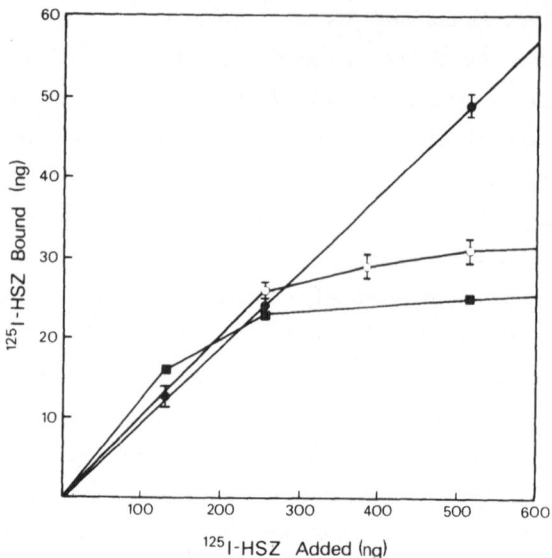

Fig. 2.   Binding of ^{125}I-heat solubilized rabbit zonae to
rabbit spermatozoa. Ejaculated spermatozoa were
capacitated _in vitro_ according to the method of
Brackett et al. (1982) which yields 63%
fertilization in our laboratory. The incubation
mixture consisted of 500µl of spermatozoa (7 x 10[6]
sperm) with increasing amounts of ^{125}I-HSRZ in
0-200µl in a total volume of one ml in Brackett's
medium. Soybean trypsin inhibitor was added at one
mg/ml for a total concentration of either 200µg or
300µg; Rabbit zonae, no STI, ●——●; Rabbit zonae,
plus 300µg STI, ○——○ ; Rabbit zonae, plus
capacitated spermatozoa fixed with 2.5%
glutaraldehyde in McCoy's 5a medium for ten
minutes, washed, blocked with 0.1M glycine and
resuspended in Brackett's medium before the
incubation with zonae, □——□; Points indicate mean
of at least two experiments and the range of
experimental values.

similarity among these species was the presence of low molecular weight
ZBP between 14,000 and 19,000 daltons. Each species also had a higher
molecular weight band or bands which were considered minor using the
blotting technique. However, the blotting technique requires a strong
[125]I label on a particular zona component and the conformational integrity
of the sperm component. Thus, in the physiological sperm-zona
interaction, these minor or weakly labeled bands may be important.

Using a similar technique, Sullivan and Bleau (1985) have identified
one or two hamster spermatozoa ZBP in the 25K-31K molecular weight range
with the major band at 26,400 daltons. The hamster zona did not appear to
react with ram, bull, or stallion spermatozoa.

To investigate further the homologous binding of rabbit zonae to rabbit spermatozoa, we have used an in vitro incubation system containing increasing amounts of ^{125}I-heat solubilized rabbit zonae (HSRZ) and 7×10^6 live motile spermatozoa. At this sperm concentration, the amount of bound zonae increases linearly with the amount of zonae added (Fig. 2). However, we also found that saturation can be achieved in this system if soybean trypsin inhibitor (STI) is added or if the spermatozoa are glutaraldehyde-fixed before incubation (Fig. 2). Under non-saturating conditions (5×10^7 sperm/ml and 0.5 ng of ^{125}I-HSRZ), the amount of bound zonae will decline over time if the experiment is not terminated by adding inhibitors or fixing the spermatozoa (Fig. 3). Figure 3 also shows that the binding and loss of bound zonae are almost the same for high ionic strength treated and in vivo capacitated spermatozoa.

Rabbit spermatozoan ZBP recognize both rabbit and pig zonae (O'Rand et al., 1985), and in vitro, rabbit spermatozoa bind to intact rabbit and pig zonae, yet rabbit spermatozoa do not penetrate pig zonae to any great

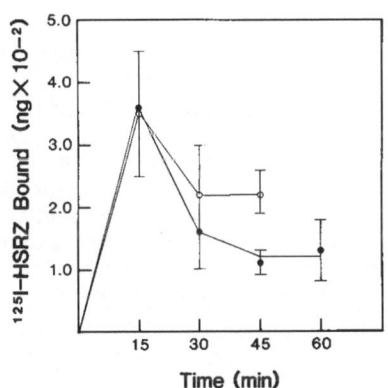

Fig. 3. Binding of ^{125}I-HSRZ to rabbit in vivo capacitated (O—O) and salt treated ejaculated spermatozoa (●—●) over time. Spermatozoa were washed in McCoy's 5a medium containing 0.3% bovine serum albumin, treated for ten minutes with McCoy's-BSA containing 3.4 mg/ml NaCl, washed and resuspended in McCoy's-BSA. The incubation mixture consisted of 500 μl of rabbit spermatozoa (5×10^7 sperm/ml) and 0.5 ng of rabbit ^{125}I-HSRZ (4.8×10^3 cpm/ng). After incubation at 37°C for each time period, 0.5ml of McCoy's-BSA containing soybean trypsin inhibitor (1 mg/ml), 5 mM PMSF and 0.2% sodium azide was added to each tube. Spermatozoa were centrifuged, washed, and counted. In vivo capacitated spermatozoa were 5.6×10^5 sperm/ml. Bars indicate range of experimental values.

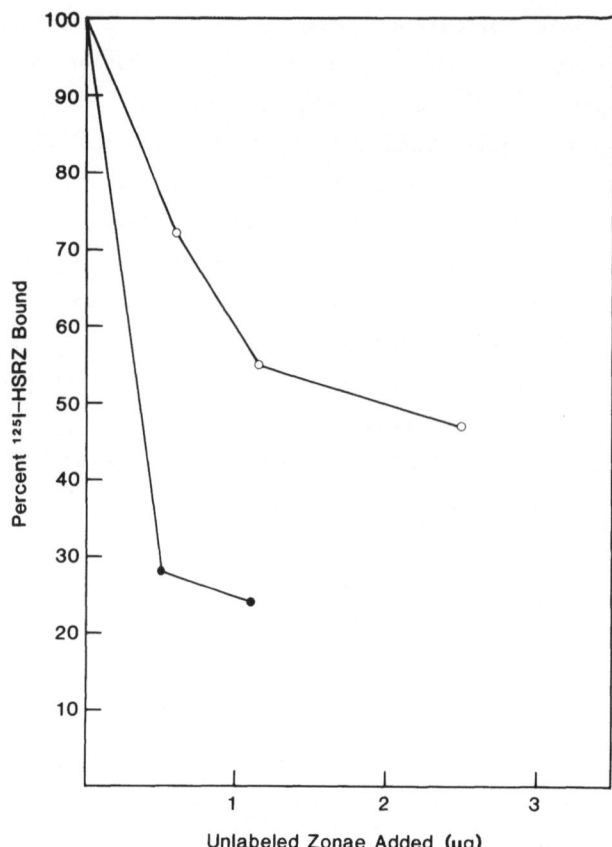

Fig. 4. Displacement of bound ^{125}I-HSRZ by unlabled rabbit
(●——●) or pig (○——○) zonae. The incubation
mixture consisted of 500 μl of 3.4 mg/ml
NaCl-treated rabbit ejaculated spermatozoa at
2 x 10^7 sperm/ml in McCoy's-BSA medium, 0.5 ng of
^{125}I-HSRZ and increasing amounts of unlabeled
zonae. At the end of 30 minutes incubation the
sperm were processed as indicated in Figure 3.

extent. To investigate this specificity further, we incubated 2×10^7 sperm/ml with 0.5 ng of ^{125}I-HSRZ and then added increasing amounts of unlabeled rabbit or pig zonae. As shown in Figure 4, unlabeled HSRZ almost completely displaced the ^{125}I-HSRZ at 1 μg of added unlabeled zonae, whereas the pig zonae was able to displace only approximately 50% of the bound ^{125}I-HSRZ at a 10-fold concentration difference of added unlabeled zonae. Thus, the rabbit ZBP appear to have a greater affinity for the homologous zonae. If the rabbit zona's sperm receptor is similar to that of the pig (Sacco et al., 1984), then the apparent Kd between the sperm receptor and the ZBP would be in the order of 10^{-12} M.

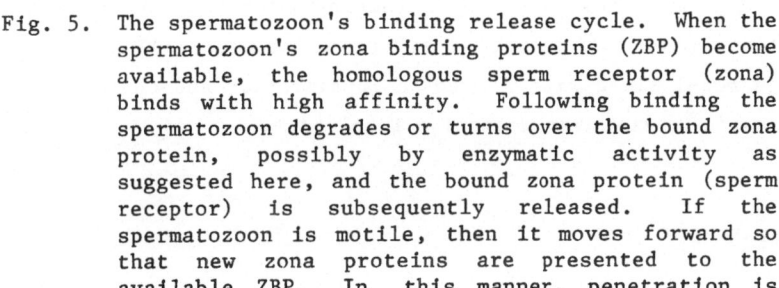

AVAILABLE
SPERM SURFACE ZONA
BINDING PROTEINS
(ZBP)

RELEASE
OF BOUND ZP
FROM ZBP

HIGH AFFINITY
BINDING OF
ZONA PROTEINS (ZP)
TO SPERM'S
ZBP

DEGRADATION
OF BOUND
ZP BY SPERM
ENZYME (S)

Fig. 5. The spermatozoon's binding release cycle. When the
spermatozoon's zona binding proteins (ZBP) become
available, the homologous sperm receptor (zona)
binds with high affinity. Following binding the
spermatozoon degrades or turns over the bound zona
protein, possibly by enzymatic activity as
suggested here, and the bound zona protein (sperm
receptor) is subsequently released. If the
spermatozoon is motile, then it moves forward so
that new zona proteins are presented to the
available ZBP. In this manner, penetration is
achieved.

THE MODEL

Based on our data and that of others, we now propose that the
specificity at the level of sperm-zona interaction lies in the ability of
the spermatozoon to penetrate the zona and that this penetration has two
requirements: 1) the high affinity of the ZBP for the sperm receptor and
2) the necessity of having bound zona protein removed or degraded from the
sperm surface. In the case of a heterologous combination, for example,
pig zonae and capacitated rabbit spermatozoa, there would be no high
affinity binding, even though the sperm's enzyme(s) might be able to
degrade the zona's components. Enzymatic dissolution of unbound zona
proteins, as might be seen in any number of heterologous combinations,
would not benefit the sperm's forward penetration progress. Penetration
would also not occur if binding did not occur or if the sperm enzyme(s)
were not available to act on the bound substrate (non-capacitated sperm;
non-acrosome reacted sperm).

Using these requirements for penetration, and given that only motile spermatozoa can penetrate the zona, we propose the model shown in Figure 5. Although there are species differences in binding to the zona (i.e. capacitated, non-capacitated, acrosome reacted, or non-reacted), this model assumes capacitated spermatozoa undergoing the acrosome reaction on the surface of the zona. The model suggests that these spermatozoa penetrate the zona by a cyclic mechanism in which there is a high affinity binding between the sperm receptor on the zona and the ZBP. After turnover or degradation of the bound components (enzymatic?), they are released from the ZBP. With the spermatozoon's forward motility, it would be able to move deeper into the zona, bind new substrate and the cycle would begin again. Without forward motility, no new substrate would be available for the next binding step, and penetration would not occur. When sperm enzymatic activity or its ability to turnover bound ligand is blocked (i.e. inhibitor added or the sperm fixed), the bound zona would not be degraded, the ZBP would remain saturated with receptor, and penetration would not occur. Further experiments will be necessary to test this model along with a precise localization and characterization of the enzymatic or turnover mechanism involved.

ACKNOWLEDGEMENTS

The authors gratefully acknowledge the generous gifts of rabbit and pig zonae from Dr. B. S. Dunbar, Baylor College of Medicine and Dr. J. L. Hedrick, University of California, Davis. We also thank Ms. J. E. Matthews for her expert technical assistance. This research was supported by NIH grant HD14232.

REFERENCES

Aarons, D., Speake, J. L., and Poirier, G. R., 1984, Evidence for a proteinase inhibitor binding component associated with murine spermatozoa, Biol. Reprod., 31:811-817.

Aarons, D., Robinson, R., Richardson, R., and Poirier, G. R., 1985, Competition between seminal and exogenous proteinase inhibitors for sites on murine epididymal sperm, Contraception, 31:177-184.

Ahuja, K. K., 1982, Fertilization studies in the hamster, Exptl. Cell Res., 140:353-362.

Bearer, E. L., and Friend, D. S., 1982, Modifications of anionic-lipid domains preceding membrane fusion in the guinea pig, J. Cell Biol., 92:604-615.

Bedford, J. M., 1977, Sperm/egg interaction: The specificity of human spermatozoa, Anat. Rec., 188:477-488.

Benau, D. A., and Storey, B. T., 1985, Mouse sperm zona-binding sites with serine protease properties, Biol. Reprod., 32:96.

Bleil, J. D., and Wassarman, P. M., 1980, Structure and function of the zona pellucida: Identification and characterization of the proteins of the mouse oocyte's zona pellucida, Devel. Biol., 76:185-202.

Brackett, B. G., Bousquet, D., and Dressel, M. A., 1982, In vitro sperm capacitation and in vitro fertilization with normal development in the rabbit, J. Androl., 3:402-411.

Conway, A. F., and Metz, C. B., 1976, Phospholipase activity of sea urchin sperm: Its possible involvement in membrane fusion, J. Exptl. Zool., 198:39-48.

Davis, B. K., Byrne, R., and Hungund, B., 1979, Studies on the mechanism of capacitation. II. Evidence for lipid transfer between plasma membrane of the rat sperm and serum albumin during capacitation in vitro, Biochim. Biophys. Acta, 558:257-266.

Dietl, J., and Czuppon, A. B., 1984, Ultrastructural studies of the porcine zona pellucida during the solubilization process by Li-3,5-diiodosalicylate, Gamete Res., 9:45-54.

Dudkiewicz, A. B., and Williams, W. L., 1976, Interaction of rabbit sperm and egg, Cell Tiss. Res., 169:277-287.

Dunbar, B. S., Wardrip, N. J., and Hedrick, J. L., 1980, Isolation, physicochemical properties, and macromolecular composition of zona pellucida from porcine oocytes, Biochem., 19:356-365.

Dunbar, B. S., and Wolgemuth, D. J., 1984, Structure and function of the mammalian zona pellucida, a unique extracellular matrix, Modern Cell Biol., 3:77-111.

Florman, H. M., and Wassarman, P. M., 1985, O-linked oligosaccharides of mouse egg ZP3 account for its sperm receptor activity, Cell, 41:313-324.

Friend, D. S., 1982, Plasma-membrane diversity in a highly polarized cell, J. Cell Biol., 93:243-249.

Friend, D. S., and Bearer, E. L., 1981, Beta-hydroxysterol distribution as determined by freeze-fracture cytochemistry, Histochem. J., 13:535-546.

Goldstein, J. L., Anderson, R. G. W., and Brown, M. S., 1979, Coated pits, coated vesicles, and receptor-mediated endocytosis, Nature, 279:679-685.

Greve, J. M., and Wassarman, P. M., 1985, Mouse egg extracellular coat is a matrix of interconnected filaments possessing a structural repeat, J. Mol. Biol., 181:253-264.

Hedrick, J. L., and Wardrip, N., 1981, Microheterogeneity in the glycoproteins of the zona pellucida is due to the carbohydrate moiety, J. Cell Biol., 91:177a.

Hedrick, J. L., and Wardrip, N., 1982, Topographical radiolabeling of zona pellucida glycoproteins, J. Cell Biol., 95:162a.

Hook, M., Kjellen, L., and Johansson, S., 1984, Cell-surface glycosaminoglycans, Ann. Rev. Biochem., 53:847-869.

Huang, T. T. F., Ohzu, E., and Yanagimachi, R., 1982, Evidence suggesting that L-fucose is part of a recognition signal for sperm-zona pellucida attachment in mammals, Gamete Res., 5:355-361.

Jacobson, K., O'Dell, D., and August, J. T., 1984, Lateral diffusion of an 80,000-dalton glycoprotein in the plasma membrane of murine fibroblasts: relationships to cell structure and function, J. Cell Biol., 99:1624-1633.

Meizel, S., 1978, The mammalian sperm acrosome reaction, a biochemical approach, in: "Development in Mammals," M. Johnson, ed., North-Holland, New York, pp.1-64.

Meizel, S., and Turner, K. O., 1984, The effects of products and inhibitors of arachidonic acid metabolism on the hamster sperm acrosome reaction, J. Exptl. Zool., 231;283-288.

Ono, K., Yanagimachi, R., and Huang, T. T. F., 1982, Phospholipase A of guinea pig spermatozoa: Its preliminary characterization and possible involvement in the acrosome reaction, Develop. Growth and Differ., 24:305-310.

O'Rand, M. G.,1979,Changes in sperm surface properties correlated with capacitation, in: "The Spermatozoon," D. Fawcett and J. M. Bedford, eds., Urban and Schwarzenherger, Baltimore-Munich, pp. 195-204.

O'Rand, M. G., 1982 Modification of the sperm membrane during capacitation, Ann. N.Y. Acad Sci., 383:392-402.

O'Rand, M. G., 1985, Steps in the fertilization Process: Understanding and control, in: "Molecular and Cellular Biology of Fertilization, J. L. Hedrick, ed., Plenum Press, New York, in press.

O'Rand, M. G., Matthews, J. E., Welch, J. E., and Fisher, S. J., 1985, Identification of zona binding proteins of rabbit, pig, human, and mouse spermatozoa on nitrocellulose blots, J. Exptl. Zool., 235:423-428.

Orci, L., Perrelet, A., and Friend, D. S., 1977, Freeze-fracture of membrane fusions during exocytosis in pancreatic islet cells, J. Cell Biol., 75:23-30.

Parks, J. E., and Hammerstedt, R. H., 1985, Developmental changes occurring in the lipids of ram epididymal spermatozoa plasma membrane, Biol. Reprod., 32;653-668.

Peterson, R. N., Russell, L. D., Bundman, D., and Freund, M., 1980, Binding of boar spermatozoa to porcine oocytes: Evidence for specific receptors on the zona pellucida, Science, 207:73-74.

Peterson, R. N., Russell, L. D., and Hunt, W. P., 1984, Evidence for specific binding of uncapacitated boar spermatozoa to porcine zonae pellucidae in vitro, J. Exptl. Zool., 231:137-147.

Peterson, R. N., Henry, L., Hunt, W., Saxena, N., and Russell, L. D., 1985, Further characterization of boar sperm plasma membrane proteins with affinity for the porcine zona pellucida, Gamete Res., 12:91-100.

Sacco, A. G., Subramanian, M. G., and Yurewicz, E. C., 1984, Association of sperm receptor activity with a purified pig zona antigen (PPZA), J. Reprod. Immunol., 6:89-103.

Saling, P. M., 1981, Involvement of trypsin-like activity in binding of mouse spermatozoa to zonae pellucidae, Proc. Natl. Acad. Sci. USA, 78:6231-6235.

Saling, P. M., 1982, Development of the ability to bind to zonae pellucidae during epididymal maturation: Reversible immobilization of mouse spermatozoa by lanthanum, Biol. Reprod., 26:429-436.

Shur, B. D., and Hall, N. G., 1982, A role for mouse sperm surface galactosyltransferase in sperm binding to the egg zona pellucida, J. Cell Biol., 95:574-579.

Singer, S. J., and Nicolson, G. L., 1972, The fluid mosaic model of the structure of cell membranes, Science, 175:720-731.

Sullivan, R., and Bleau, G., 1985, Interaction of isolated components from mammalian sperm and egg, Gamete Res., 12:101-116.

Swenson, C. E., and Dunbar, B. S., 1982, Specificity of sperm-zona interaction, J. Exptl. Zool., 219:97-104.

Wassarman, P. M., Florman, H. M., and Greve, J.M., 1985, Receptor-mediated sperm-egg interactions in mammals, in: "Biology of Fertilization," vol. 2, C. B. Metz and A. Monroy, eds., Academic Press, New York pp. 341-360.

Wolf, D. E., and Voglmayr, J. K., 1984, Diffusion and regionalization in membranes of maturing ram spermatozoa, J. Cell Biol., 98:1678-1684.

Yanagimachi, R., 1977, Specificity of sperm-egg interaction, in: "Immunobiology of Gametes," M. Edidin and M.H. Johnson, eds., Cambridge University Press, Cambridge, pp. 255-295.

Yanagimachi, R., 1981, Mechanisms of fertilization in mammals, in: "Fertilization and Embryonic Development in vitro," L. Mastroianni and J. D. Biggers, eds., Plenum Press, New York, pp. 81-182.

Yanagimachi, R., and Philips, D. M., 1984, The status of acrosomal caps of hamster spermatozoa immediately before fertilization in vivo, Gamete Res., 9:1-19.

THE INTERACTION OF EGG PEPTIDES WITH SPERMATOZOA

David L. Garbers, Thomas D. Noland, Lawrence J. Dangott,
Chodavarapu S. Ramarao, and J. Kelley Bentley

The Howard Hughes Medical Institute and the Departments of
Pharmacology and Molecular Physiology and Biophysics
Vanderbilt University Medical Center
Nashville, Tennessee 37232

INTRODUCTION

Mechanisms by which eggs (includes acellular matrices as well as
other attached cells, such as cumulus oophorus cells) and/or other
components of the female reproductive tract communicate with spermatozoa
are beginning to be understood. Research has progressed to the molecular
level in both invertebrates and vertebrates, and it can be anticipated
that the genes responsible for the synthesis of some of the regulatory
components will be soon identified and their function (or fate) in animals
separated by large evolutionary distances determined.

Three different sites where gamete communication demonstrates some
degree of species-specificity have been pursued with some vigor (Fig. 1).
Although different groups of animals will have their own variations of the
schemes presented, the sea urchin and the mouse are depicted in the figure
as representative of invertebrates and mammals, respectively.

One site of communication is the species-specific adhesion of
spermatozoa to the egg. In the sea urchin, a protein with a molecular
weight of about 30,000, named bindin, which exists as a part of the
acrosome vesicle (Vacquier and Moy, 1977; Vacquier, 1983) binds to a
receptor on the egg vitelline layer (Glabe and Vacquier, 1978; Glabe and
Lennarz, 1979; Glabe et al., 1982; Rossignol et al., 1984). It has been
suggested that the binding is mediated by a lectin-polysaccharide type of
interaction (Glabe et al., 1982). In the mammal, species-specific binding

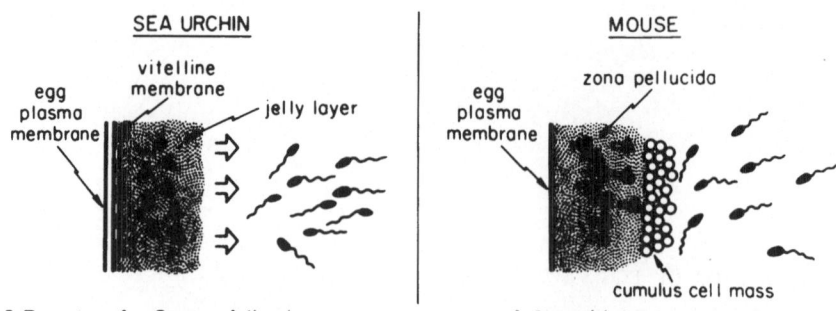

Fig. 1. Scheme which depicts egg/spermatozoa interactions in the sea urchin and the mouse.

occurs at the level of the zona pellucida, and in the mouse, the zona pellucida protein responsible for sperm binding has been called ZP3 (Bleil and Wassarman, 1980, 1983; Florman and Wassarman, 1985). Recent work has shown that removal of O-linked but not N-linked oligosaccharides from ZP3 destroys sperm receptor activity (Florman and Wassarman, 1985). Therefore, some similarities between the receptors for sperm adhesion on the mouse zona pellucida and the sea urchin vitelline layer may exist.

The functions of these receptors for spermatozoa remain to be studied in greater detail. Aside from relative species-specific binding in the sea urchin and the mammal (Glabe and Lennarz, 1979; Bedford, 1981; Wassarman, 1983), the receptor appears to induce an acrosome reaction in mouse spermatozoa (Bleil and Wassarman, 1983). In addition, Glabe (1985) has shown that bindin promotes membrane fusion in artificial membranes, and has suggested a potential role of bindin in membrane fusion at fertilization.

A second site of study is the induction of an acrosome reaction. In sea urchins and various other animals, it is clear that a specific induction occurs due to factors associated with the egg (SeGall and Lennarz, 1979; Garbers and Kopf, 1980), but in the mammal this has been less clear. Spermatozoa from Eutherian mammals must first undergo an ill-defined process known as capacitation. Once capacitated (definitions of capacitation vary), the acrosome reaction occurs in response to an apparent Ca^{2+} influx. The Ca^{2+} influx is common to mammalian and other animal spermatozoa where an acrosome reaction occurs (Singh et al., 1978; Garbers and Kopf, 1980). Mammalian spermatozoa can be capacitated and induced to undergo acrosome reactions in the absence of eggs or female reproductive tract fluids in vitro (Bedford, 1983). These observations, however, do not lend strong support to the occasionally given hypothesis that specific egg-related substances which induce an acrosome reaction therefore do not exist. In fact, by variation of extracellular Ca^{2+} and/or pH it is also possible to easily induce acrosome reactions in the absence of eggs in the sea urchin.

In the sea urchin, the substance responsible for induction of an acrosome reaction is a highly sulfated molecule or group of molecules containing fucose and protein (SeGall and Lennarz, 1979, 1981; Garbers and Kopf, 1980; Garbers et al., 1983). The complex acts in a species-specific manner and is a component of the jelly layer (SeGall and Lennarz, 1979). The fucose-sulfate-rich material diffuses from the jelly layer, but it is doubtful that the released factor is of physiological significance since sea urchin spermatozoan fertility decreases rapidly after induction of an acrosome reaction and is essentially zero by 40 seconds (Vacquier, 1979). Therefore, the acrosome reaction of the fertilizing sperm cell probably occurs within the jelly layer or at the level of of the vitelline layer. The fucose-sulfate-rich material, in addition to causing the induction of an acrosome reaction, causes marked increases in $^{45}Ca^{2+}$ uptake, and up to 300-fold elevations of cyclic AMP (Garbers and Kopf, 1980; Garbers et al., 1983). The elevations of cyclic AMP, however, appear to be dependent on the increased Ca^{2+} influx and not on direct receptor-mediated modulation of adenylate cyclase activity.

In the mouse, it has been reported that spermatozoa undergo an acrosome reaction after binding to the zona pellucida (Saling et al., 1978; Storey et al., 1984), and ZP3 appears to be responsible for the acrosome reaction in sperm bound to the zona pellucida (Bleil and Wassarman, 1983). A number of recent reports also have suggested that glycosaminoglycans can induce an acrosome reaction in bovine or hamster spermatozoa (Lenz et al., 1982, 1983; Meizel and Turner, 1984). Such glycosaminoglycans could function under normal fertilization conditions since some of them appear to be associated with the zona pellucida (Talbot, 1984; Ball et al., 1982); the sulfated forms resemble the fucose-sulfate-rich material from sea urchin egg jelly, but a species-specificity of these molecules seems doubtful.

The third area of investigation involves the regulation of sperm metabolism and motility. Again, it is clear that egg substances modify the metabolism and motility (velocity, pattern) in many different animals (Garbers and Kopf, 1980). In the mammal, a hyperactivated type of motility has been described (Yanagimachi, 1981), but this type of motility can be observed in the absence of eggs.

In the sea urchin, small peptides have been isolated from eggs which can stimulate sperm metabolism and motility under appropriate conditions (Ohtake, 1976; Hansbrough and Garbers, 1981; Suzuki et al., 1981; Nomura et al., 1983; Suzuki et al., 1984); at least one of these peptides is also a sperm chemoattractant (Ward et al., 1985a). In the mammal, specific motility stimulants which might function under normal fertilization conditions have yet to be rigorously defined. Two substances of some recent interest have been hypotaurine/taurine and bicarbonate (Mrsny et al., 1979; Garbers et al., 1982a; Okamura et al., 1985). Additionally, a number of manuscripts have suggested that f-Met-Leu-Phe-related peptides can act as chemoattractants in mammalian spermatozoa (Iqbal et al., 1980; Vijayasarathy et al., 1980.; Gnessi et al., 1985), although at least one report has raised serious questions with respect to the methods used (Miller, 1982).

In this manuscript, the research on the sea urchin egg peptides will be described in detail followed by a short description of recent studies on the effects of bicarbonate on mammalian spermatozoa.

EGG PEPTIDES IN THE SEA URCHIN

Peptide Structure

The two peptides studied in greatest detail have been speract (Gly-Phe-Asp-Leu-Asn-Gly-Gly-Gly-Val-Gly) isolated from Strongylocentrotus purpuratus (Hansbrough and Garbers, 1981; Garbers et al., 1982b) or Hemicentrotus pulcherrimus (Suzuki et al., 1981) and resact (Cys-Val-Thr-Gly-Ala-Pro-Gly-Cys-Val-Gly-Gly-Gly-Arg-LeuNH$_2$) obtained from Arbacia punctulata eggs (Suzuki et al., 1984). Two peptides isolated from the sea urchin, Anthocidaris crassispina, have the structure Gly-Phe-Asp-Leu-Thr-Gly-Gly-Gly-Val-Gly and Gly-Phe-Asp-Leu-Ser-Gly-Gly-Gly-Val-Gly (Nomura et al., 1983). Resact and speract do not crossreact in a detectable manner with spermatozoa of the species containing the opposite peptide.

There is general agreement with respect to structure/activity relationships of speract when respiration rates are determined (Garbers et al., 1982b; Nomura and Isaka, 1985), the most detailed study being that of Nomura and Isaka (1985). Considerable substitution is possible in the

NH_2-terminal portion of speract with retention of biological activity, but deletion of the CO_2H-terminal Gly or Val-Gly results in large or total losses of respiration-stimulating activity (Smith and Garbers, 1983). Most synthesized analogues have had an equal or decreased potency relative to speract, but one analogue (Gly-Phe-Asp-Leu-Ser-Gly-Gly-Gly-Val-Pro) appears to be 500-times more potent (Nomura and Isaka, 1985).

For the most part, structure/activity studies have concentrated on the relationship of structure to respiration-stimulation; for a number of speract analogues, respiration-stimulating activity has coincided with cyclic nucleotide-elevating activity (Garbers et al., 1982b). Recently, however, Shimomura and Garbers (submitted) prepared various analogues of resact and found that relative potencies varied dependent on the physiological parameter measured.

Modification of the CO_2H-terminal leucine-NH_2 of resact did not alter biological activity, but substitution of the two cysteinyl residues by Ser or Tyr or methylation of the cystenyl residues resulted in divergent relative potencies dependent on whether respiration rates or cyclic nucleotide concentrations were measured (Table I). [Ser[1], Tyr[8]]resact was approximately 20 percent as potent as resact when respiration rates were measured, but was 0.1 percent as potent as resact when cyclic GMP elevations were determined. An NH_2-terminal fragment (Cys-Val-Thr-Gly-Ala-Pro-Gly) neither stimulated respiration nor elevated cyclic nucleotide levels at concentrations up to 10μM, whereas a CO_2H-terminal fragment (Cys-Val-Gly- Gly-Gly-Arg-LeuNH_2) had approximately the same respiration-activity but 0.1 percent of the cyclic GMP-elevating activity of resact. It appears, then, that cyclic nucleotide elevations can be separated from the stimulation of respiration suggestive that a single receptor can cause multiple primary events dependent on the agonist bound, or that multiple receptors exist.

Peptide Receptor

The receptors for speract and resact exist on the plasma membrane of spermatozoa, but the regional distribution remains to be determined.

In initial studies on the receptor, the Bolton/Hunter analogue of speract was synthesized and shown to possess equivalent respiration-stimulating activity to speract (Smith and Garbers, 1983). The radioiodinated Bolton/Hunter adduct of speract was bound to intact S.

Table I. Concentrations of resact and of resact analogues required to one half-maximally increase sperm respiration rates or cyclic GMP concentrations.

Peptides	Approximate Concentration of Peptide Required to One Half Maximally Increase.	
	Respiration Rate	Cyclic GMP
Resact	12 nM	30 nM
$[Tyr^1, Ser^8]$ Resact	60 nM	100 nM
$[Ser^1, Tyr^8]$ Resact	60 nM	3000 nM
$[Tyr^1, Cys(Me)^8]$ Resact	60 nM	500 nM
$[Cys(Me)^1, Tyr^8]$ Resact	60 nM	500 nM
$[Cys(Me)^1, Cys(Me)^8]$ Resact	60 nM	80 nM
Resact (1-7)[a]	>>10,000 nM	>>20,000 nM
Resact (8-14)[a]	40 nM	20,000 nM

[a]Resact (1-7) represents the NH_2-terminal heptapeptide and resact (8-14) represents the CO_2H-terminal heptapeptide amino acids.

purpuratus but not to A. punctulata spermatozoa, and speract analogues with over 10^6-fold differences in their relative ability to stimulate respiration rates showed the same relative potencies in competition binding assays with ^{125}I-Bolton/Hunter speract. The binding of the $[^{125}I]$-analogue was highly dependent on the presence of extracellular Na^+, but binding was decreased only slightly in the absence of added Ca^{2+} or Mg^{2+} (Smith and Garbers, 1983).

Although the ^{125}I-Bolton/Hunter adduct was valuable, the lack of a free amino group negated potential crosslinking studies to identify the receptor. Therefore, an analogue (Gly-Gly-Gly-Gly-Tyr-Asp-Leu-Asn-Gly-Gly-Gly-Val-Gly) was synthesized which retained respiration-stimulating activity equivalent to speract (Dangott and Garbers, 1984). The analogue also competed with ^{125}I-Bolton/Hunter speract for receptor binding with equivalent potency to speract. The $GGG[Y^2]$speract was radiolabeled with ^{125}I and subsequently crosslinked to the apparent receptor with dissuccinimidyl suberate (Dangott and Garbers, 1984). The receptor was found to be a glycoprotein with an apparent molecular weight of 77,000

(Na.dodecyl.SO$_4$.gels, reducing conditions). Under non-reducing conditions, the receptor migrated with an apparent molecular weight of approximately 55,000. That the association of peptide with the protein was specific was determined by competition experiments with various speract analogues, and by the failure of the radiolabeled peptide to crosslink to spermatozoa from species which do not crossreact with speract.

These studies which clearly identified only a single radioactive band need to be viewed with some caution, however, since the amount of added radioactive peptide was generally in the 1 nM range. Cyclic nucleotide concentrations are elevated half-maximally near 100 nM and, therefore, it remains possible that a higher affinity receptor has been identified but that a lower affinity receptor, not observed under these crosslinking conditions, still exists.

Responses of the Intact Cell to Receptor/Peptide Interaction

When added to intact spermatozoa, speract causes a stimulation of respiration and motility, especially at acidic extracellular pH values (Ohtake, 1976; Hansbrough and Garbers, 1981; Suzuki et al., 1981), a net efflux of H$^+$ (Repaske and Garbers, 1983), elevations of cyclic AMP and cyclic GMP (Hansbrough and Garbers, 1981), and apparent decreases in guanylate cyclase activity [estimated after homogenization of cells and after the addition of detergent (Ramarao and Garbers, 1985)]. Resact causes the same changes in $\underline{A}$. punctulata spermatozoa, but in addition has been shown to act as a chemoattractant (Ward et al., 1985a). Whether or not the various other egg peptides serve as chemoattractants has yet to be determined. In addition, resact has been shown to cause a shift in the apparent molecular weight of a plasma membrane protein (Suzuki et al., 1984), a protein whose apparent molecular weight was first observed to change in response to egg jelly (Ward and Vacquier, 1983). The plasma membrane protein has been identified as guanylate cyclase (Ward et al., 1985b), an enzyme found in high activity in sea urchin spermatozoa (Gray et al., 1976). Upon addition of resact, guanylate cyclase, originally in a 160,000 molecular weight form, rapidly converts to an apparent molecular weight of 150,000 (Suzuki et al., 1984). The mechanism responsible for the apparent molecular weight change has yet to be established. The 160,000 molecular weight protein becomes labeled with ^{32}P when intact cells are incubated with ^{32}P, and the addition of resact also causes the rapid loss of ^{32}P; therefore, the apparent change in molecular weight could be due to dephosphorylation (Ward and Vacquier, 1983) or to a significant decrease in mass due to a loss of protein or carbohydrate.

To answer this question, the enzyme responsible for the 160,000/150,000 conversion is now being purified. This enzyme, tentatively referred to as guanylate cyclase converting enzyme, is present in the cytosolic and membrane fractions of homogenized cells, but the majority of the activity appears to be in the soluble fractions. When spermatozoa solubilized with detergent are incubated with partially purified converting enzyme at ambient temperature, the guanylate cyclase activity in homogenates of spermatozoa is decreased by 70 to 80 percent within 20 minutes of incubation (Fig. 2). In addition, converting enzyme also causes a shift in the apparent molecular weight of guanylate cyclase (160,000 to 150,000). Concomitantly, ^{32}P label on the 160,000 dalton form of guanylate cyclase, obtained from spermatozoa which were previously labeled with ^{32}P$_i$, is lost upon incubation of the detergent-solubilized spermatozoa with converting enzyme. The specificity of the enzyme

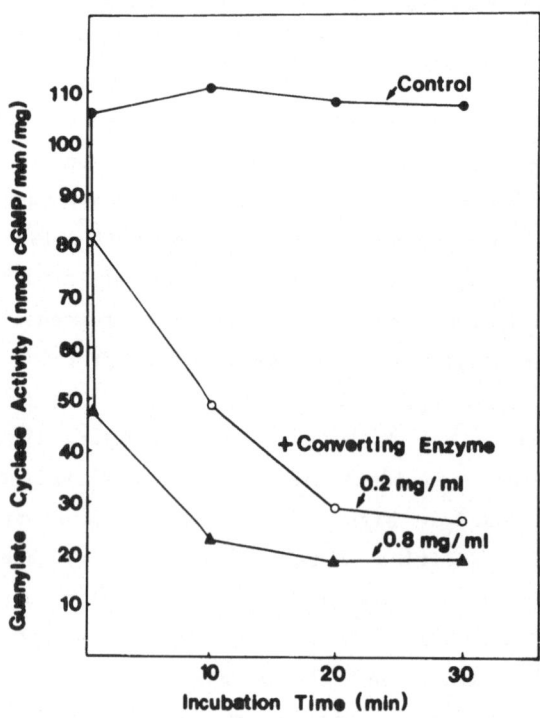

Fig. 2. Decrease in guanylate cyclase activity due to the action of another enzyme (guanylate cyclase converting enzyme).

preparation is apparent from studies where sperm membranes have been phosphorylated with $(\gamma^{32}P)ATP$. In these experiments, both the 150,000 and 160,000 dalton proteins are radiolabelled. However, ^{32}P is removed only from the 160,000 molecular weight band upon addition of partially purified enzyme. A wide variety of added proteolytic and phosphatase inhibitors have not protected guanylate cyclase from the action of the converting enzyme, and therefore its enzymatic activity remains uncertain.

A summary of our knowledge with intact cells is given schematically in Figure 3. Peptides interact with a specific plasma membrane receptor whose molecular weight appears to be about 77,000. Within 3 seconds, a membrane phosphoprotein identified as guanylate cyclase shifts in apparent molecular weight (160,000 to 150,000) and loses all associated ^{32}P. Although peptide appears to bind to receptor and to stimulate respiration rates in the absence of extracellular Ca^{2+}, a chemotactic response of A. punctulata spermatozoa is not observed unless extracellular Ca^{2+} is present. Another early response to the addition of peptide is a net efflux of H^{+}, although peak net release appears to occur later (at ∿ 7 seconds) than the shift in apparent molecular weight of the plasma membrane protein.

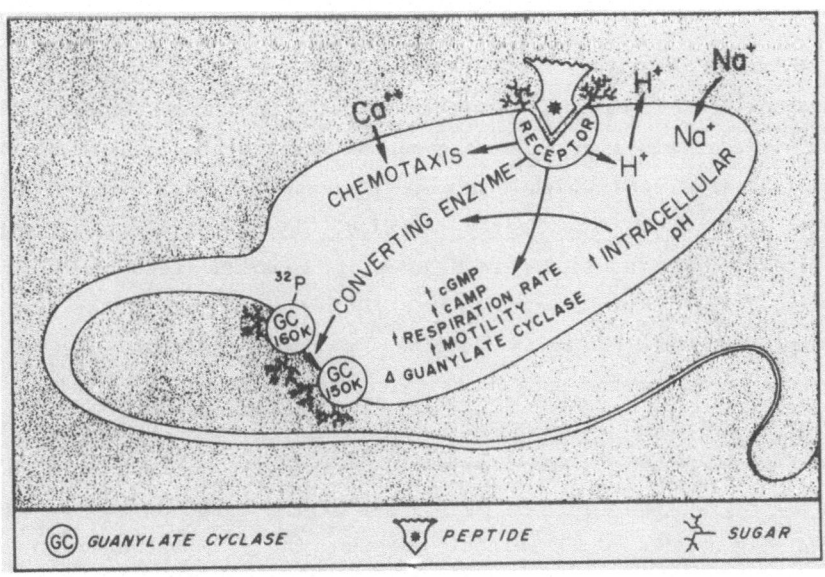

Fig. 3. Summary of known events which occur upon an egg peptide interacting with its receptor on the sperm cell.

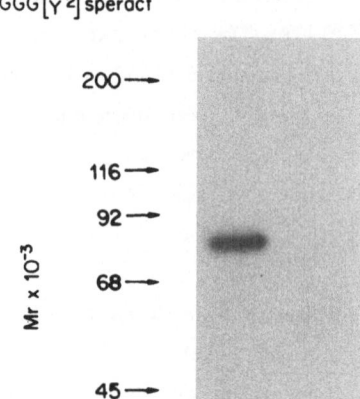

GGG[Υ 2] speract

Fig. 4. Crosslinking of ^{125}I-GGG[Υ^2]speract to membrane vesicles prepared from S. purpuratus spermatozoa. The reaction was performed in the absence or presence of 100 nM non-labeled peptide (as indicated) and the samples were prepared for Na.dodecyl.SO$_4$- polyacrylamide gel electrophoresis in the presence of β-mercaptoethanol.

Receptor-mediated Events in Isolated Membranes

In order to dissect the early events which occur after receptor occupation by the peptide, attempts were made to prepare membrane vesicles which would retain a peptide receptor as well as one or more of the same physiological events which occur in response to peptide binding to the intact cell. In the early studies, hypotonic homogenization of spermatozoa had resulted in a loss of apparent receptor (Smith and Garbers, 1983). Bentley and Garbers (1986a) succeeded in the preparation of membrane vesicles which retained a speract receptor by releasing membranes from spermatozoa with nitrogen cavitation techniques. The membranes retained the capacity to bind ^{125}I-Bolton/Hunter speract. Gly-Phe-Asp-Leu-Asn-Gly-Gly-Gly-Val-Gly, Tyr-Asp-Leu-Asn-Gly-Gly-Gly-Val-Gly, Tyr-Asp-Leu-Thr-Thr-Gly-Gly-Gly-Val-Gly, and Gly-Phe-Ala-Leu-Gly-Gly-Gly-Val-Gly caused a 50 percent decrease in ^{125}I-Bolton/Hunter speract binding at 10, 600, 1,260 and 3,160 nM, respectively, and these coincide with relative binding observed in the intact cell. Phe-Asp-Leu-Asn-Gly-Gly-Gly, which has no biological activity in the intact cell, failed to compete with ^{125}I-Bolton/Hunter speract at concentrations of 10 μM. In

154

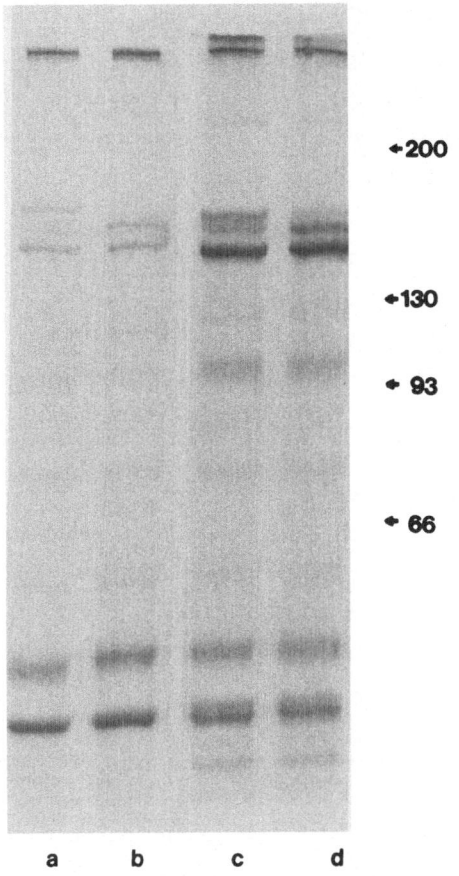

a   b   c   d

Fig. 5. Apparent molecular weight conversion of the 160,000 molecular weight protein in A. punctulata spermatozoa and spermatozoan membranes in response to resact. Spermatozoa were collected and washed once in pH 6.0 artificial seawater. Spermatozoan membranes were prepared as described (Bentley and Garbers, submitted) and resuspended in a solution containing 0.5 M NaCl, 0.1 M NaF, and 0.02 M [N-morpholino] ethane sulfonic acid (MES) at pH 6.0. At time zero, intact spermatozoa or membranes were diluted with 9 volumes of artificial seawater (pH 8.0) containing 0 or 450 nM resact (pH 7.8 final). After 1 min at 21°C, the reaction was stopped by the addition of three volumes of buffer (pH 6.5) containing 1 M tris(hydroxymethyl)-aminomethane, 0.3 M ethylenediaminetetraacetic acid, 0.1 M NaF, 0.1 M benzamidine, and 1 percent (w/v) Na.dodecyl.SO$_4$. Shown are the Coomassie blue-stained gels (30 μg protein/lane). Lane a: intact cells treated at pH 7.8 without resact; Lane b: intact cells treated at pH 7.8 with resact, 450 nM; Lane c: spermatozoan membranes treated without resact; Lane d: spermatozoan membranes treated with 450 nM resact. Numbers on the right represent molecular weights (x10^{-3}) of standard proteins.

final studies, the (GGG[Y^2]-speract) was crosslinked to the apparent receptor with disuccinimidyl suberate. A radiolabeled protein was identified on Na.dodecyl.SO$_4$ gels whose apparent molecular weight was 77,000 (reducing conditions), the same apparent molecular weight of the receptor observed in intact cells (Fig. 4).

Since a physiological response to receptor binding also was a goal of these studies, attempts were then made to prepare membrane vesicles from A. punctulata spermatozoa where a receptor would be retained and the plasma membrane protein of apparent molecular weight 160,000/150,000 would be retained in the 160,000 molecular weight form. Bentley and Garbers (1986b) succeeded in preparing such vesicles (Fig. 5). In addition, when such membranes were treated with resact, the same apparent molecular weight change seen in intact cells could now be observed in the isolated membranes (Fig. 5). In the intact cell, resact causes a rapid and substantial decrease in guanylate cyclase activity if enzyme activity is measured in the presence of Mn^{2+} and detergent (Ramarao and Garbers,

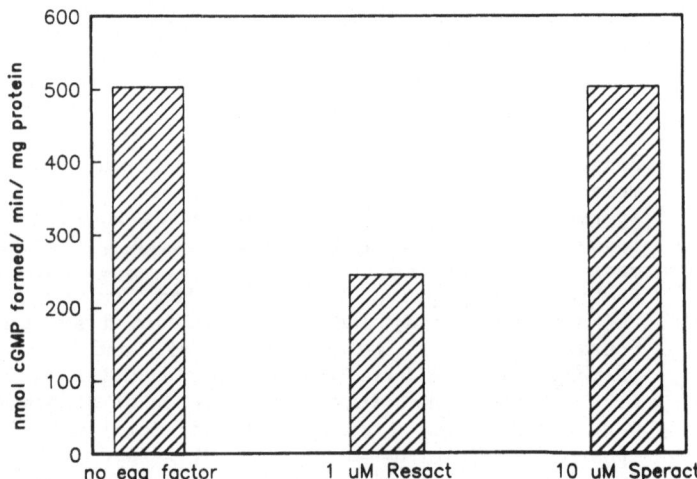

Fig. 6. Specificity of peptide-induced inactivation of guanylate cyclase. A. punctulata spermatozoan membranes were prepared and suspended in a solution containing 0.5 M NaCl, 0.1 M NaF, and 0.02 M MES at pH 6.0 on ice. At time zero, they were diluted into artificial seawater (pH 8.0) at 21°C with no peptide, 1 μM resact, or 10 μM speract. At 1 minute of incubation at a final pH of 7.8, the membranes were diluted with 9 volumes of a buffer consisting of 0.5 percent (w/v) Lubrol PX, 25 mM MES pH 6.0, 10 mM KF, and 100 μM Na$_3$VO$_4$ at 0-4°C. After standing for 1 hour on ice, 800 ng of protein were assayed for guanylate cyclase activity as described (Ramarao and Garbers, 1985).

1985). The membrane vesicles showed the same response; the inactivation was dependent on receptor occupancy since speract failed to cause any activity change (Fig. 6).

Although the mechanisms by which these events occur need to be carefully evaluated, it will now be possible to study the receptor-mediated responses in a greatly simplified system.

Since we also have shown that a protein with characteristics of N_i or N_o exists in sea urchin and other animal spermatozoa (VanDop et al., unpublished), it is possible that this protein is coupled to the receptor. These data are in disagreement with a report of Hildebrandt et al. (1985).

THE STIMULATION OF MAMMALIAN SPERMATOZOA BY BICARBONATE

HCO_3^- has been reported to stimulate sperm metabolism and motility in various animals (Hamner and Williams, 1964; Foley and Williams, 1967; Murdoch and Davis, 1978; Lodge and Salisbury, 1962), and a positive effect on caput epididymal spermatozoan motility has been reported (Vijayaraghavan et al., 1985). Fixation of CO_2 into the tricarboxylic acid cyclic could explain some or all of these positive effects (Garbers et al., 1982a). In addition, an alteration of intracellular pH by HCO_3^- could represent its general mechanism of action; this has been suggested by Hoskins and coworkers (Vijayaraghavan et al., 1985) in bovine caput epididymal spermatozoa and by Hyne (1984) in guinea pig sperm cells.

In 1982, Garbers et al. reported that Ca^{2+}-induced elevations of cyclic AMP in guinea pig spermatozoa required the presence of HCO_3^-. Although cyclic AMP could be elevated slightly (up to 2-fold) by HCO_3^- in the absence of added Ca^{2+}, the presence of both agents together resulted in up to 10- to 20-fold elevations of cyclic AMP (Garbers et al., 1982a). Adenylate cyclase activity, assayed in the presence of Mg^{2+}, Ca^{2+}, or HCO_3^-, did not appear to be further increased by HCO_3^- and Ca^{2+} in combination.

Recently, Okamura and Sugita (1983) reported that seminal plasma contained an activator of sperm adenylate cyclase; this activator was subsequently purified and was shown to be bicarbonate (Okamura et al., 1985). Adenylate cyclase could be activated in broken cell preparations by as much as 3.5-fold; the cyclic AMP concentration of intact cells,

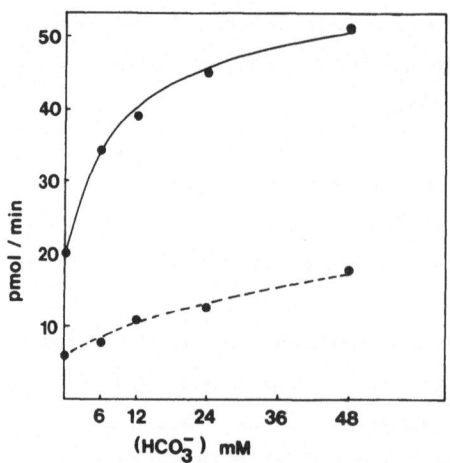

Fig. 7. Effects of bicarbonate on bovine sperm adenylate cyclase activity. Bovine caudal epididymal spermatozoa were homogenized and the heads were removed by centrifugation at 1500 x g for 10 minutes. The supernatant fluid was collected and a crude membrane fraction was prepared by centrifugation at 100,000 x g for 1 hour. The pellet was then resuspended in a solution containing 10 mM Hepes and 150 mM NaCl at pH 7.5. The membrane fractions were assayed for adenylate cyclase activity at 37°C for 20 minutes in the presence of 50 mM Tris, pH 7.4, 2.2 mM cyclic AMP, 7 mM creatine phosphate, 305 μg/ml creatine phosphate kinase, 1 mM ATP (specific activity [^{3}H]ATP = 20 cpm/pmol), and 10 mM MgCl$_2$. Product (cyclic AMP) accumulation was determined in the absence (o--o) or presence (o—o) of 0.05 percent Lubrol PX.

however, were elevated less than 2-fold, similar to the previous studies in the guinea pig (Garbers et al., 1982).

The mechanism of the HCO$^-_3$ effect on sperm cells remains to be clarified (direct, indirect, activation, deinhibition), but we (Noland and Garbers, unpublished) have recently confirmed the studies of Okamura et al. (Fig. 7), using bovine caudal epididymal spermatozoa. In addition, the apparent stimulation of adenylate cyclase activity was observed in the presence of detergent (Fig. 7).

It is of interest to point out that the bacterial adenylate cyclase from Brevibacterium liquefaciens is activated up to 100-fold by α-keto acids (Takai et al., 1974). Whether or not the bicarbonate response represents a modification of the same binding site remains to be determined.

ACKNOWLEDGEMENTS

This work was supported by grant HD10254 from the National Institute of
Child Health and Human Development.

REFERENCES

Ball, G. D., Bellin, M. E., Ax, R. L., and First, N. L., 1982,
    Glycosaminoglycans in bovine cumulus-oocyte complexes: Morphology
    and chemistry, Mol. Cell Endocrin., 28:113-122.

Bedford, J. M., 1981, Why mammalian gametes don't mix, Nature,
    291:286-288.

Bedford, J. M., 1983, Significance of the need for sperm capacitation
    before fertilization in Eutherian mammals, Biol. Reprod.,
    28:108-120.

Bentley, J. K., and Garbers, D. L., 1986a, Retention of the speract
    receptor by isolated plasma membranes of sea urchin spermatozoa,
    Biol. Reprod., 34:413-421.

Bentley, J. K., and Garbers, D. L., 1986b, Isolation of spermatozoan
    plasma membranes which retain a functional receptor for resact,
    Cell, in press.

Bentley, J. K., Garbers, D. L., Domino, S. E., Noland, T. D., VanDop, C.,
    1985, personal communication.

Bleil, J. D., and Wassarman, P. M., 1980, Mammalian sperm-egg interaction:
    identification of a glycoprotein in mouse egg zonae pellucidae
    possessing receptor activity for sperm, Cell, 20:873-882.

Bleil, J. D., and Wassarman, P. M., 1983, Sperm-egg interactions in the
    mouse: Sequence of events and induction of the acrosome reaction
    by a zona pellucida glycoprotein, Dev. Biol., 95:317-324.

Dangott, L. J., and Garbers, D. L., 1984, Identification and partial
    characterization of the receptor for speract, J. Biol. Chem.,
    259:13712-13716.

Florman, H. M., and Wassarman, P. M., 1985, O-linked oligosaccharides of
    mouse egg ZP3 account for its sperm receptor activity, Cell,
    41:313-324.

Foley, C. W., and Williams, W. L., 1967, Effect of bicarbonate and oviduct
    fluid on respiration of spermatozoa, Proc. Soc. Exp. Biol. Med.,
    126:634-637.

Garbers, D. L., and Kopf, G. S., 1980, The regulation of spermatozoa by
    calcium and cyclic nucleotides, Adv. Cyclic Nucleotide Res.,
    13:251-306.

Garbers, D. L., Kopf, G. S., Tubb, D. J., and Olson, G., 1983, Elevation of sperm adenosine 3':5'-monophosphate concentrations by a fucose-sulfate-rich complex associated with eggs: I. Structural characterization, Biol. Reprod., 29:1211-1220.

Garbers, D. L., Tubb, D. J., and Hyne, R. V., 1982a, A requirement of bicarbonate for Ca^{2+}-induced elevations of cyclic AMP in guinea pig spermatozoa, J. Biol. Chem., 257:8980-8984.

Garbers, D. L., Watkins, H. D., Hansbrough, J. R.. Smith, A., and Misono, K.S., 1982b, The amino acid sequence and chemical synthesis of speract and of speract analogues, J. Biol. Chem., 257:2734-2737.

Glabe, C. G., 1985, Interaction of the sperm adhesive protein, bindin, with phospholipid vesicles. II. Bindin induces the fusion of mixed-phase vesicles that contain phosphatidylcholine and phosphatidylserine in vitro, J. Cell Biol., 100:800-806.

Glabe, C. G., Grabel, L. B., Vacquier, V. D., and Rosen, S. D., 1982, Carbohydrate specificity of sea urchin sperm bindin: A cell surface lectin mediating sperm-egg adhesion, J. Cell Biol., 94:123-128.

Glabe, C. G., and Lennarz, W. J., 1979, Species-specific sperm adhesion in sea urchins: A quantitative investigation of bindin mediated egg agglutination, J. Cell Biol., 83:595-604.

Glabe, C. G., and Vacquier, V. D., 1978, Egg surface glycoprotein receptor for sea urchin sperm bindin, Proc. Natl. Acad. Sci. USA, 75:881-885.

Gnessi, L., Ruff, M. R., Fraioli, F., and Pert, C.B., 1985, Demonstration of receptor-mediated chemotaxis by human spermatozoa: A novel quantitative bioassay, Exptl. Cell Res., 161:219-230.

Gray, J. P., Drummond, G. I., Luk, D. W., Hardman, J. G., and Sutherland, E. W., 1976, Enzymes of vertebrate sperm, Arch. Biochem. Biophys., 172:20-30.

Hamner, C. E., and Williams, W. L., 1964, Identification of a sperm stimulating factor of rabbit oviduct fluid, Proc. Soc. Exp. Biol. Med., 117:240-243.

Hansbrough, J. R., and Garbers, D. L., 1981, Purification and characterization of a peptide associtated with eggs that activates spermatozoa, J. Biol. Chem., 256:1447-1452.

Hildebrandt, J. D., Codina, J., Tash, J. S., Kirchick, H. J., Lipschultz, L., Sekura, R. D., and Birnbaumer, L., 1985, The membrane-bound spermatozoal adenylyl cyclase system does not share coupling characteristics with somatic cell adenylyl cyclases, Endocrinology, 116:1357-1366.

Hyne, R. V., 1984, Bicarbonate- and calcium-dependent induction of rapid guinea pig sperm acrosome reactions by monovalent ionophores, Biol. Reprod., 31:312-323.

Iqbal, M., Shivaji, S., Vijayasarathy, S., and Balaram, P., 1980, Synthetic peptides as chemoattractants for bull spermatozoa structure activity correlations, Biochem. Biophys. Res. Commun., 96:235-242.

Lenz, R. W., Ax, R. L., Grimek, H. J., and First, N. L., 1982, Proteoglycan from bovine follicular fluid enhances an acrosome reaction in bovine spermatozoa, Biochem. Biophys. Res. Commun., 106:1092-1098.

Lenz, R. W., Ball, G. D., Lohse, J. K., First, N. L., and Ax, R. L., 1983, Chondroitin sulfate facilitates an acrosome reaction in bovine spermatozoa as evidenced by light microscopy, electron microscopy, and in vitro fertilization, Biol. Reprod., 28:683-690.

Lodge, J. R., and Salisbury, G. W., 1962, Initiation of anaerobic metabolism of mammalian spermatozoa by carbon dioxide, Nature, 195:293-294.

Meizel, S., and Turner, K. O., 1984, Glycosaminoglycans stimulate the acrosome reaction of previously capacitated hamster sperm, J. Cell Biol., 99:261a.

Miller, R. L., 1982, Synthetic peptides are not chemoattractants for bull sperm, Gamete Res., 5:395-402.

Mrsny, R. J., Waxman, L., and Meizel, S., 1979, Taurine maintains and stimulates motility of hamster sperm during capacitation in vitro, J. Exp. Zool., 210:123-128.

Murdoch, R. N., and Davis, W. D., 1978, Effects of bicarbonate on the respiration and glycolytic activity of boar spermatozoa, Aust. J. Biol. Sci., 31:385-394.

Nomura, K.. and Isaka, S., 1985, Synthetic study on the structure-activity relationship of sperm activating peptides from the jelly coat of sea urchin eggs, Biochem. Biophys. Res. Commun., 126:974-982.

Nomura, K., Suzuki, N., Ohtake, H., and Isaka, S., 1983, Structure and action of sperm activity peptides from the egg jelly of a sea urchin, Anthocidaris crassispina, Biochem. Biophys. Res. Commun., 117:147-153.

Ohtake, H., 1976, Respiratory behavior of sea urchin spermatozoa. I. Effect of pH and egg water on the respiratory rate, J. Exp. Zool., 198:303-312.

Okamura, N., and Sugita, Y., 1983, Activation of spermatozoan adenylate cyclase by a low molecular weight factor in porcine seminal plasma,

J. Biol. Chem., 258:13056-13062.

Okamura, N., Tajima, Y., Soejima, A., Masuda, H., and Sugita, Y., 1985, Sodium bicarbonate in seminal plasma stimulates the motility of mammalian spermatozoa through direct activation of adenylate cyclase, J. Biol. Chem., 260:9699-9705.

Ramarao, C. S., and Garbers, D. L., 1985, Receptor-mediated regulation of guanylate cyclase activity in spermatozoa, J. Biol. Chem., 260:8390-8396.

Repaske, D. R., and Garbers, D, L., 1983, A hydrogen ion flux mediates stimulation of respiratory activity by speract in sea urchin spermatozoa, J. Biol. Chem., 258:6025-6029.

Rossignol, D. P., Earles, B. J., Decker, G. L., and Lennarz, W. J., 1984, Characterization of the sperm receptor on the surface of the eggs of Strongylocentrotus purpuratus, Dev. Biol., 104:308-321.

Saling, P. M., Storey, B. T., Wolf, D. P., 1978, Calcium-dependent binding of mouse epididymal spermatozoa to the zona pellucida, Dev. Biol., 65:515-525.

SeGall, G. K., and Lennarz, W. J., 1979, Chemical characterization of the component of the jelly coat from sea urchin eggs responsible for induction of the acrosome reaction, Dev. Biol., 71:33-48.

SeGall, G. K., and Lennarz, W. J., 1981, Jelly coat and induction of the acrosome reaction in echinoid sperm, Dev. Biol., 86:87-93.

Shimomura, H., and Garbers, D. L., 1985, Differential effects of resact analogues on sperm respiration rates and cyclic nucleotide concentrations, submitted.

Singh, J. P., Babcock, D. F., and Lardy, H. A., 1978, Increased calcium-ion influx is a component of capacitation of spermatozoa, Biochem. J., 172:549-556.

Smith, A. C., and Garbers, D. L., 1983, The binding of an ^{125}I-speract analogue to spermatozoa, in: "Biochemistry of Metabolic Processes," D. L. F. Lennon, F. W. Stratman, and R. N. Zahlten, eds., Elsevier Science Pub. Co., New York, pp. 15-28.

Storey, B. T., Lee, M. A., Muller, C., Ward, C. R., and Wirtshafter, D. G., 1984, Binding of mouse spermatozoa to the zonae pellucidae of mouse eggs in cumulus: Evidence that the acrosomes remain substantially intact, Biol. Reprod., 31:1119-1128.

Suzuki, N., Nomura, K., Ohtake, H., and Isaka, S., 1981, Purification and the primary structure of sperm-activating peptides from jelly coat of sea urchin eggs, Biochem. Biophys. Res. Commun., 99:1238-1244.

Suzuki, N., Shimomura, H., Radany, E. W., Ramarao, C. S., Ward, G. E., Bentley, J. K., and Garbers, D. L., 1984, A peptide associated with eggs causes a mobility shift in a major plasma membrane protein of spermatozoa, J. Biol. Chem., 259:14874-14879.

Takai, K., Kurashina, Y., Suzuki-Hori, C., Okamota, H., and Hayaishi, O., 1974, Adenylate cyclase from Brevibacterium liquefaciens, I. Purification crystallization, and some properties, J. Biol. Chem., 249:1965-1972.

Talbot, P., 1984, Hyaluronidase dissolves a component in the hamster zona pellucida, J. Exp. Zool., 229:309-316.

Vacquier, V. D., 1979, The fertilizing capacity of sea urchin sperm rapidly decreases after induction of the acrosome reaction, Dev. Growth Differ., 21:61-60.

Vacquier, V. D., 1983, Purification of sea urchin sperm bindin by DEAE-cellulose chromatography, Anal. Biochem., 129:497-501.

Vacquier, V. D., and Moy, G. W., 1977, Isolation of bindin: The protein responsible for adhesion of sperm to sea urchin eggs, Proc. Natl. Acad. Sci. USA, 74:2456-2460.

Vijayaraghavan, S., Critchlow, L. M., and Hoskins, D. D., 1985, Evidence for a role for cellular alkalinization in the cyclic adenosine 3',5'-monophosphate-mediated initiation of motility in bovine caput spermatozoa, Biol. Reprod., 32:489-500.

Vijayasarathy, S., Shivaji, S., Iqbal, M., and Balaram, P., 1980, Formyl-Met-Leu-Phe induces chemotaxis and acrosomal enzyme release in bull sperm, FEBS Lett., 115:178-180.

Ward, G. E., Brokaw, C. J., Garbers, D. L., and Vacquier, V. D., 1985a, Chemotaxis of Arbacia punctulata spermatozoa to resact, a peptide from the egg jelly layer, J. Cell Biol., 101:2324-2329.

Ward, G. E., Garbers, D. L., and Vacquier, V. D., 1985b, Effects of extracellular egg factors on sperm guanylate cyclase, Science, 227:768-770.

Ward, G. E., and Vacquier, V. D., 1983, Dephosphorylation of a major sperm membrane protein is induced by egg jelly during sea urchin fertilization, Proc. Natl. Acad. Sci., 80:5578-5582.

Wassarman, P. M., 1983, Fertilization in: "Cell Interactions and Development: Molecular Mechanisms," K. Yamada, ed., Wiley, New York, pp. 1-27.

Yanagimachi, R., 1981, Mechanisms of fertilization in mammals, in: "Fertilization and Embryonic Development In vitro," L. M. Mastroianni, and J. D. Biggers, eds., Plenum, New York, pp. 81-182.

THE ROLE OF TYROSINE PHOSPHORYLATION IN THE REGULATION OF INSULIN AND INSULIN-LIKE GROWTH FACTOR-I RECEPTOR KINASE ACTIVITIES

Kin-Tak Yu and Michael P. Czech

Department of Biochemistry
University of Massachusetts Medical School
Worcester, Massachusetts 01605

INTRODUCTION

Insulin is a polypeptide hormone which plays a crucial role in regulating cellular metabolism and growth. Although both the acute and long-term effects of insulin in a variety of cell types have been widely examined and reported, the molecular mechanism or mechanisms mediating the actions of this hormone are still unknown. Due to the diverse effects of insulin on cellular functions, it is quite possible that a multitude of pathways may be involved in the signalling mechanism of this hormone. In spite of the possible existence of such an intricate network of signalling pathways, it is reasonable to expect that they may originate from the interaction of insulin with its cell surface receptor. Thus, a logical approach to elucidate the molecular mechanism of insulin action is to investigate events which may occur immediately after the binding of insulin to its receptor.

The aim of this chapter is to focus on one of these chemical events, the phosphorylation of the insulin receptor on tyrosine residues upon binding of insulin. The insulin-stimulated tyrosine phosphorylation of the insulin receptor β-subunit was first demonstrated by Kahn and co-workers (Kasuga et al., 1982a, 1982b). This area of research has subsequently become the subject of intense investigation by a number of laboratories. The goal of our studies is to determine whether the functional characteristics of the receptor is altered following ligand-induced phosphorylation. It is hoped that these studies will yield important clues on the role of insulin receptor phosphorylation in transmitting the signal of the hormone.

STRUCTURAL AND ENZYMATIC ASPECTS OF THE INSULIN RECEPTOR

The general structure of the insulin receptor was assessed by a number of laboratories using three major approaches: affinity labeling with ^{125}I-insulin (Pilch and Czech, 1979, 1980; Massague et al., 1980, 1981; Massague and Czech, 1982a), isotopic labeling of the receptor with [124]I, ^{3}H, and [35]S (Jacobs et al., 1980; Kasuga et al., 1982c), and affinity purification of the receptor (Jacobs et al., 1977, 1980; Harrison and Itin, 1980; Fujita-Yamaguchi et al., 1983, 1984). The insulin receptor is postulated to consist of two disulfide linked heterodimers (Fig. 1). Each dimer is composed of two polypeptides designated α- and β-subunits also linked by disulfide bonds. The α- and β-subunits exhibit apparent molecular weights of 130,000 and 95,000 daltons, respectively (Massague and Czech, 1982a) on SDS-polyacrylamide gel. It has been suggested that the α-subunit contains the hormone binding domain in light of the relative ease in affinity cross-linking of this subunit to ^{125}I-insulin (Pilch and Czech, 1979, 1980). Based on the relative

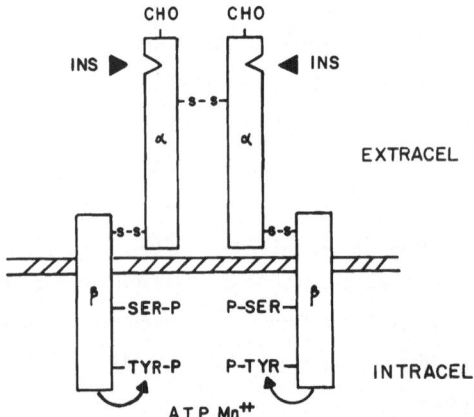

Fig. 1. Current model of the insulin receptor. The insulin receptor consists of two identical disulfide-linked heterodimers. Each dimer is composed of an α- and β-subunit. The α-subunit resides on the cell surface and is anchored to the β-subunit by thiol bonds. The α-subunit contains the hormone binding domain. The β-subunit is an integral protein. A large portion of the β-subunit is on the cytoplasmic side of the membrane. Following interaction of insulin with the α-subunit, the β-subunit undergoes phosphorylation both on tyrosine and serine residues. The tyrosine phosphorylation of the insulin receptor β-subunit is an autocatalytic reaction, while the serine phosphorylation reaction is probably mediated through other exogenous kinases.

accessibilities of the two receptor subunits to ^{125}I-labeling in intact cells (Hedo and Simpson, 1984) and subcellular fractions, it has been suggested that the α-subunit resides on the surface of the cell while the β-subunit transverses the plasma membrane into the cytosol.

The validity of the insulin receptor model is substantiated by the recent isolation of the human insulin receptor cDNA (Ullrich et al., 1985; Ebina et al., 1985). The deduced primary structure of the insulin receptor consists of a single polypeptide of 1370-1382 amino acids. It is predicted that cleavage at a protease-preferred site along the receptor precursor polypeptide leads to the generation of the α- and β-subunits. Hydropathy analysis of the receptor structure indicates that only the β-subunit contains a stretch of hydrophobic amino acids long enough to serve as a transmembrane domain. This information predicts that the α-subunit is entirely extracellular and is anchored to the plasma membrane through disulfide linkage and probably also by hydrophobic bonding to the transmembrane β-subunit. Thus, the predicted topographical arrangement of the receptor α- and β-subunits appears to be consistent with that based on experimental data (Hedo and Simpson, 1984).

A key finding of the analysis of the cytoplasmic domain of the insulin receptor β-subunit is that a region of the receptor subunit exhibits a high degree of homology to the kinase domain of a number of tyrosine kinases in the src kinase family (Ullrich et al., 1985; Ebina et al., 1985). These tyrosine kinases include the human epidermal growth factor receptor and the products of v-able, v-src, v-fes, and v-fms oncogenes. This finding indicates that the insulin receptor is a tyrosine kinase and confirms the conclusion based on earlier studies using highly purified insulin receptor (Kasuga et al., 1983a) and affinity labeling with radioactive ATP analogues (Roth and Cassell, 1983; Shia and Pilch, 1983; Van Obberghen et al., 1983).

The enzymatic characteristics of the insulin receptor have been investigated by a number of laboratories (Petruzzelli et al., 1982; Avruch et al., 1982; Kasuga et al., 1983b; Zick et al., 1983a, 1983b; Stadtmauer and Rosen, 1983; Nemenoff et al., 1984; White et al., 1984; Yu and Czech, 1984b, 1985, 1986). The tyrosine kinase activity of the insulin receptor is stimulated several fold upon binding of insulin. The receptor is capable of self phosphorylating its β-subunit as well as phosphorylating a number of artificial substrates such as histone, casein, angiotensin II, and copolymers of glutamic acid and tyrosine. The Km for ATP of the

167

receptor kinase is approximately 100–200 µM. At low concentrations of ATP, manganese ion is the preferred ion while magnesium ion is also effective at higher concentrations of ATP (Nemenoff et al., 1984; White et al., 1984). The insulin-stimulated kinase activity of the receptor is markedly enhanced in the presence of reductant such as dithiothreitol. The effects of dithiothreitol appear to result from the reduction of intramolecular thiol groups instead of intermolecular disulfide bonds between the two receptor subunits (Shia et al., 1983).

Under in vitro conditions, the insulin-stimulated tyrosine phosphorylation of the insulin receptor β-subunit occurs on multiple sites (White et al, 1984; Yu and Czech, 1984b; Petruzzelli et al., 1984). The identities of these phosphorylation sites have not yet been elucidated. However, the question regarding the role of receptor β-subunit tyrosine phosphorylation in the regulation of the receptor activity has been extensively examined by Rosen and colleagues (Rosen et al., 1983) and by our laboratory (Yu and Czech, 1984b; Yu et al., 1985, 1986). The results of these investigations are presented in the following section and strongly indicate that the tyrosine kinase activity of the insulin receptor towards exogenous substrates is markedly activated following tyrosine phosphorylation of the receptor β-subunit.

IN VITRO ACTIVATION OF INSULIN AND INSULIN-LIKE GROWTH FACTOR I RECEPTOR KINASES BY TYROSINE PHOSPHORYLATION

A common occurrence among all the tyrosine kinases examined to date is that they readily undergo autophosphorylation on tyrosine residues (Hunter and Cooper, 1985). The significance of this phenomenon has yet to be fully evaluated. It is generally felt that the phosphorylation state of these kinases may play a regulatory role in the activity of the enzyme. Studies performed by us (Yu and Czech, 1984b; Yu et al., 1986) and those by Rosen and co-workers (Rosen et al., 1983) indicated that the insulin receptor tyrosine kinase activity towards exogenous substrates is markedly stimulated following tyrosine phosphorylation of the receptor β-subunit.

Our experiments were conducted using a novel approach as described in Figure 2. Our strategy is to purify the insulin receptor by adsorption to insulin-agarose following wheat germ agglutinin-agarose chromatography. While immobilized on insulin-agarose, the insulin receptor kinase activity was assayed using histone as substrate. The insulin receptor dependent

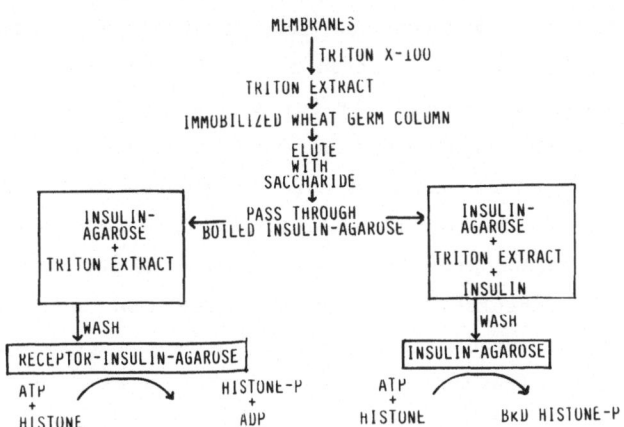

Fig. 2. Affinity purification of the insulin receptor and
assay of its kinase activity. Placental membranes
were solubilized in 1% Triton X-100. The detergent
extracts were applied to a wheat germ agglutinin-
agarose column. The sugar-containing insulin
receptor was eluted from the lectin column with
0.3 M N-acetylglucosamine. The receptor preparation was further affinity purified by adsorption
to insulin-agarose in the presence and absence of
100 μg/ml free insulin. Insulin receptor bearing
and deficient insulin-agarose preparations were
then generated. The kinase activities of these
insulin receptor preparations were measured using
histone H2B as substrate and [γ-^{32}P] ATP as
phosphate donor. The receptor dependent kinase
activity was calculated as the difference between
the kinase activity of the receptor bearing
insulin-agarose and that of the receptor deficient
insulin-agarose.

kinase activity was calculated as the difference between the kinase
activity of the receptor bearing insulin-agarose and that of the insulin-
agarose preparation specifically devoid of insulin receptor. The insulin
receptor deficient insulin-agarose preparation was generated by including
excess free insulin in the adsorption mixture in order to competitively
inhibit the receptor from binding to the agarose matrix.

When this purified preparation of insulin receptor was first
incubated with increasing concentrations of unlabeled ATP in the presence
of Mn^{++} and Mg^{++} and subsequently washed to remove the unreacted ATP, the
receptor dependent kinase activity towards histone was progressively
increased (Fig. 3). Significantly, the amount of phosphate incorporated
into the β-subunit of the insulin receptor also increased as the
concentrations of ATP used for incubation were raised. Thus, a close

parallel is observed between the incorporation of phosphate into the insulin receptor and the activation of its histone kinase activity. The half-maximal concentrations of ATP for activation of the receptor histone kinase and β-subunit phosphorylation are between 50-100 μM (Yu and Czech, 1984b). This range of values appears to correspond to the reported Km value for ATP of the insulin receptor kinase (Nemenoff et al., 1984; White et al., 1984). Phosphoamino acid analysis of the receptor-phosphorylated histone and the receptor β-subunit indicated that both phosphorylation reactions occurred predominately on tyrosine residues (Yu and Czech, 1984b). These results strongly suggest that tyrosine phosphorylation of the insulin receptor β-subunit enhances the tyrosine kinase activity of the receptor towards exogenous substrate.

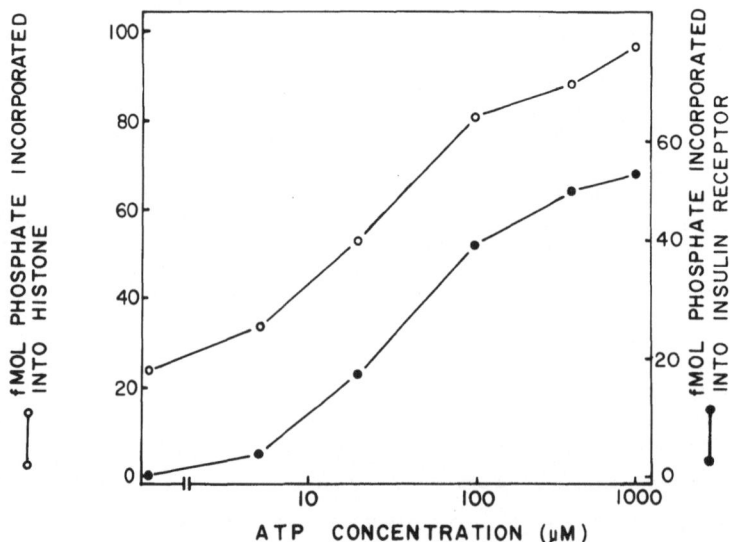

Fig. 3. Relationship between the phosphorylation of the insulin receptor β-subunit and its associated kinase activity. Insulin receptor-bearing insulin-agarose preparations were incubated with different concentrations of unlabeled or [γ-32P] ATP for 1 h. at 22°C. The preparations were washed extensively to remove the unreacted ATP. Samples incubated with [γ-32P] ATP were boiled in 0.2 ml of electrophoresis buffer containing SDS and DTT and electrophoresed on an 8% gel. The 32P-labeled band corresponding to the β-subunit of the insulin receptor was excised and radioactivity was determined by liquid scintillation counting. Samples incubated with unlabeled ATP were used to assay the receptor-associated kinase activity using histone as substrate and 5 μM [γ-32P] ATP as phosphate donor in the presence of Mn^{++} and Mg^{++} (Yu and Czech, 1984b).

The correlation between the phosphotyrosine content of the insulin receptor β-subunit and the receptor kinase activity towards exogenous substrates is further substantiated by the observation that the ATP-activated receptor histone kinase activity could be readily reversed by treatment of the receptor with alkaline phosphatase. Most important, the deactivation of the elevated receptor kinase activity was accompanied by a similar decrease in phosphate content of the receptor β-subunit (Table I). Because the action of alkaline phosphatase was directed specifically on phosphotyrosine residues of the insulin receptor β-subunit (Yu and Czech, 1984b), these studies provide further support to the concept that tyrosine phosphorylation of the receptor β-subunit leads to the activation of the receptor tyrosine kinase.

Using high pressure liquid chromatography, we had extensively characterized the phosphorylation patterns of the insulin receptor subunit (Yu and Czech, 1984b). Our results indicated that receptor tyrosine phosphorylation occurred on three tryptic peptide fractions (Fig. 4). One of these three phosphopeptide fractions (peak 2 in Fig. 4) may contain the

Table I. Effect of alkaline phosphatase treatment on the phosphate content of the insulin receptor and its associated kinase activity[a].

Alkaline Phosphatase	Temperature	^{32}P Remaining in Receptor	Decrease	Receptor-mediated phosphorylation of histone	Decrease
	°C	cpm	%	cpm	%
−	15	292		1643	
+	15	142	51	538	67
−	22	282		691	
+	22	144	49	253	64

Insulin receptor-bearing insulin-agarose preparation were phosphorylated with either 400 μM unlabeled or ^{32}P-labeled ATP (specific activity, 10 μCi/nmol) at 22°C for 1 h. After extensive washing, the resin was incubated in the presence and absence of 12 units/ml of alkaline Phosphatase for 1 h. at 15 or 22°C. Samples phosphorylated with [γ-^{32}P] ATP were used for the measurement of phosphate incorporation into the subunit of the insulin receptor and sample treated with unlabeled ATP were used for the assays of receptor kinase activity. [a]Reprinted with permission from Yu and Czech, 1984b.

tyrosine phosphorylation site which is involved in the activation of the receptor kinase. This postulate is based on the analysis that the dephosphorylating activity of alkaline phosphatase was directed primarily on peak 2 (Fig. 4). Significantly, the percentage decrease in phosphotyrosine content in this particular receptor phosphopeptide fraction appears to correlate with that in the kinase activity of the insulin receptor following phosphatase treatment (Yu and Czech, 1984b). As discussed above, the exact location of tyrosine phosphorylation sites on the receptor β-subunit is still unknown. The identification of this

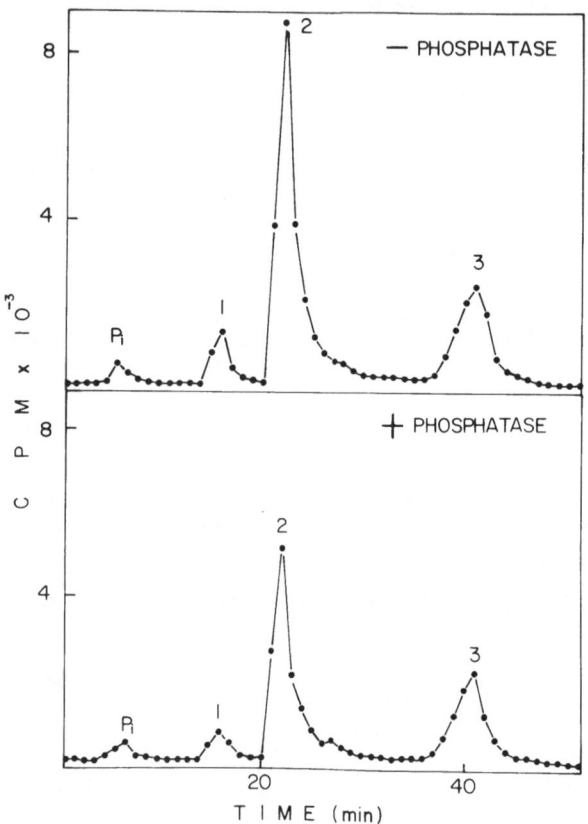

Fig. 4. HPLC tryptic peptide mapping of the ^{32}P-labeled β-subunit of the insulin receptor with or without alkaline phosphatase treatment. Insulin receptor preparations were phosphorylated with 400 μM of [γ-^{32}P] ATP (specific activity; 100 μCi/nmol) at 22°C for 1 h. The samples were then treated with or without alkaline phosphatase at 15°C followed by SDS-polyacrylamide gel electrophoresis and exhaustive tryptic digestion of the ^{32}P-labeled β-subunit. Resolution of the ^{32}P-labeled tryptic peptides was performed on a HPLC reverse-phase C-18 column. (Yu and Czech, 1984b).

phosphatase sensitive site on the primary structure of the receptor β-subunit will provide interesting insight into the mechanism through which the insulin receptor kinase is activated.

We have recently investigated the enzymatic properties of the insulin-like growth factor I receptor (IGF-I receptor). Earlier studies in this and other laboratories (Massague and Czech, 1982b; Kasuga et al., 1981, 1982d; Chernausek et al., 1981; Bhaumick et al., 1981; Kull et al., 1983) indicated that the general structure of the IGF-I receptor was very similar to that of insulin receptor. The IGF-I receptor is postulated to consist of two identical disulfide-linked heterodimers. Each dimer is composed of a 130,000 dalton α- and a 95,000 dalton β-subunits (Massague and Czech, 1982b). These two receptor subunits are also disulfide linked. The IGF-I receptor exhibits high affinity to IGF-I and low affinity to insulin while the reverse is true for the insulin receptor (Massague et al., 1982b; Kasuga et al., 1982d). Apart from its resemblance in gross receptor structure to that of the insulin receptor, the IGF-I receptor is also capable of mediating ligand-induced biological responses similar to those of insulin receptor. These insulin-like responses include stimulation of hexose and amino acid transport (Yu and Czech, 1984a), glycogen synthesis (Versphol et al., 1984) and cell proliferation (Zapf et al., 1978; King et al., 1980; Schmid et al., 1983; Massague et al., 1982c). In addition, the IGF-I receptor undergoes phosphorylation on tyrosine residues following binding of its ligand (Jacobs et al., 1983; Rubin et al., 1983; Sasaki et al., 1985). It has long been suspected that the IGF-I receptor, like the insulin receptor, contains intrinsic tyrosine kinase activity, although this possibility has yet to be proven.

We reason that if the IGF-I receptor is a true tyrosine kinase, perhaps the receptor kinase activity may also be activated upon tyrosine phosphorylation of its β-subunit in a fashion similar to that of the insulin receptor. In order to examine this issue, we have used the affinity chromatography strategy similar to that employed for the insulin receptor to study the kinase characteristics of the IGF-I receptor immobilized on IGF-I-agarose. As shown in Figure 5, the immobilized IGF-I receptor derived from human placenta exhibited a marked 50-fold enhancement in histone tyrosine kinase activity following treatment with 1 mM unlabeled ATP. The nucleotide dependent activation was fully reversible when the phosphorylated IGF-I receptor was incubated with alkaline phosphatase to effect dephosphorylation of the receptor (Yu et al., 1986). Detailed investigation of this receptor kinase indicated that

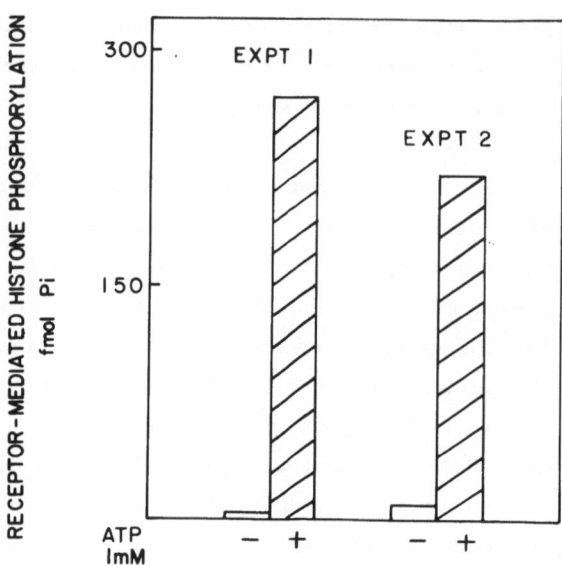

Fig. 5. Activation of IGF-I receptor dependent histone kinase activity by prior treatment with ATP. IGF-I receptor-bearing and deficient IGF-I-agarose preparations were incubated in 0.05 ml of 50 mM Hepes containing 0.1% Triton X-100, 1 mM PMSF, 10 µg/ml leupeptin, 10 mM $MgCl_2$ and 3 mM $MnCl_2$ in the presence and absence of 1 mM unlabeled ATP at 22°C for 30 min. The preparations were washed extensively to remove the unreacted ATP and were then suspended in 0.05 ml of phosphorylation buffer containing 0.4 mg/ml of histone H2B and 10 µM reaction was allowed to continue for 1 h. at 22°C and was then terminated by the addition of 0.1 ml of electrophoresis sample buffer. The samples were electrophoresed on a 6-16% SDS-polyacrylamide gradient gel. The gel was stained and destained. The histone bands were excised and the incorporated ^{32}P-radioactivity was determined by Cerenkov counting. Shown here are the results of two experiments. Each bar represents the IGF-I receptor-mediated histone phosphorylation as determined by subtracting histone phosphorylation using receptor deficient preparation from that by the receptor-bearing preparation.

its enzymology is essentially identical to that of the insulin receptor kinase. Thus, the IGF-I receptor kinase was progressively activated following incubation with increasing concentrations of ATP. Accompanying the elevation in receptor histone kinase activity, the phosphate content of the IGF-I receptor β-subunit was also increased in parallel. Half maximal activation of the IGF-I receptor kinase was achieved with a similar concentration range of 30-50 µM of ATP. The nucleotide specificity of activation for these two receptor systems appears to be identical with ATP being most effective. Furthermore, the tyrosine kinase activity of the IGF-I receptor was also markedly enhanced in the presence of reductant. The list of similarities between the two receptor kinases even extends to the tryptic phosphopeptide maps of their β-subunits. Thus, the major phosphopeptide fractions derived from the tryptic hydrolysates of the ^{32}P-labeled IGF-I receptor β-subunit were chromatographically indistinguishable from those of the insulin receptor subunit when resolved by reverse phase C-18 high pressure liquid chromatography column (Yu et al., 1986). Taken together, these data strongly support the notion that the IGF-I receptor is a tyrosine kinase which is enzymatically and structurally very similar to the insulin receptor kinase.

IN VIVO ACTIVATION OF THE INSULIN RECEPTOR KINASE BY TYROSINE PHOSPHORYLATION

Our results presented above indicate that both the insulin and IGF-I receptor kinases are activated in vitro following tyrosine phosphorylation of their β-subunits. A major question remains as to whether the same regulatory mechanism of these two receptor kinases also applies under in vivo conditions. This consideration is particularly important in view of the complex and dynamic cellular environment around the insulin and IGF-I receptors in intact cells. Unlike the in vitro situation where receptor phosphorylation occurs primarily on tyrosine residues, significant degrees of serine and threonine phosphorylation were also evident when the phosphorylation of insulin receptor was examined in control and insulin-treated cells (Kasuga et al., 1982b; White et al., 1985; Yu and Czech, 1986). It is quite possible that the role of tyrosine phosphorylation in regulating the insulin receptor kinase activity in intact cells may be altered in the presence of serine and threonine bound phosphate in the receptor β-subunit. However, attempts to resolve this issue had been hampered by the lack of knowledge on the identity of the putative substrate and substrates for the insulin receptor kinase in intact cells. Thus, a direct measure of the receptor kinase activity in vivo is not feasible.

In order to circumvent this problem, we recently developed an experimental procedure to assay the kinase activity of the insulin receptor under in vitro conditions following purification of the receptor from cells which were treated with or without insulin. The insulin receptor was purified by sequential affinity chromatography of Triton X-100 solubilized H-35 hepatoma cell extracts through wheat germ agglutinin- and insulin-agaroses. We included protease and phosphatase inhibitors in all the buffers used in order to preserve the structural and phosphorylation features of the insulin receptor. This precautionary measure is essential so that the physical and chemical states of the purified insulin receptor will resemble those in intact cells. The kinase activity of the insulin receptor was assayed using histone as substrate while immobilized on insulin-agarose. This step ensures that the insulin receptor preparations from control and insulin-treated cells are equally saturated with insulin. Thus, any difference in the insulin receptor kinase activity between the receptor preparations from control and insulin-treated cells is not due to the presence of prebound insulin but rather is due to changes that occur to the receptor while still in the cell.

Using the protocol described above, we detected a marked 20-fold increase in the histone tyrosine kinase activity of immobilized insulin receptors purified from insulin-treated H-35 cells when compared to control (Fig. 6). Kinetic analyses of the receptor kinase preparations from control and insulin-treated cells indicated that treatment of cells with insulin resulted in a 22-fold increase in the V_{max} of the receptor histone kinase while the Km value for histone remained unaltered. These results are in agreement with our earlier in vitro findings (Yu and Czech, 1984b). Significantly, the elevated insulin receptor kinase activity from insulin-treated cells was reversed by incubating the receptor preparation with alkaline phosphatase (Yu and Czech, 1986). When the gross phospho-amino acid contents of the insulin receptor from control and insulin-treated cells were analyzed, it became clear that insulin stimulated both tyrosine and serine phosphorylation of the receptor β-subunit (Yu and Czech, 1986). Subsequent incubation of the ^{32}P-labeled insulin receptor preparation from insulin-treated cells with alkaline phosphatase resulted in a marked decrease in the phosphotyrosine and phosphoserine contents of the receptor β-subunit (Yu and Czech, 1986). Thus, the insulin receptor kinase activity is directly related to the phosphate content of the receptor β-subunit.

The question whether tyrosine or serine phosphorylation or both may be responsible for activating the receptor kinase in vivo was addressed by detailed analyses of the tryptic phosphopeptide maps of the insulin receptor β-subunit ^{32}P-labeled in intact H-35 cells. Our results

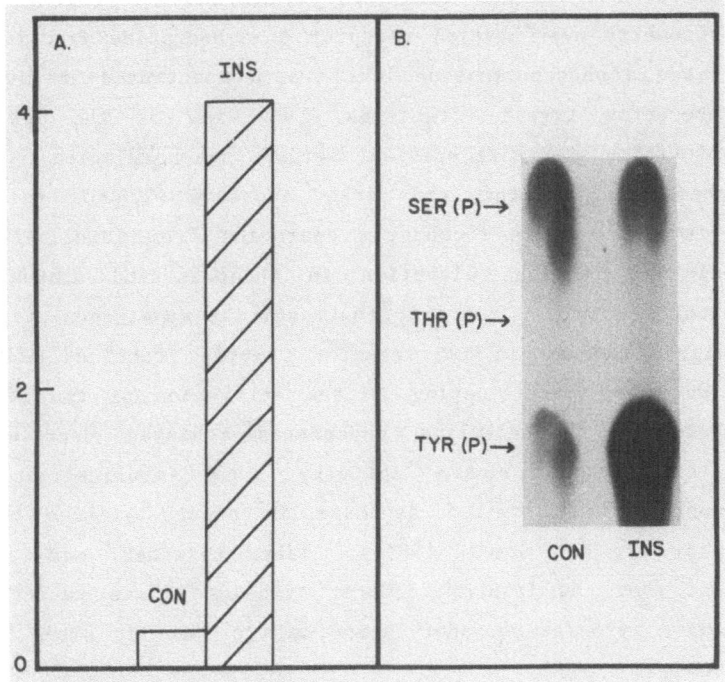

Fig. 6. Kinase activities of insulin receptor purified from control and insulin-treated H-35 hepatoma cells. Two groups of subconfluent H-35 cells were serum-starved for 24 h. and incubated further in phosphate-free DME for 17 h. at 37°C. Insulin (100 nM) was added to one group of cells and the incubation of these two groups of cells was continued for another 10 min. The cultures were lysed in 1% Triton X-100 containing phosphatase and protease inhibitors. Insulin receptor-bearing and -deficient insulin-agarose beads were prepared from each of these two groups of detergent hepatoma cell extracts by sequential affinity chromatographies through wheat germ agglutinin- and insulin-agaroses according to the procedures described in Figure 1. While immobilized on insulin-agarose, the insulin receptor was assayed for its kinase activity by using histone (0.4 mg/ml) and $[\gamma-^{32}P]$ ATP (0.3μM) over a period of 15 min at 25°C. The values of histone phosphorylation shown in Panel A represent the insulin receptor dependent kinase activity calculated by subtracting the histone associated ^{32}P-radioactivity in the receptor deficient preparation from that in the receptor-bearing preparation. The magnitudes of histone phosphorylation by the receptor deficient preparations from control and insulin-treated cells are 5,200 and 6,000 cpm respectively. Shown in Panel B is the phosphoamino acid analyses of histone ^{32}P-labeled by insulin receptor-bearing-agarose prepared from control and insulin-treated hepatoma cells. (Yu and Czech, 1986).

indicated that insulin-stimulated serine phosphorylation occurred in low stoichiometry over several receptor phosphopeptide fractions whereas the increase in phosphotyrosine level was concentrated in high abundance in two receptor tryptic fractions. In view of the low abundance of insulin-stimulated site specific serine phosphorylation (< 10% of basal), it seems unlikely that the marked increase (2,000%) in histone kinase activity of insulin receptor preparation from insulin-treated cell is associated with the elevation in phosphoserine contents in certain receptor β-subunit sites. In contrast, the appearance of phosphotyrosine in high abundance in two receptor tryptic fractions made it a likely candidate for participating in the activation of the receptor kinase. Furthermore, the alkaline phosphatase-mediated decrease in insulin receptor histone kinase activity from insulin-treated cells was accompanied by a marked decrease in phosphotyrosine content of the receptor (Yu and Czech, 1985b). Taken together, our results strongly suggest that the insulin receptor tyrosine kinase is activated in vivo following tyrosine phosphorylation on two specific sites on the receptor β-subunit.

CONCLUSIONS AND FUTURE DIRECTION

Studies presented above have allowed us to formulate a model on the regulation of the insulin and insulin-like growth factor I receptor kinases by tyrosine phosphorylation (Fig. 7). According to this model, the insulin and IGF-I receptor kinases are in a high V_{max} state when the receptor β-subunits are tyrosine phosphorylated. Dephosphorylation of the tyrosine phosphorylated receptors subunit returns the receptor kinase to a low V_{max} state. This mode of regulation of insulin and IGF-I receptor tyrosine kinases by phospho- and dephosphorylation also appears to be utilized by the epidermal growth factor receptor (Bertics and Gill, 1985). It will be interesting to determine whether the regulatory mechanism of the above receptor kinases by tyrosine phosphorylation also applies to other tyrosine kinases.

Clearly an important issue needed to be addressed is the identification and characterization of the putative substrate and substrates for the insulin and IGF-I receptor kinases in intact cells. Studies directed at this issue will form the basis for advancing our understanding on the molecular mechanism of insulin action.

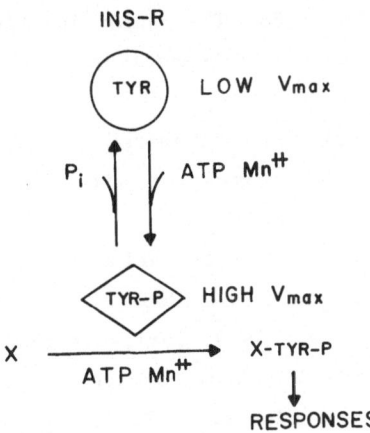

Fig. 7. Schematic model of regulation of insulin receptor kinase by tyrosine phosphorylation. Upon binding of insulin, the intrinsic kinase of the insulin receptor catalyzes tyrosine phosphorylation of its β-subunit. In the tyrosine phosphorylated form, the insulin receptor exhibits elevated tyrosine kinase activity towards exogenous substrate. The generation of insulin responses in the cell may be initiated as a result of tyrosine phosphorylation of some putative cellular substrates by the insulin receptor kinase. Termination of the signal may be achieved by lowering the receptor kinase activity following dephosphorylation of the receptor β-subunit by exogenous phosphatases.

REFERENCES

Avruch, J., Nemenoff, R. A., Blackshear, P. J., Pierce, M. W., and Osathanondh, R., 1982, Insulin-stimulated tyrosine phosphorylation of the insulin receptor in detergent extracts of human placental membranes, J. Biol. Chem., 257:15162-15166.

Bertics, P. J., and Gill, G. N., 1985, Self-phosphorylation enhances the protein-tyrosine kinase activity of the epidermal growth factor receptor, J. Biol. Chem., 260:14642-14647.

Bhaumick, B., Bala, R. M., and Hollenberg, M. D., 1981, Somatomedin receptor of human placenta: Solubilization, photolabeling, partial purification and comparison with insulin receptor, Proc. Natl. Acad. Sci. USA, 78:4279-4283.

Chernausek, S. D., Jacobs, S., and Van Wyk, J. J., 1981, Structural similarities between human receptors for somatomedin C and insulin. Analysis by affinity labeling, Biochemistry, 20:7345-7550.

Ebina, Y., Ellis, L., Jarnagin, K. G., Edery, M., Graf, L., Clauser, E., Ou, J. H., Mariarz, F., Kan, Y. W., Goldfine, I. D., Roth, R. A., and Rutter, W. J., 1985, The human insulin receptor cDNA: The structural basis for hormone-activated transmembrane signalling, Cell, 40:747-758.

Fujita-Yamaguchi, Y., Choi, S., Sakamoto, Y., and Itakura, K., 1983, Purification of insulin receptor with full binding activity, J. Biol. Chem., 258:5045-5049.

Fujita-Yamaguchi, Y., 1984, Characterization of purified insulin receptor subunits, J. Biol. Chem., 259:1206-1211.

Harrison, L. C., and Itin, A., 1980, Purification of the insulin receptor from human placenta by chromatography on immobilized wheat germ lectin and receptor antibody, J. Biol. Chem., 255:12066-12072.

Hedo, J. A., and Simpson, I. A., 1984, Internalization of insulin receptors in the isolated rat adipose cell, J. Biol. Chem., 259:11083-11089.

Hunter, T., and Cooper, J. A., 1985, Protein tyrosine kinase, Ann. Rev. Biochem., 54:897-930.

Jacobs, S., Shechter, Y., Bissell, K. and Cuatrecasas, P., 1977, Purification and properties of insulin receptors from rat liver membranes, Biochem. Biophys. Res. Commun., 77:981-988.

Jacobs, S., Hazum, E., and Cuatrecasas, P., 1980, The subunit structure of rat liver insulin receptor, J. Biol. Chem., 255:6937-6940.

Jacobs, S., Kull, F. C., Jr., Earp, H. S., Svoboda, M. E., Van Wyk, J. J., and Cuatrecasas, P., 1983, Somatomedin C stimulates the phosphorylation of the β subunit of its own receptor, J. Biol. Chem., 258:9581-9584.

Kasuga, M., Van Obberghen, E., Nissley, S. P., and Rechler, M. M., 1981, Demonstration of the subtypes of insulin-like growth factor receptors by affinity cross-linking, J. Biol. Chem., 256:5305-5308.

Kasuga, M., Karlsson, F. A., and Kahn, C. R., 1982a, Insulin stimulates the phosphorylation of the 95,000-dalton subunit of its own receptor, Science, 215:185-187.

Kasuga, M., Zick, Y., Blithe, D. L., Karlsson, F. A. Haring, H. U., and Kahn, C. R., 1982b, Insulin stimulation of phosphorylation of the subunit of the insulin receptor, J. Biol. Chem., 257:9891-9894.

Kasuga, M., Hedo, J. A., Yamada, K. M., and Kahn, C. R., 1982c, The structure of insulin receptor and its subunits, J. Biol. Chem., 257:10392-10399.

Kasuga, M., Van Obberghen, E., Nissley, S. P. and Rechler, M. M., 1982d, Structure of the insulin-like growth factor receptor in chicken embryo fibroblast, Proc. Natl. Acad. Sci. USA, 79:1864-1868.

Kasuga, M., Fujita-Yamaguchi, Y., Blithe, D. L., and Kahn, C. R., 1983a, Tyrosine-specific protein kinase activity is associated with the purified insulin receptor, Proc. Natl. Acad. Sci. USA, 80: 2137-2141.

Kasuga, M., Fujita-Yamaguchi, Y., Blithe, D. L., White, M. F., and Kahn, C. R., 1983b, Characterization of the insulin receptor kinase purified from human placental membranes, J. Biol. Chem., 258:10973-10980.

King, G. L., Kahn, C. R., Rechler, M. M., and Nissley, S. P., 1980, Direct demonstration of separate receptors for growth and metabolic activities of insulin and multiplication-stimulating activity (an insulin-like growth factor) using antibodies to the insulin receptor, J. Clin. Invest., 66:130-140.

Kull, F. C., Jacobs, S., Su, Y. F., Svoboda, M. E., Van Wyk, J. J., and Cuatrecasas, P., 1983, Monoclonal antibodies to receptors for insulin and somatomedin-C, J. Biol. Chem., 258:6561-6566.

Massague, J., Pilch, P. F., and Czech, 1980, Electrophoretic resolution of three major insulin receptor structures with unique stoichiometries, Proc. Natl. Acad. Sci. USA, 77:7137-7141.

Massague, J., Pilch, P. F., and Czech, 1981, A unique proteolytic cleavage site on the β subunit of the insulin receptor, J. Biol. Chem., 256:3182-3190.

Massague, J., and Czech, M. P., 1982a, Role of disulfides in the subunit structures of the insulin receptor, J. Biol. Chem., 257:6729-6738.

Massague, J., and Czech, M. P., 1982b, The subunit structures of two distinct receptors for insulin-like growth factor I and II and their relationship to the insulin receptor, J. Biol. Chem., 257:5038-5045.

Massague, J., Blinderman, L. A., and Czech, M. P., 1982c, The high affinity insulin receptor mediates growth stimulation in rat hepatoma cells, J. Biol. Chem., 257:13958-13963.

Nemenoff, R. A., Kwok, Y. C., Shulman, G. I., Blackshear, P. J., Osathanondh, R., and Avruch, J., 1984, Insulin-stimulated tyrosine protein kinase: Characterization and relation to the insulin receptor, J. Biol. Chem., 259:5058-5065.

Petruzzelli, L. M., Ganguly, S., Smith, C. J., Cobb, M. H., Rubin, C. S., and Rosen, O. M., 1982, Insulin activates a tyrosine-specific protein kinase in extracts of 3T3-L1 adipocytes and human placenta, Proc. Natl. Acad. Sci. USA 79:6792-6796.

Petruzzelli, L. M., Herrera, R., and Rosen, O. M., 1984, Insulin receptor is an insulin-dependent tyrosine protein kinase: Copurification of insulin-binding activity and protein kinase activity to homogeneity from human placenta, Proc. Natl. Acad. Sci. USA, 81:3327-3331.

Pilch, P. F., and Czech, M. P., 1979, Interaction of cross-linking agents with the insulin effector system isolated fat cells: Covalent linkage of ^{125}I-insulin to a plasma membrane receptor protein of 140,000 daltons, J. Biol. Chem., 254:3375-3381.

Pilch, P. F., and Czech, M. P., 1980, The subunit structure of the high affinity insulin receptor, J. Biol. Chem., 255:1722-1731.

Rosen, O. M., Herrera, R., Olowe, Y., Petruzzelli, L. M., and Cobb, M., 1983, Phosphorylation activates the insulin receptor tyrosine protein kinase, Proc. Natl. Acad. Sci. USA, 80:3237-3240.

Roth, R. A., and Cassell, D. J., 1983, Insulin receptor: Evidence that it is a protein kinase, Science, 219:299-301.

Rubin, J. B., Shia, M. A., and Pilch, P. F., 1983, Stimulation of tyrosine specific phosphorylation in vitro by insulin-like growth factor, Nature (London), 305:438-440.

Sasaki, N., Rees-Jones, R. W., Zick, Y., Nissley, S. P., and Rechler, M. M., 1985, Characterization of insulin-like growth factor I-stimulated tyrosine kinase activity associated with the β-subunit of type I insulin-like growth factor receptors of rat liver cells, J. Biol. Chem., 260:9793-9804.

Schmid, C. H., Steiner, T. H., and Froesch, E. R., 1983, Preferential enhancement of myoblast differentiation by insulin-like growth factors, FEBS Lett., 161:117-121.

Shia, M. A., and Pilch, P. F., 1983, The β subunit of the insulin receptor is an insulin-activated protein kinase, Biochemistry, 22:717-721.

Shia, M. A., Rubin, J. B., and Pilch, P. F., 1983, The insulin receptor protein kinase. J. Biol. Chem. 258:14450-14455.

Stadtmauer, L. A., and Rosen, O. M., 1983, Phosphorylation of exogenous substrates by the insulin receptor-associated protein kinase, J. Biol. Chem., 258:6682-6685.

Ullrich, A., Bell, J. R., Chen, E. Y., Herrera, R., Petruzzelli, L. M., Dull, T. J., Gray, A., Coussens, L., Liao, Y. C., Tsubokawa, M., Mason, A., Seeburg, P. H., Grunfeld, C., Rosen, O. M., and Ramachandran, J., 1985, Human insulin receptor and its relationship to the tyrosine kinase family of oncogenes, Nature, 313:756-761.

Van Obberghen, E., Rossi, B., Kowalski, A., Gazzano, H., and Ponzio, G., 1983, Receptor-mediated phosphorylation of the hepatic insulin receptor: Evidence that the Mr-95,000 receptor subunit is its own kinase, Proc. Natl. Acad. Sci. USA, 80:945-949.

Verspohl, E. J., Roth, R. A., Vigneri, R., and Goldfine, I. D., 1984, Dual regulation of glycogen metabolism by insulin and insulin-like growth factors in human hepatoma cells (HEP-G2), J. Clin. Invest., 74:1436-1443.

White, M. F., Haring, H. U., Kasuga, M., and Kahn, C. R., 1984, Kinetic properties and sites of autophosphorylation of the partially purified insulin receptor from hepatoma cells, J. Biol. Chem., 259:255-264.

White, M. F., Takayama, S., and Kahn, C. R., 1985, Differences in the sites of phosphorylation of the insulin receptor in vivo and in vitro, J. Biol. Chem., 260:9470-9478.

Yu, K. T., and Czech, M. P., 1984a, The type I insulin-like growth factor receptor mediates the rapid effects of multiplication-stimulating activity on membrane transport systems in rat soleus muscle, J. Biol. Chem., 259:3090-3095.

Yu, K. T., and Czech, M. P., 1984b, Tyrosine phosphorylation of the insulin receptor subunit activates the receptor-associated tyrosine kinase activity, J. Biol. Chem., 259:5277-5286.

Yu, K. T., Werth, D. K., Pastan, I. H., and Czech, M. P., 1985, Src kinase catalyzes the phosphorylation and activation of the insulin receptor kinase, J. Biol. Chem., 260:5838-5846.

Yu, K. T., and Czech, M. P., 1986, Tyrosine phosphorylation of insulin receptor β subunit activates the receptor tyrosine kinase in intact H-35 hepatoma cells, J. Biol. Chem., in press.

Yu, K. T., Peters, M. A., and Czech, M. P., 1986, Similar control mechanisms regulate the insulin and type I-insulin-like growth factor receptor kinase, J. Biol. Chem., in press.

Zapf, J., Schoenle, E., and Froesch, E. R., 1978, Insulin-like growth factor I and II: Some biological actions and receptor binding characteristics of two purified constituents of nonsuppressible insulin-like activity of human serum, Eur. J. Biochem., 87:285-296.

Zick, Y., Kasuga, M., Kahn, C. R., and Roth, J., 1983a, Characterization of of insulin-mediated phosphorylation of the insulin receptor in a cell free system, J. Biol. Chem., 258:75-80.

Zick, Y., Rees-Jones, R. W., Grunberger, G., Taylor, S. I., Moncada, V., Gorden, P., and Roth, J., 1983b, The insulin-stimulated receptor kinase is a tyrosine-specific casein kinase, Eur. J. Biochem., 137:631-637.

INTERNALIZATION OF HORMONE RECEPTOR COMPLEXES: ROUTE AND SIGNIFICANCE

Barry I. Posner, Masood N. Khan, Denis G. Kay,
and John J. M. Bergeron

Departments of Medicine and Anatomy
McGill University
Montreal, Quebec   H3A 1A1

INTRODUCTION

Receptor-mediated endocytosis is a general process by which a wide variety of ligands are taken up with high specificity and affinity into cells. In this review we shall primarily discuss data in regard to the internalization of peptide hormones, especially insulin, but it is noteworthy that there are similarities as well as interesting differences by which membrane-bound lectins, immune complexes, viruses, toxins, etc. are handled by cells (Bergeron et al., 1985).

The initial event consists of the binding of ligand to specific receptors on the cell surface. In some instances, as in the case of LDL endocytosis, this initial binding event is followed by aggregation of ligand-receptor complexes in coated pits (Goldstein et al., 1985). In the case of insulin and other peptide hormones, ligand-receptor complexes appear to be dispersed over the entire cell surface, and the role of coated pits in their subsequent aggregation and internalization remains uncertain (Posner et al, 1982).

Internalization of ligands is a rapid energy-dependent process which has been studied in detail by a large number of investigators using subcellular fractionation, morphologic methods, radioautography, and immunocytochemistry (Posner et al., 1981; Bergeron et al., 1985). Our early studies employed both electron microscope radioautography and cell fractionation methods to demonstrate receptor-mediated uptake of

185

[125]I-labeled hormones into Golgi-enriched fractions of rat liver (Posner et al., 1981). Golgi vesicular elements in rat liver were defined on the basis of two criteria: the morphological appearance of intraluminal lipoprotein particles, and enrichment in the biochemical marker galactosyltransferase (Bergeron et al., 1973; Ehrenreich et al., 1973). We interpreted our data to indicate uptake into Golgi elements since, by electron microscope radioautography, the radiolabeled hormone was in lipoprotein-containing structures and, by subcellular fractionation, ^{125}I-hormone was especially concentrated in elements co-sedimenting with galactosyltransferase (Posner et al., 1981).

It was clear that concentration of internalized hormone into lysosomes was not a significant process. Thus, when we examined the uptake of ^{125}I-insulin into rat hepatocytes _in vivo_, the vesicular elements accumulating ligand were acid phosphatase negative (Bergeron et al., 1979). Furthermore, when we compared uptake into Golgi-enriched and lysosomal fractions (Table I) with or without chloroquine treatment, ^{125}I-insulin accumulated in Golgi-enriched fractions to a much greater degree (5- to 20-fold) than in the lysosomal fractions. Nor did chloroquine, though concentrating to a greater extent in lysosomes, influence the relative extent of ^{125}I-insulin accumulation in the fractions (Table I). Furthermore, virtually all the radiolabel in the lysosomal fractions was found in lipoprotein-containing vesicles and never over secondary lysosomes (Posner et al., 1982a).

Though the structures involved in the early phase of peptide hormone internalization are not lysosomes, there is evidence for degradation of peptide hormones in lysosomes. For example, studies of gold and ferritin labeled ligands have shown ultimate accumulation of these particles in lysosomes (Dickson et al., 1981; McKanna et al., 1979). The extent to which this is the exclusive fate of internalized protein ligands has not been established, and other routes of cellular clearance are possible.

THE ENDOSOMAL SYSTEM (ENDOSOMES)

Differentiation from Known Organelles

The recognition that lipoprotein-containing vesicles, co-sedimenting in lysosomal fractions, were the structures internalizing ^{125}I-insulin ("unique" vesicles) in these fractions (Khan et al., 1981) led to studies

Table I. Chloroquine concentration and its effect on ^{125}I-insulin accumulation in Golgi and lysosomal fractions.

Cell Fraction	^{125}I-insulin (cpm/mg protein)[a]		Relative Chloroquine Concentration
	Control	+Chloroquine	$\dfrac{\text{Fraction[b]}}{\text{Homogenate}}$
Golgi light	18,035 ± 5,321	104,154 ± 22,416	1.9 ± 0.3
intermediate	32,387 ± 3,615	209,712 ± 39,332	3.2 ± 0.5
heavy	8,580 ± 1,237	24,586 ± 2,252	2.5 ± 0.3
Lysosomal L_2	1,804 ± 490	10,767 ± 4,672	18.0 ± 0.7
Fractions L_3	577 ± 97	3,637 ± 315	7.9 ± 0.2

[a]Each animal received 12 x 106 cpm of ^{125}I-insulin 20 min before sacrifice. Chloroquine treatment was 10 mg/100 body wt. 2 hrs and 1 hr, respectively, before ^{125}I-insulin injection. All values are mean ± SE of four fractionations. See Posner et al. (1982a) for details.

[b]Subcellular fractions, prepared 60 min after chloroquine, were extracted and measured as described elsewhere (Posner et al., 1982a). Each value is the mean ± SE of three fractionations.

demonstrating that lipoprotein-containing vesicles in Golgi fractions were heterogeneous. Thus, it was possible to resolve Golgi-enriched fractions into two receptor-rich vesicle populations by Percoll gradient centrifugation: a low density population, co-sedimenting with galactosyltransferase, and a high density population devoid of this Golgi marker (Fig. 1; Khan et al., 1982).

On reexamining ^{125}I-insulin internalization into these fractions, it was seen that at 1 minute postinjection, radiolabel was largely in elements of low density, but by 10 minutes it was found predominantly in elements of high density (Fig. 1). The integrity of ^{125}I-insulin was substantially retained across the gradient and, on morphological analysis, was exclusively over lipoprotein-containing structures (Khan et al., 1982).

Although these observations suggested initial uptake into low density structures with subsequent transfer to high density ("unique") vesicles, other data have raised the possibility of the independent uptake of peptide hormones, at two different rates, into these two populations of

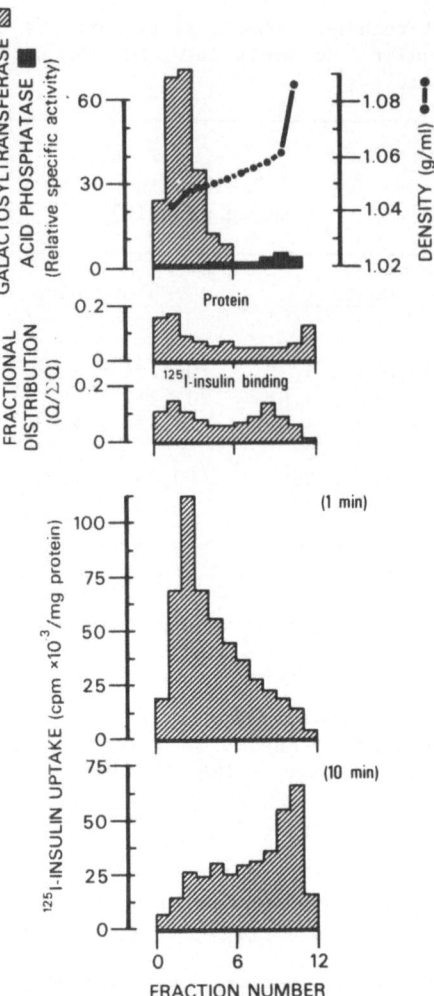

Fig. 1. Distribution of marker enzymes, protein, ^{125}I-insulin binding, and radioactivity following ^{125}I-insulin administration in Golgi intermediate subfractions generated by Percoll gradient centrifugation. Animals received 12 x 10[6] cpm of ^{125}I-insulin and were studied at 1 minute and 10 minutes after radiolabel injection. Galactosyltransferase acid phosphatase, ^{125}I-insulin binding, and protein were assayed as described by Khan et al. (1982) wherein further details may be obtained.

lipoprotein-containing vesicles (Posner et al., 1982b; Bergeron et al., 1983). In addition, these observations raised further doubts as to the homogeneity of the low density lipoprotein-containing vesicles. Perhaps there was one pool involved in a biogenetic route and another engaged in endocytosis and catabolism. The possibility that a substantial proportion of hepatic lipoprotein is derived via endocytosis has been given strong credence by the appreciation that the liver in most animals probably accounts for the bulk of low density lipoprotein catabolism (Attie et al., 1982; Chao et al., 1981).

The advent of a powerful new technique for resolving endocytic structures has permitted us to examine critically the extent to which internalized ^{125}I-insulin co-sedimenting with galactosyltransferase (Fig. 1) is actually located in the same structures containing this enzyme. The technique is based upon the polymerization of diaminobenzidine (DAB) to a dense complex by horseradish peroxidase in the presence of H_2O_2. Courtoy and his colleagues coupled horseradish peroxidase to galactosylated bovine serum albumin and allowed this ligand for the asialoglycoprotein receptor to be internalized by rat liver. Following this, they prepared cell fractions containing the ligand and showed that on adding DAB and H_2O_2 the endocytic structures were shifted

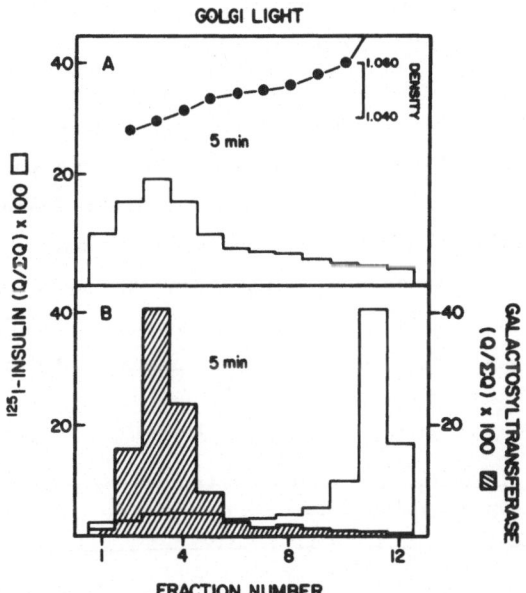

Fig. 2. Internalized insulin is not in Golgi components. At 5 min. following the co-injection of ^{125}I-insulin and galactose-bovine serum albumin horseradish peroxidase, the Golgi light fraction was isolated and incubated in the absence (A) or presence (B) of diaminobenzidine and H_2O_2. In the control (A) the peak of radioactive insulin is coincident with GT the marker enzyme of the (shaded in B). In the treated sample (B), the containing endosome has been clearly shifted away from the GT. Comparable data have been seen with the Golgi intermediate fraction as well (Kay et al., 1984).

to a higher density on subsequent centrifugation (Courtoy et al., 1984). This density technique has been applied by us to study ^{125}I-insulin uptake (Kay et al., 1984) and it has become clear that the bulk of endocytic elements containing ^{125}I-insulin are devoid of galactosyltransferase (Fig. 2).

Characteristics

Various terms have been employed to describe the structures involved in concentrating internalized substances within the cell (Table II).

It is now clear that these structures can be distinguished from plasma membrane (Posner et al., 1981), lysosomes (Posner et al., 1982a) and, in large part, from Golgi elements as well (Kay et al., 1984). The term "endosome", first defined by Stockem (1969) and more recently by Helenius et al. (1983) is a convenient term for describing the intracellular nonlysosomal components involved in the uptake and concentration of exogenous substances by cells.

There is substantial evidence that endosomes are heterogeneous. In rat liver, at least three endosomal components can be distinguished by subcellular fractionation (Bergeron et al., 1985; Posner et al., 1982): The first is a small vesicular fraction which probably originates from a series of uncoated vesicles and tubules near the cell surface (Posner et al., 1980). A second major component consists of larger structures (150–500 nm) which are vesicular but often associated with tubules. These

Table II. Terminology for structures involved in internalization.

Term	Reference
Lysosomes	Gorden et a., 1980
Golgi Fractions (elements)	Posner et al., 1981
Ligandosomes	Smith and Peters, 1982
Compartment for Uncoupling Receptor and Ligand (CURL)	Geuze et al., 1983
Receptosomes	Willingham and Pastan, 1980
Multivesicular Bodies	McKanna et al., 1979; Walsh et al., 1984)
Tubular Vesicles	Wall et al., 1980
Lipoprotein-containing vesicles	Posner et al., 1981;
"Unique" vesicles	Khan et al., 1981; Posner et al., 1982
Endosomes	Helenius et al., 1983

components are low density on Percoll gradients and may not be low-pH entities since, in our experience (Posner et al., 1982a; Khan et al., 1985b), they do not appear to concentrate the acidotropic agent chloroquine. The third class of endosomal structures is heterogeneous and is lysosome-like in that it contains low concentrations of acid phosphatase (Khan et al., 1982). These correspond to heavier density components resolved on Percoll gradients (Fig. 1) and to those entities previously called "unique" vesicles which co-sediment with lysosomes (Khan et al., 1981). Since chloroquine concentrates significantly in these structures they probably constitute a low-pH compartment (deDuve et al., 1974). There is substantial agreement that at least a portion of the endosomal system is at low pH, likely maintained as such by an ATP-dependent proton pump (Galloway et al., 1983). It has been suggested that the segregation of ligand from receptor occurs in this compartment (Tycko et al., 1982); a hypothesis favored by recent direct visualization of such segregation (Geuze et al., 1983). The acronym CURL (Compartment for Uncoupling Receptor and Ligand) has been proposed for this endosomal component (Geuze et al., 1983).

It is presently unclear as to what extent the endosomal system represents a unique organelle as opposed to portions of plasma membrane involved in pinocytic activity. For example, how much endosomal structure is determined by membrane donation from the Golgi apparatus as well as plasma membrane?

Control of Endocytosis

Agents which raise the pH of endocytic structures cause significant accumulation of ligand in endosomes (Posner et al., 1982a). Chloroquine treatment leads to significant insulin but not prolactin accumulation in rat liver endosomes (Khan et al., 1985a). This corresponds to the more marked effect of low pH on insulin versus prolactin binding (Posner et al., 1978; Posner et al., 1979). Thus, raising endosomal pH from acid to near-neutral has only minimal effects on the state of prolactin dissociation and hence clearance (Khan et al., 1985a), whereas the effects on insulin dissociation are marked and hence clearance is strongly influenced (Posner et al., 1982a). The acidotropic agents may not only influence ligand dissociation but may prevent membrane fusion events between endosomal components and lysosomes.

Microtubule polymerization is of importance in regulating endosomal structures (Posner et al., 1982b; Bergeron et al., 1983). Temperature-sensitive steps have been described at the cell surface and within the endosomal system. Thus, internalization is greatly reduced at temperatures less than 5°C (Marsh and Helenius, 1980) and transit from the endosomal compartment is inhibited at temperatures less than 20°C (Dunn et al., 1980). The temperature effect may reflect a unique composition of the phospholipid bilayer of "late" endosomes which must undergo a specific perturbation to permit membrane coalescence, presumably with lysosomes.

INTERNALIZATION AND RECYCLING OF RECEPTORS: A ROLE FOR ENDOSOMES

The above discussion has emphasized the internalization of insulin and other peptide hormones, but it is now quite clear that it is hormone-receptor complexes which are internalized. This was suggested by early observations that ligand uptake is receptor-mediated and that internalized ligand concentrates in receptor-rich cell fractions (Posner et al., 1981). This view is further supported by the recognition that ligands are metabolized by cells far more rapidly than their receptors. Thus, insulin is metabolized rapidly (t1/2 ≅ 30 minutes) and the insulin receptor much more slowly (t1/2 ≅ 10 hours) by target cells (Krupp and Lane, 1981).

Using photoaffinity labeling techniques, Fehlmann et al. (1982) have provided strong evidence for insulin-receptor internalization and recycling. Thus, photoaffinity-labeled insulin receptors of cultured hepatocytes become trypsin-insensitive on incubation at 37°C but, on more extended incubation (up to 6 hours), become partially sensitive to trypsin, indicating an initial loss from and subsequent return to enzyme accessibility at the cell surface (Fehlmann et al., 1982a). Similar studies of rat adipocytes confirmed the internalization of photolabeled receptors (Berhanu et al., 1982). These studies can be criticized on the basis that an unnatural linkage between receptor and ligand was formed. Nevertheless, they are supported by a range of observations. Thus, exposing chick liver cells to insulin resulted in a loss (downregulation) of cell surface receptors with a corresponding increase in intracellular receptor content (Krupp and Lane, 1981). The injection of insulin has been shown to effect a loss of insulin receptors from liver cell plasma membranes (Desbuquois et al., 1982; Pezzino et al., 1980) along with a corresponding increase in endosomal receptors (Desbuquois et al., 1982; Khan et al., 1985). Comparable observations have been made made on exposing isolated adipocytes to insulin (Green and Olefsky, 1982).

The use of antibodies prepared against various receptors has contributed greatly to defining more precisely the fate of the receptor. Thus, several groups using antibody to the LDL receptor (Basu et al., 1981), the transferrin receptor (Enns et al., 1983), and the Fc receptor on macrophages (Mellman et al., 1983) have shown that receptor is internalized along with ligand. In elegant studies Geuze, Schwartz, and their colleagues (1983) have used double-label immunoelectron microscopy to co-localize asialoglycoprotein and its receptor. They showed that ligand and receptor were closely associated at the plasma membrane and in clathrin-coated vesicles close to the surface. However an endosomal compartment was identified wherein they could demonstrate the dissociation of ligand from receptor, with the former found in the lumen of vesicles and the latter concentrated in tubular elements free of ligand. They interpreted these observations to indicate the existence of a cell compartment which they called CURL (see above and Table II) in which ligand-receptor complexes dissociated.

Other studies have shown that there are different routes for the handling of ligands depending, presumably at least in part, on the ease with which they are dissociated from receptor. Thus, Hopkins (1983), studying transferrin-receptor complex internalization, has shown the existence of both a peripheral endosomal system and a juxtanuclear area involved in the handling of ligand-receptor complexes. He suggested that each compartment plays a specific role in directing internalized receptors to particular cellular domains. In a study comparing the uptake of the asialoglycoprotein receptor with that for polymeric IgA (membrane secretory component) Geuze, Schwartz, and colleagues (1984) showed that the intracellular routes for these two receptors corresponded at an early stage but subsequently diverged with each entering distinctive endosomal regions. The IgA receptor and its ligand were directed to the bile canalicular surface and secreted into the bile (Kuhn and Kraehenbuhl, 1982).

The above noted studies attest to the complexity of the endosomal apparatus but especially the role of this system in directing the flow of ligand-receptor complexes and the dissociated species as well. Figure 3 attempts to schematize the major intracellular routes traversed by hormones and their receptors. The endosomal apparatus (EN) is depicted as playing a major role in directing traffic to Golgi and lysosomes. A key question is how receptor is recycled to the plasma membrane. In this scheme, receptor is visualized as recycling from endosomes and/or via the

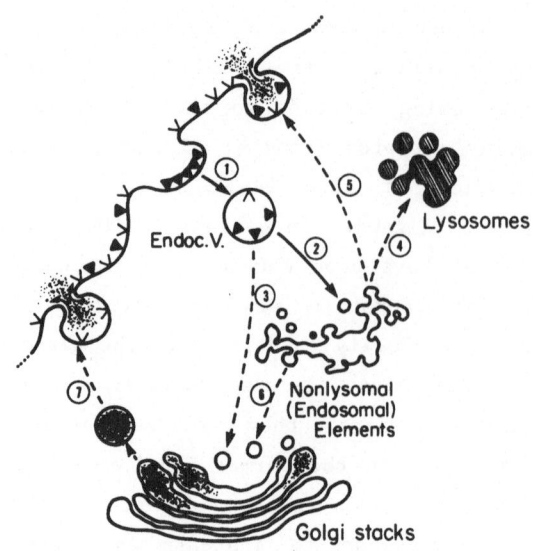

Fig. 3. Scheme of possible intracellular routes traversed by hormone-receptor complexes and individual hormone and receptor. 1) internalization to early endosome; 2) accumulation in later endosomal elements (EN); 3) possible limited internalization to Golgi; 4) and 5) dissociation of hormone-receptor complex with sorting and distribution of hormone to lysosomes (Ly) and hormone receptor to cell surface; 6) transfer of receptors from EN to Golgi; 7) recycling of receptors to cell surface via the secretory pathway and secretory granules.

Golgi apparatus and secretory granules (Sg). The demonstration that ^{125}I-insulin is internalized into secretory granules of the pancreas (Cruz et al., 1984), as well as the appearance of peptide hormones in secretions from various cells, is consistent with such a role for this route. The high concentration of several different receptors in Golgi saccules, as defined by quantitative immunocytochemistry, has indicated a role for this organelle not only in receptor biogenesis, but also in receptor recycling to the cell surface (Geuze et al., 1984).

In concluding this section, it is worthwhile to note that the endosomal apparatus plays a role in facilitating the entry of agents such as enveloped viruses and bacterial toxins into the cytosol (Helenius et al., 1983). Here low pH appears to play a key role in the process by which the agent penetrates endosomal membranes. In summary, the endosomal apparatus appears to be a distinctive entity engaged in sorting intracellular traffic of hormones and their receptors and other penetrating agents as well. In the next few years, the exact composition of this apparatus should become clearer as receptor antibodies, ultrathin

cryoelectron microscopy, and the DAB density shift procedure, among other
new techniques, are applied to its study.

SIGNIFICANCE OF INTERNALIZATION INTO THE ENDOSOMAL SYSTEM

Internalization of hormone-receptor complexes can be linked to three
different processes. It is clear that in most target cells, internalized
hormone is degraded (Mock and Niswender, 1983; Pastan and Willingham,
1981). This could conceivably be a mechanism for ensuring that once bound
to receptor, the hormone cannot dissociate and rebind to another receptor.
Internalization could be a mechanism for clearing ligand from the
circulation. In short, one can see this process as contributing to the
termination of hormone action.

A second process which interlocks with internalization is that of
receptor regulation. Ligand-induced loss or down-regulation of receptors
has been documented by a number of investigators (Gavin et al., 1974; Catt
et al., 1979), as has ligand induced up-regulation of receptors (Posner et
al., 1975). Temporary down-regulation of receptors might ensure that,
following the acute elevation of circulating hormone concentration, target
cells remain in a state of reduced sensitivity until hormone levels
decline. Thus internalization-dependent receptor regulation probably
plays a role in regulating cell sensitivity to hormone.

An interesting and important question is the role of hormone-receptor
complex internalization in hormone action. In the case of insulin, most,
if not all, of its biological effects are mimicked by antibodies to the
insulin receptor (Jacobs, 1985). Though there is evidence indicating that
insulin degradation products may play a role in augmenting insulin action
(Kikuchi et al., 1981), the studies with receptor antibodies suggest that
the ligand as such is not critical for promoting bioeffects. Some studies
have suggested that a second messenger mediating insulin's actions may be
generated from plasma membranes following insulin-receptor complex
formation (Larner et al., 1979). It is possible that such molecules might
derive from endosomal elements as well. Of great interest is the recent
demonstration that a variety of receptors function as protein kinases.
Thus insulin, insulin-like growth factor (IGF)-1, epidermal growth factor
(EGF), and platelet-derived growth factor (PDGF) are examples of peptide
ligands whose receptors possess intrinsic tyrosine-kinase activity (Hunter
and Cooper, 1985). It has been suggested that these kinases effect

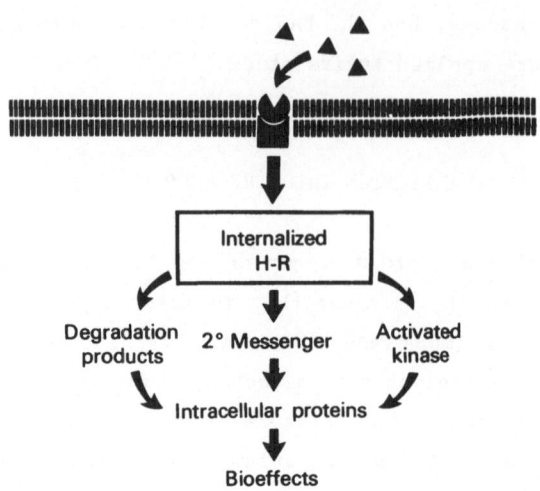

Fig. 4. A model of insulin action following internalization
of the hormone-receptor complex. For discussion
see text.

hormone and growth factor action by phosphorylating key intracellular
protein substrates entraining a phosphorylation cascade. Recent studies
from our laboratories (Khan et al., 1985; Kay et al., 1985) as well as
those of Cohen and Fava (1985) indicate that functional receptor kinases
are internalized to endosomes. In the case of insulin, the receptor
kinase appears to be in an activated state and, with or without insulin,
is maintained in this activated state in the endosomal system (Khan et
al., 1985). Thus internalization of active receptor kinases may
facilitate their interaction with key intracellular substrates whose
phosphorylation is needed for the realization of the bioresponse.

Figure 4 depicts this possibility and that of other modalities of
signal transmission. It is clear that further work is needed to establish
the biological significance of kinase function, the nature of the
physiologic substrates of the kinase, and the role of internalized kinase
in hormone action.

REFERENCES

Attie, A. D., Pittman, R. C., and Steinberg, D., 1982, Hepatic catabolism
    of low density lipoprotein: Mechanisms and metabolic consequences,
    Hepatology, 2:269-281.

Basu, S. K., Goldstein, J. K., Anderson, R. G. W., and Brown, M. S., 1981,

Monensin interrupts the recycling of low density lipoprotein receptors in human fibroblasts, Cell, 24:493-502.

Bergeron, J. J. M., Ehrenreich, J. H., Siekevitz, P., and Palade, G. E., 1973, Golgi fractions prepared from rat liver homogenates: Isolation procedure and morphological characterization, J. Cell Biol., 59:73-88.

Bergeron, J. J. M., Sikstrom, R., Hand, A. R., and Posner, B. I., 1979, Binding and uptake of ^{125}I-insulin into rat liver hepatocytes and endothelium, J. Cell. Biol., 80:427-443.

Bergeron, J. J. M., Resch, L., Rachubinski, R., Patel, B., and Posner, B. I., 1983 , Effect of colchicine on internalization of prolactin in female rat liver: An in vivo radioautographic study, J. Cell Biol., 96:875-886.

Bergeron, J. J. M., Cruz, J., Khan, M. N., and Posner, B. I., 1985, Uptake of insulin and other ligands into receptor-rich endocytic components of target cells: The endosomal apparatus, Ann. Rev. Physiol., 47:383-403.

Berhanu, P., Olefsky, J. M., Tsai, F., Thamm, P., Saunders, D., and Brandenburg, D., 1982, Internalization and molecular processing of insulin receptors in isolated rat adipocytes, Proc. Natl. Acad. Sci. USA, 79:4069-4073.

Catt, K. J., Harwood, J. P., Aquilera, G., and Dufau, M. L., 1979, Hormonal regulation of peptide receptors and target cell responses, Nature, 280: 109.

Chao, Y. S., Jones, A. L., Hradek, G. I., Windler, E. E. T., and Havel, R. J., 1981, Autoradiographic localization of low density lipoproteins in the liver of normal and estrogen-treated rats, Proc. Natl. Acad. Sci. USA, 78:597-601.

Cohen, S., and Fava, R. A., 1985, Internalization of functional epidermal growth factor: Receptor kinase complexes in A-431 cells, J. Biol. Chem., 260:12351-12358.

Courtoy, P. J., Quintart, J., and Baudhuin, F., 1984, Shift of equilibrium density induced by 3, 3-diaminobenzidine cytochemistry: A new procedure for the analysis and purification of peroxidase-containing organelles, J. Cell Biol., 98:870-876.

Cruz, J., Posner, B. I., and Bergeron, J. J. M., 1984, Receptor-mediated endocytosis of ^{125}I-insulin into pancreatic acinar cells in vivo, Endocrinology, 115:1996-2008.

de Duve, C., de Bousy, T., Poole, B., Trouet, A., Tulkens, P., and Van Hoof, F., 1974, Lysosomotropic agents, Biochem. Pharmacol., 23:2495-2531.

Desbuquois, B., Lopez, S., and Burlet, H., 1982, Ligand-induced translocation of insulin receptors in intact rat liver, J. Biol. Chem., 257:10852-10860.

Dickson, R. B., Willingham, M. C., and Pastan, I., 1981, $_2$-macroglobulin absorbed to colloidal gold: A new probe in the study of receptor-mediated endocytosis, J. Cell Biol., 89:29-34.

Dunn, W. A., Hubbard, A. L., and Aronson, A. L., 1980, Low temperature selectively inhibits fusion between pinocytic vesicles and lysosomes during heterophagy of ^{125}I-asialofetuin by the perfused rat liver, J. Biol. Chem., 255:5971-5978.

Ehrenreich, J. H., Bergeron, J. J. M., Siekevitz, P., and Palade, G. E., 1973, Golgi fractions prepared from rat liver homogenates: Isolation procedure and morphological characterization, J. Cell Biol., 59:45-72.

Enns, C. A., Larrick J. W., Suomalainen, H., Schroder, J., and Sussman, H. H., 1983, Co-migration and internalization of transferrin and its receptor on K562 cells, J. Cell Biol., 97:579-585.

Fehlmann, M., Carpentier, J.-L., Le Cam, A., Thamm, P., Saunders, D., Brandenburg, D., Orci, L., and Freychet, P., 1982, Biochemical and morphological evidence that the insulin receptor is internalized with insulin in hepatocytes, J. Cell Biol., 93:82-87.

Fehlmann, M., Carpentier, J.-L., Van Obberghen, E., Freychet, P., Thamm, P., Saunders, D., Brandenburg, D., and Orci, L., 1982a, Internalized insulin receptors are recycled to the cell surface in rat hepatocytes, Proc. Natl. Acad. Sci. USA, 79:5921-5925.

Galloway, C. J., Dean, C. E., Marsh, M., Rudnick, G., and Mellman, I., 1983, Acidification of macrophage and fibroblast endocytic vesicles in vitro, Proc. Natl. Acad. Sci. USA, 80:3334-3339.

Gavin, J. R. III, Roth, J., Neville, D. M. Jr., De Meyts, F., and Buell, D. N., 1974, Insulin-dependent regulation of insulin receptor concentrations: A direct demonstration in cell culture, Proc. Natl. Acad. Sci. USA, 71:84-88.

Geuze, H. J., Slot, J. W., Strous, G. J., Lodish, H. F., and Schwartz, A. A. L., 1983, Intracellular site of asialoglycoprotein receptor-ligand uncoupling: Double-label immunoelectron microscopy during receptor-mediated endocytosis, Cell, 32:277-287.

Geuze, H. J., Slot, J. W., and Schwartz, A. L., 1984, Intracellular receptor sorting during endocytosis: Comparative immunoelectron microscopy of multiple receptors in rat liver, Cell, 37:195-204.

Goldstein, J. L., Brown, M. S., Anderson, R. G. W., Russell, D. W., and Schneider, W. J., 1985, Receptor-mediated endocytosis: Concepts emerging from the LDL receptor system, <u>Ann. Rev. Cell Biol.</u>, 1:1-39.

Gorden, P., Carpentier, J.-L., Freychett, P., and Orci, L., 1980, Internalization of polypeptide hormones: Mechanism, intracellular localization and significance, <u>Diabetologia</u>, 18:263-274.

Green, A., and Olefsky, J. M., 1982, Evidence for insulin-induced internalization and degradation of insulin receptors in rat adipocytes, <u>Proc. Natl. Acad. Sci. USA</u>, 79:427-431.

Helenius, A., Mellman, J. I., Wall, D., and Hubbard, A., 1983, Endosomes, <u>Trends Biochem. Sci.</u>, 7:245-250,

Hopkins, C. R., 1983, Intracellular routing of transferrin and transferrin receptors in epidermoid carcinoma A431 cells, <u>Cell</u>, 32:321-330.

Hunter, I., and Cooper, J. A., 1985, Protein-tyrosine kinases, <u>Ann. Rev. Biochem.</u>, 54:897-930.

Jacobs, S., 1985, Immunochemical characterization of receptors for insulin and insulin-like growth factor-1, <u>in</u>: "Molecular Basis of Insulin Action", M. P. Czech, ed., Plenum Press, New York, pp.31-43.

Kay, D. G., Khan, M. N., Posner, B. I., and Bergeron, J. J. M., 1984, ^{125}I-insulin in hepatic Golgi fractions: Application of the diaminobenzidine (DAB)-shift protocol, <u>Biochem. Biophys. Res. Commun.</u>, 123:1144-1148.

Kay, D. G., Lai, W. H., Uchihashi, M., Posner, B. I., and Bergeron, J. J. M., 1985, In vivo internalization of active EGF receptor kinase into rat liver endosomes, <u>J. Cell. Biol.</u>, 101:298a (abstr. 1124).

Khan, M. N., Posner, B. I., Verma, A. K., Khan, R. J., and Bergeron, J. J. M., 1981, Intracellular hormone receptors: Evidence for insulin and lactogen receptors in a unique vesicle sedimenting in lysosome fractions of rat liver, <u>Proc. Natl. Acad. Sci. USA</u>, 78:4980-4984.

Khan, M. N., Posner, B. I., Khan, R. J., and Bergeron, J. J. M., 1982, Internalization of insulin into rat liver Golgi elements: Evidence for vesicle heterogeneity and the path of intracellular processing, <u>J. Biol. Chem.</u>, 257:5969-5976.

Khan, M. N., Bergeron, J. J. M., and Posner, B. I., 1985, In vivo activation of the insulin receptor protein kinase in rat liver, <u>J. Cell. Biol.</u>, 101:50a (abstr. 185).

Khan, R. J., Khan, M. N., Bergeron, J. J. M. and Posner, B. I., 1985a, Prolactin internalization into rat liver Golgi fractions: Differential effects of chloroquine, <u>Biochim. Biophys. Acta</u>, 838:77-83.

Khan, M. N., Savoie, S., Bergeron, J. J. M., and Posner, B. I., 1985b, Insulin and insulin receptor uptake into Golgi fractions: Effect and possible site of chloroquine action, Diabetes, 34:1025-1030.

Kikuchi, K., Larner, J., Freer, R. J., and Day, A. R., 1981, Effect of insulin fragments on biological activity of insulin and desoctapeptide insulin, I. Potentiation of biological activities, J. Biol. Chem., 256:9441-9444.

Krupp, M., and Lane, M. D., 1981, On the mechanism of ligand-induced down-regulation of insulin receptor level in the liver cell, J. Biol. Chem., 256:1689-1694.

Kuhn, L. C., and Kraehenbuhl, J. P., 1982, The sacrificial receptor-translocation of polymeric IgA across epithelia, Trends Biochem. Sci., 7:299-302.

Larner, J., Galasko, G., Cheng, K., Depaoli-Roach, A. A., Huang, L., Daggy, P., and Kellogg, J., 1979, Generation by insulin of a chemical mediator that controls protein phosphorylation and dephosphorylation, Science, 206;1408-1410.

Marsh, M., and Helenius, A., 1980, Adsorptive endocytosis of Semliki Forest Virus, J. Mol. Biol., 142:439-454.

McKanna, J. A., Haigler, H. T., and Cohen, S., 1979, Hormone receptor topology and dynamics: Morphological analysis using ferritin-labeled epidermal growth factor, Proc. Natl. Acad. Sci. USA, 76:5689-5693.

Mellman, I. S., Plutner, H., Steinman, R. M., Unkeless, J. C., and Cohn, Z. A., 1983, Internalization and degradation of macrophage Fc receptors during receptor-mediated phagocytosis, J. Cell Biol., 96:887-895.

Mock, E. J., and Niswender, G. D., 1983, Differences in the rate of internalization of ^{125}I-labeled human chorionic gonadotropin, luteinizing hormone and epidermal growth factor in ovine luteal cells, Endocrinology, 113:259-.

Pastan, I. H., and Willingham, M. C., 1981, Receptor-mediated endocytosis of hormones in cultured cells, Ann. Rev. Physiol., 43:239-250.

Pezzino, V., Vigneri, R., Piliam, N. B., and Goldfine, I. D., 1980, Rapid regulation of plasma membrane insulin receptors, Diabetologia, 19:211-215.

Posner, B. I., Kelly, P. A., and Friesen, H. G., 1975, Prolactin receptors in rat liver: Possible induction by prolactin, Science, 187:57-59.

Posner, B. I., Bergeron, J. J. M., and Josefsberg, Z., 1978, Intracellular polypeptide hormone receptors: Characterization of insulin binding sites on rat liver Golgi membranes, J. Biol. Chem., 253:4067-4073.

Posner, B. I., Josefsberg, Z., and Bergeron, J. J. M., 1979, Intracellular polypeptide hormone receptors: Characterization and induction of prolactin receptors in the Golgi apparatus of rat liver, J. Biol. Chem., 254:12494-12499.

Posner, B. I., Patel, B., Verma, A. K., and Bergeron, J. J. M., 1980, Uptake of insulin by plasmalemma and Golgi subcellular fractions of rat liver, J. Biol. Chem., 255:735-741.

Posner, B. I., Bergeron, J. J. M., Josefsberg, Z., Khan, M. N., Khan, R. J., Patel, B. A., Sikstrom, R. A., and Verma, A. K., 1981, Polypeptide hormones: Intracellular receptors and internalization, Rec. Prog. Horm. Res., 37:539-582.

Posner, B. I., Khan, M. N., and Bergeron, J. J. M., 1982, Endocytosis of peptide hormones and other ligands, Endocrine Rev., 3:280-298.

Posner, B. I., Patel, B., Khan., M. N., Bergeron, J. J. M., 1982a, Effect of chloroquine on the internalization of ^{125}I-insulin into subcellular fractions of rat liver: Evidence for an effect of chloroquine on Golgi elements, J. Biol. Chem., 257:5789-5799.

Posner B. I., Verma, A. K., Patel, B., and Bergeron, J. J. M., 1982b, Effect of colchicine on the uptake of prolactin and insulin into Golgi fractions of rat liver, J. Cell. Biol., 93:560-567.

Smith, G. D., and Peters, I. J., 1982, The localization in rat liver of alkaline phosphodiesterase to a discrete organelle implicated in ligand internalization, Biochim. Biophys. Acta 716:24-30.

Stockem, W., and Wohlfarth-Botterman, K., 1969, Pinocytosis (endocytosis), in: "Handbook of Molecular Cytology", A. Lima-de-Faria, ed., North Holland, Amsterdam, pp. 1373-1400

Tycko, B., and Maxfield, F.R., 1982, Rapid acidification of endocytic vesicles containing $_2$-macroglobulin, Cell, 28:643-651.

Wall, D. A., Wilson, G., and Hubbard, B., 1980, The galactose specific recognition system of mammalian liver: The route of ligand internalization in rat hepatocytes, Cell, 21:79-93.

Walsh, R. J., Posner, B. I., and Patel, B., 1984, Binding and uptake of ^{125}I-iodoprolactin by epithelial cells of the rat choroid plexus: An in vivo autoradiographic analysis, Endocrinology, 114:1496-1505.

Willingham, M. C., and Pastan, I., 1980, The receptosome: An intermediate organelle of receptor-mediated endocytosis in cultured fibroblasts, Cell, 21:67-77.

PART II

GENE EXPRESSION AND REPRODUCTION

CHROMATIN STRUCTURE AND GENE EXPRESSION

IN GERM LINE AND SOMATIC CELLS

Mark Groudine and Maxine Linial

Fred Hutchinson Cancer Research Center
Division of Basic Sciences
Seattle, Washington 98104

University of Washington School of Medicine
Department of Radiation Oncology
Seattle, Washington 98195

INTRODUCTION

The paternal genome is, according to recent reports, somehow marked for selective expression during early development. For example, the male pronucleus is essential for the formation of the extraembryonic layers during embryogenesis of the mouse (Surani et al., 1984; McGrath and Solter, 1984). In addition, the paternal X chromosome is selectively inactivted in the extraembryonic tissues of various species (West et al., 1977; Takagi et al., 1978; Harper et al., 1982). Thus, while it is formally possible that paternal DNA is completely reprogrammed on fertilization by components of the egg, there is evidence for the templating of information for differential gene expression by some component of sperm. We have initiated a series of experiments to determine whether the structure of sperm chromatin or the pattern of methylation of sperm DNA (or both) might provide a basis for the templating of information important in the selective expression of patenal genes early in development. The rationale for these experiments is based on observations concerning the relationship between chromatin structure and gene expression in somatic cells.

Changes in chromatin structure are associated with the transcriptional activation of eukaryotic genes. For example, active genes are preferentially sensitive to digestion with deoxyribonuclease I (DNase

I; Weintraub and Groudine, 1976) and contain sites hypersensitive to digestion with several nucleases including DNase I, nuclease S1 (S1), and restriction endonucleases (Elgin, 1981; Conklin and Groudine, 1984). In addition to the structural features of chromatin revealed by nucleases, actively transcribed genes are relatively undermethylated compared to inactive genes (Razin and Riggs, 1980; Riggs, 1984). In attempting to understand the developmental significance of the appearance of changes in chromatin structure during embryogenesis, we have examined various molecular aspects of globin gene expression during avian embryogenesis.

During the early development of chicken embryos, the posterior portion of a particular region of the embryo, the area opaca, will give rise to most of the primitive line of avian erythroid cells that first appear as blood islands in the area vasculosa at about 35 hr of development (Lillie, 1952). Figure 1 shows embryos at two stages of early chicken development. We have focused on the head fold stage present at about 20-23 hr of incubation, and have removed by dissection the presumptive red-cell-forming region from embryos of this stage (shown by the white line in Fig. 1A). While there are about 10^6 cells in the total

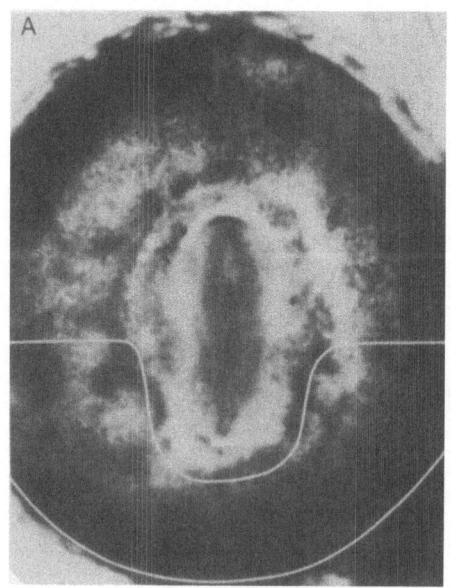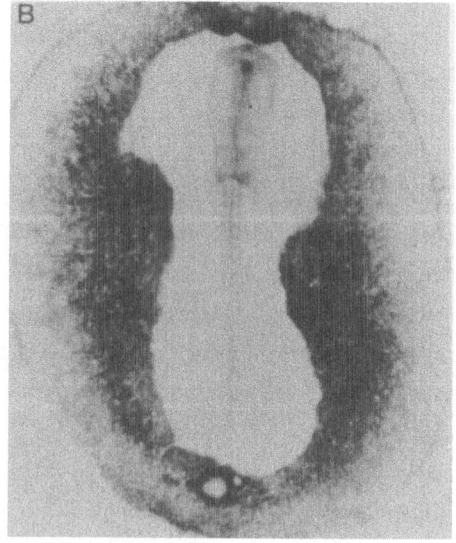

Fig. 1. Chicken blastoderms at various stages of incubation. (A) Head fold; stage 5; 23 hr. (B) Stage 11; 40-45 hr, containing blood islands. Area outlined in white (A) shows the precursor region that is cut out for the biochemical analysis. Most analysis is done with 20-22 hr embryos.

embryo at this time, this precursor region contains about 250,000 cells. These numbers suggest that the primary determination for erythropoiesis occurs very early during chicken development, perhaps as early as the 4- to 10-cell stage. The average generation time of cells in these precursor region is about 8 hr. Thus, these cells are about two divisions away from terminal differentiation into Hb-producing erythroblasts. When this occurs at 35 hr, about 10^6 red cells appear per embryo (Weintraub et al., 1971). While we realize that the cell division kinetics that convert a precursor population of 250,000 cells (at 20-23 hr) to a red cell population of 10^6 (at 35 hr) could be very complex (for example, one must consider differential rates of division and cell death, differential rates of overt differentiation, and symmetric vs. asymmetric divisions), it is nevertheless probable that these numbers suggest that the majority of this population of cells is erythroid percursors (Hagopian et al., 1972).

To test this possibility, we dissociated the presumptive precursor population into single cells and plated them in tissue culture. After several days, colonies grew, and these were stained for Hb with benzidine. The percentage of erythroid cells in each clone increases with time in culture. In many colonies, so-called bursts of red cells can be seen. These presumably represent founder cells that give rise to at least two cells that have the potential to differentiate into individual erythroid clones. Most important, however, is the fact that of the total number of colonies, 85-90% are erythroid. As controls, explantation or single cell cultures from other regions of the 23 hr embryo fail to yield significant numbers of red cells or red cell colonies (our unpublished observations; Samarut et al., 1979). These results, together with the numerical calculations presented above, suggest that most of the cells in the presumptive erythropoietic region of a 20-23 hr chick blastoderm are erythroid progenitor cells.

These experiments, raise at least two fundamental questions regarding the control of gene expression during embryogenesis: 1) Are tissue-specific genes differentially repressed or activted during differentiation (e.g., are globin genes transcribed in precursor cells prior to the appearance of hemoglobin)? and 2) How can specific determinative events induced in precursor cells at one time in development be expressed in progeny cells later, independent of the concurrent actin of the original inducing influence?

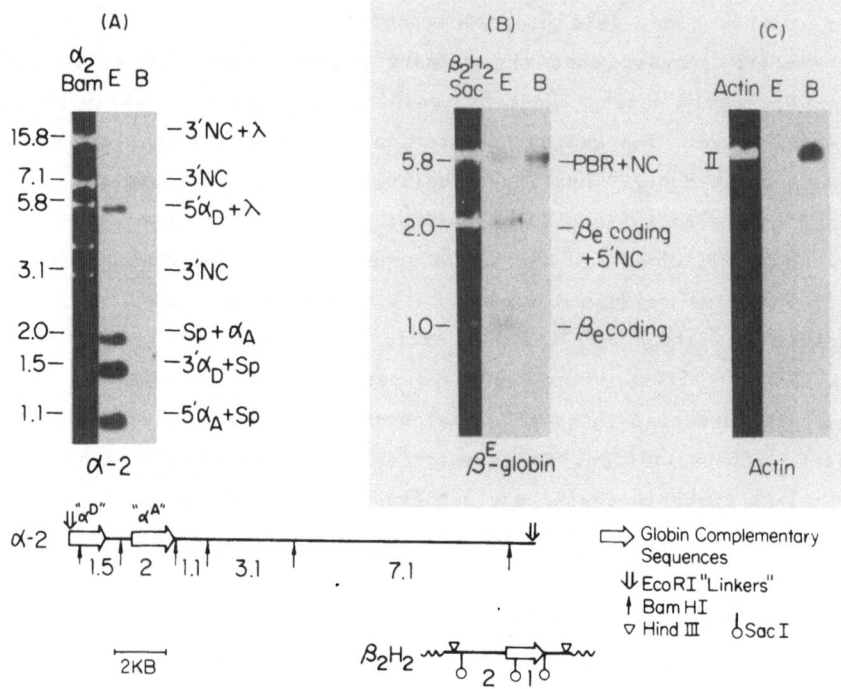

Fig. 2. Absence of initiated polymerases on globin genes in precursor cells. Nuclei from primitive erythroblasts (from 5-day embryos) or precursor cells (500 embryos) were isolated. Runoff transcription in the presence of ^{32}P-UTP and subsequent purification of RNA were performed as previously described (Groudine et al., 1981). The RNA was then hybridized to blots containing (A) Bam-digested α-2, (B) Sac-digested β2H2 (an embryonic β-globin-containing clone), or (C) an actin cDNA clone. E represents RNA from immature erythrocytes; B, RNA from blastoderms or precursors.

ABSENCE OF TRANSCRIPTION OF GLOBIN GENES IN PRECURSOR CELLS

Previously, it was shown that whole chicken blastoderms at 20-23 hr contain less than one copy of stable globin RNA sequence per cell (Groudine et al., 1974). Using a different type of assay (Groudine et al., 1981), we determined whether bound RNA polymerases could be detected on the globin genes. Nuclei were isolated from precursor red cells and from primitive erythroblasts synthesizing embryonic globin chains. Radioactive triphosphates were added to the nuclei, and the bound RNA polymerase molecules were allowed to elongate for 50-100 bases (1-2 min of synthesis under these conditions). The synthesized RNA was isolated and then hybridized to a southern blot containing the separated restriction fragments from a genomic α-globin clone, a genomic embryonic β-globin clone, or an actin cDNA clone. Figure 2 shows that under comparable conditions there is marked hybridization to the coding regions of the α- and β-globin clones from RNA synthesized by the red cell nuclei, but no detectable hybridization (even at high exposures) to the globin-coding regions from RNA synthesized in precursor red cell nuclei. At high exposures, signals are observed from some flanking region fragments of both the α and β clones. These signals are seen, however, in all cell types, and, by blot hybridization to genomic DNA, these regions are known to contain DNA sequences present elsewhere in the genome. We therefore believe that they do not necessarily reflect transcription from nonglobin regions of the globin-gene clusters.

As an internal control for the red cell precursor transcripts, Figure 3C shows that this RNA hybridizes to a chicken β-actin cDNA clone (provided by D. Cleveland and M. Kirschner), whereas there is very little hybridization of the RNA from these dividing red cells to the actin clone. These results reinforce the conclusion that the control of globin-gene activation during development is dictated by the transcriptional activation of these genes.

PROPAGATION OF CHROMATIN STRUCTURE AND GENE EXPRESSION

In attempting to understand the developmental significance of the appearance of changes in chromatin structure during embryogenesis, we (Groudine and Weintraub, 1982) attempted to investigate a specific feature of the differentiation process common to many multicellular organisms. Certain determinative events appear to be induced in precursor cells at one time in development and, independent of the concurrent action of the original inducing influence, the effects of these events are expressed in progeny cells that begin overt differentiation some time later (Alberts et al., 1983). Thus, it would seem that a specific event occurs at one time in development, and, in some way, the consequences of that event can be propagated to progeny cells in the absence of the original stimulus. Such a situation appears to exist for the maintenance in progenitor cells of competency to generate progeny erythroblasts when removed from the embryonic mileu and explanted in culture (see above). One possible mechanism for the transmission of early developmental signals to progeny cells could be the setting up of certain structural features of chromatin that have the ability to self-propagate. In this viewpoint, these structural features, once established in progenitor cells, might then provide a substrate for the products of genes expressed later in a specific lineage; or, they might respond to other embryonic cues present at a later developmental time. The formation of local perturbations in DNA structure assayed as nuclease hypersensitive sites has been reported to be influenced by salt, pH, and other ionic variables. In addition, it has been reported that the induction of specific puffs in polytene chromosomes isolated from unfixed salivary glands of Chironomous is ion- and pH-dependent (Lezzi and Robert, 1976). Since gradients and/or transient temporal changes in such variables have been described in many developmental systems (Alberts et al., 1983), we wondered if hypersensitive sites might have the property of self-propagation, and hence, a property that could serve as a structural analog for the transmission of early developmental signals to progeny cells. In an attempt to test this hypothesis, we investigated the possibility that nuclease hypersensitive sites 5' to chicken globin genes might have the ability to form in cells that are removed from the stimuli that led to the formation of such hypersensitive sites in progenitor cells.

Since the transitions from progenitor stem cell to precursor red blood cell to erythroblast occur so rapidly and asynchronously in the developing chicken embryo, we were unable to address this question during the course of hematopoesis in normal development. Instead, we used a cell culture system that we felt might harbor the potential to address this question. It was reported by Garry et al. (1981) that growth of normal chick embryo fibroblasts (CEF) in medium containing high NaCl (0.2-0.25 mol) resulted in the acquisition by these cells of many of the phenotypic properties of CEF that are transformed by various oncogenic retroviruses, including Rous sarcoma virus (RSV). It was also shown that the actual level of NaCl within these cells increased commensurately as a result of this treatment, mimicking the increase in internal NaCl observed after oncogenic transformation. On returning these cells to normal medium, the intracellular NaCl concentration reverted to the levels present before exposing the cells to NaCl shock, and these cells no longer displayed any of the phenotypic characteristics of transformed CEF. Thus, given the possible influence of ionic conditions on hypersensitive site formation, the acquisition of a transformed phenotype by CEF exposed to high NaCl, and the reported activation of globin genes by CEF transformed by RSV (Groudine and Weintraub, 1975; 1980), we wondered if high salt could induce DNase I hypersensitive sites within the globin containing chromatin in CEF.

These experiments (Fig. 3) revealed that after treatment of CEF with

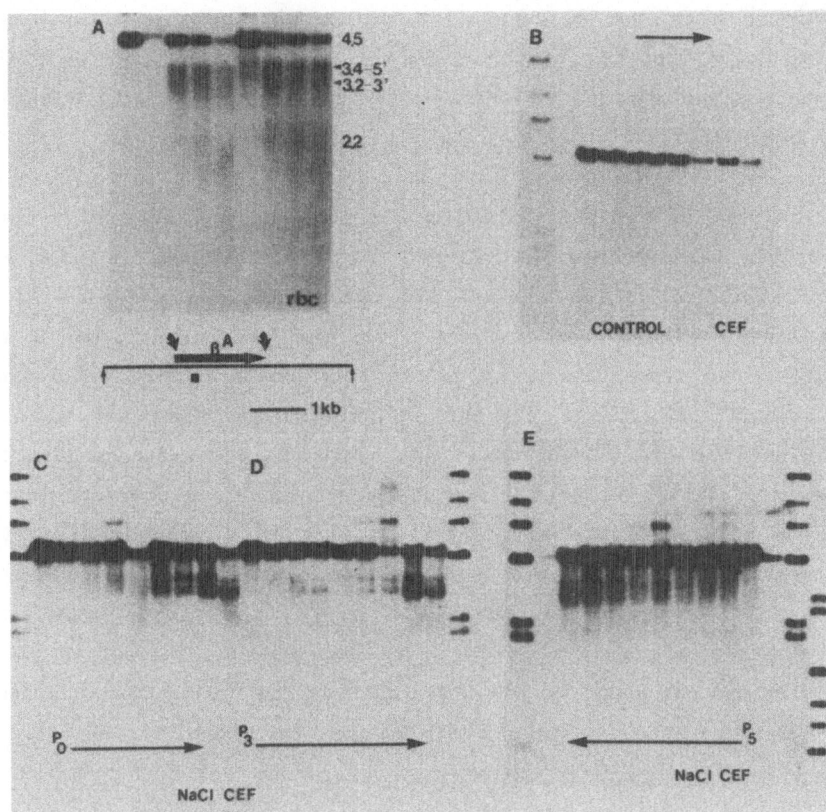

Fig. 3. Induction and propagation of β^A-globin hypersensitive sites after salt shock. Normal CEF were grown in F-10 medium (Gibco) with an additional 0.1 M NaCl for 12 hrs. Nuclei were either prepared immediately (P_0) or the cells were allowed to divide through 3 passages (P_3), roughly 12 divisions; or 5 passages (P_5), roughly 20 divisions. The 3 sets of nuclei were digested with DNase I (increasing concentrations indicated by the direction of the arrows) and the purified DNA digested with Bam HI and hybridized to a β^A-globin intervening sequence probe [dark square in the map shown in (A)]. As shown in digests from normal red cells (A), this probe detects a 5' and a 3' hypersensitive site. These sites are not present in normal CEF (B). However, they appear after NaCl shock of CEF (C) and they persist after return to normal medium and continued passages (D and E).

high NaCl for 12 hours, the normal DNase I hypersensitive sites usually observed in red blood cells appeared at the 5' and 3' side of the gene. In addition, these sites persisted after CEF were treated with high concentrations of NaCl for 12 hours, were returned to medium containing physiologic concentrations of NaCl (0.15 mol), and then were allowed to grow logarithmically for about 20 generations in the normal medium. Thus, hypersensitive sites in globin chromatin induced by NaCl were propagated to progeny cells over 20 generations in the absence of continued exposure to high NaCl; that is, in the absence of the activity that was initially responsible for generating the altered chromosomal structure assayed by DNase hypersensitivity.

Analysis of run-off transcription products and of steady-state RNA from these salt-shocked CEF (either during incubation in high NaCl or after return to normal medium) did not reveal the presence of globin transcripts. Thus, while changes in the ionic environment may be sufficient for the initiation (and stablization) of a hypersensitive site, this inheritable change in chromatin structure does not seem to be sufficient for the initiation of RNA synthesis. This uncoupling between the hypersensitive structure and transcription is a characteristic that would be compatible with a molecular analog of a determinative event in development. These experiments suggest that at least some hypersensitive sites might serve as structural analogs for the transmission of early developmental signals to progeny cells, perhaps through a mechanism involving protein bound to the DNA sequences comprising the hypersensitive site (Groudine and Weintraub, 1982; Emerson and Felsenfeld, 1984).

Clearly, not all hypersensitive sites are stably propagated from cell generation to cell generation (e.g., some of the vitellogenin and metallothionein sites). One possibility is that since these unstable sites are associated with inducible genes, these sites may represent footprints of the receptor complexed to DNA. In this viewpoint, these sites would be dependent on the continued presence or absence of the inductive event (e.g., hormone introduction or withdrawal).

CHROMATIN STRUCTURE IN SPERM

Several groups of investigators have shown that significant changes in structure and expression of chromatin occur during vertebrate spermatogenesis. For example, prior to the condensation of the nucleoprotein complex in mature sperm of various species, protamines replace some classes of histones and transcriptional activity ceases (Bellve, 1979; Bellve and O'Brian 1983). In addition, changes in the level of methylation of sperm DNA as compared to somatic DNA have been found for a number of genes in most vertebrates thus far examined (Razin and Riggs, 1980; Riggs, 1984). Thus, these phenomena present an interesting problem. If signals provided by hypersensitive sites or

212

undermethylation or specific regions of DNA are important in providing information for gene expression, how is this information maintained in the hypermethylated, hypercondensed sperm chromatin in a form that might be important in permitting the expression of specific genes on the paternal chromosomes during early development?

As an initial approach to investigate the possibility that information important in determining the developmental activation of specific paternal genes might be encoded in some aspect of sperm chromatin, we analyzed the chromatin structure and methylated state in sperm DNA of constitutively expressed, tissue-specific, and inactive genes of the chicken. We thought that if chromatin structural elements are involved in the propagation of information from sperm to developing embryo, then housekeeping genes like thymidine kinase (TK), which are transcribed early in spermatogenesis (that is, in spermatogonial cells), would maintain the hypersensitive sites associated with the expression of these genes in the transcriptionally inert mature sperm chromatin, whereas inactive genes such as the avian endogenous virus ev-1 and tissue-specific genes like the globin gene, which are not active in spermatogonia, would be propagated in an inactive chromatin structure in sperm nuclei.

In order to determine whether an active chromatin structure in genes destined to be expressed in the early embryo is maintained in mature sperm chromatin, we compared the chromatin structure, as assayed by the pattern of nuclease hypersensitive sites, of chicken thymidine kinase gene, the adult β^A-globin gene, and an inactive endogenous avian retrovirus, ev-1 (Astrin et al., 1980; Hayward et al., 1980), in mature sperm to the structure of these genes in somatic tissues and in the testes of 4-week old chickens. By histological studies, these early testes contain a minimum of 40 percent replicating spermatogonial cells, but few more mature spermatogenic cells. For all three genes examined (Fig. 4), the pattern of nuclease hypersensitive sites detected in sperm chromatin is distinctly different from those patterns found in chromatin analyzed from cells which express these genes, or from cells in which these genes are inactive. For example, no hypersensitive site is present at the 5' end of the TK gene in sperm chromatin, whereas in other tissues, including red blood cells (RBC) and the early testes, the 5' hypersensitive site is the dominant site (Fig. 4A; Groudine and Casimir, 1984). Similarly, our analysis of the location of the hypersensitive sites of the adult β^A-globin gene reveals that in sperm chromatin the dominant site near the

5' end of β^A (Fig. 4B, site c) is different from the dominant hypersensitive site in the red blood cells in which this gene is expressed (Fig. 4B, site a; Larsen and Weintraub, 1982). Our analysis also reveals that the inactive ev-1 provirus displays several nuclease hypersensitive sites in sperm chromatin, whereas no such sites are typically evident in any other somatic tissue examined (Fig. 4C; Groudine et al., 1981; Conklin et al., 1982).

In attempting to understand the basis of these different hypersensitive sites in sperm chromatin, we considered that such sites might result from the strain induced by the extreme packaging of DNA during spermatogenesis. It has been shown that supercoiled, but not relaxed, DNA will relieve some of the supercoil-induced strain by forming alternative DNA conformations that are preferentially sensitive to S1 (Larsen and Weintraub, 1982; Lilley, 1980, 1981; Panayotatos and Wells, 1981). Thus, we compared the locations of hypersensitive sites in sperm

Fig. 4. Hypersensitive sites in sperm chromatin resemble those observed in supercoiled plasmid DNA, rather than those characteristic of active chromatin. Nuclei from designated tissues were digested with increasing concentrations (as designated by the direction of the arrow) of DNase I or S1. DNA was isolated, digested with Hind III (A,B) or Hind III + Eco RI (C), and subjected to blot hybridization (Stalder et al., 1980a, b) with ^{32}P-labeled nick-translated probe (indicated on line drawing) after electrophoresis in 1 or 1.5% agarose gels. Plasmid DNA was digested with S1 (Larsen and Weintraub, 1982), and processed similarly. Subbands indicative of hypersensitive sites are indicated by arrows and labeled with letters a through e. The locations of the corresponding hypersensitive sites are shown on the line drawings; H, Hind III; R, Eco RI. (A) Analysis of hypersensitive sites of the thymidine kinase (TK) gene. (B) β^A-globin hypersensitive sites. In the analysis of the plasmid S1 sites, 0 is the undigested plasmid DNA, S1 is S1-digested plasmid or nuclear DNA. The two 5' Hind III sites shown on the line drawing, HP and HN, correspond to a sequence polymorphism in the plasmid (HP) compared to genomic/nuclear (HN) DNA. (C) ev-1 hypersensitive sites. The bold line in the drawing represents proviral sequences, filled-in boxes represent the viral LTR's, the thin line indicates host DNA flanking ev-1. The plasmid used for the S1 analysis is defined by the Hind III (H) and Eco RI (R) sites on the line drawing.

214

A. Thymidine Kinase (TK)

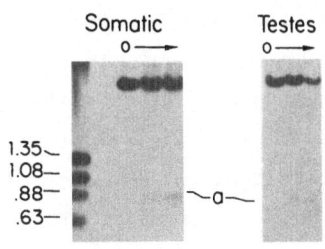

Somatic   Testes

o ⟶   o ⟶

1.35
1.08
.88   a
.63

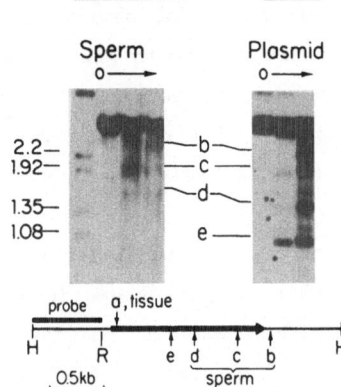

Sperm   Plasmid

o ⟶   o ⟶

2.2   b
1.92   c

1.35   d
1.08   e

probe   a, tissue

H   R   e d c b   H

0.5kb   sperm

plasmid

B. β^A Globin

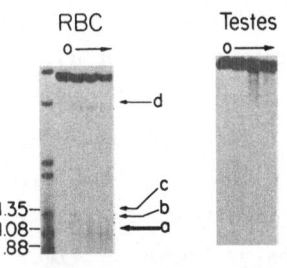

RBC   Testes

o ⟶   o ⟶

d

c
b
1.35   a
1.08
.88

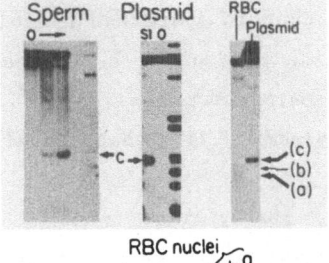

Sperm   Plasmid   RBC
                   Plasmid
o ⟶   S1 O

c   c   (c)
                   (b)
                   (a)

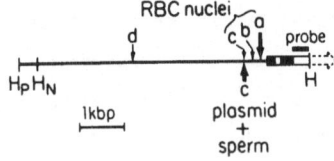

RBC nuclei

d   c b a   probe

H_P H_N   H

1kbp   c

plasmid
+
sperm

C. ev-1

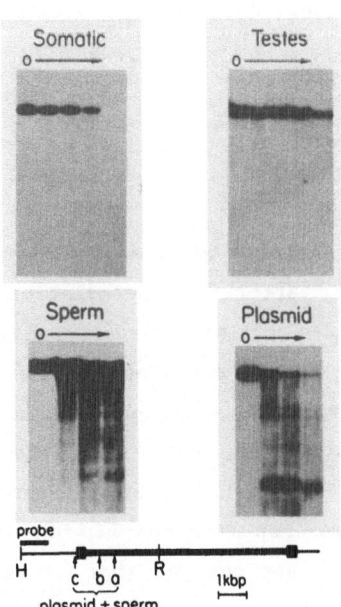

Somatic   Testes

o ⟶   o ⟶

Sperm   Plasmid

o ⟶   o ⟶

probe

H   c b a   R   1kbp

plasmid + sperm

chromatin to the location of S1 sites in supercoiled plasmid DNA
containing genomic inserts of identical sequence. As shown in Figure 4A,
S1 digestion of supercoiled plasmid DNA that contains the chicken TK gene
reveals a number of hypersensitive sites, none of which map to the 5' end
of the gene, although several correspond to those sites observed in mature
sperm DNA (sites b, c, and d). Similarly, the prominent S1 site in the
supercoiled plasmid containing the adult β^A-globin gene is the most 3' of
the cluster of sites at the 5' end of this gene (Fig. 4B, site c; Larsen
and Weintraub, 1982), the same site that is dominant in sperm but not RBC
chromatin. The difference in location between the major RBC and plasmid
or sperm sites is shown in Figure 4B. In addition, several minor sites
are coincident in sperm and plasmid, but are absent in RBC nuclei (Figure
4B). In the case of ev-1, at least three of the nuclease hypersensitive
sites in sperm chromatin (Fig. 4C, sites a, b, and c) are coincident with
those in plasmid DNA's that contain a portion of this provirus.

Thus, the similar plasmid and sperm S1 sites suggest that the
hypersensitive sites that can be assayed in sperm chromatin may be the
consequence of the fashion in which DNA is packaged during
spermatogenesis. In addition, these results suggest that when sperm
nuclei enter the egg, the chromatin structural characteristics of
inactive and active genes are not present and therefore cannot provide the
proposed structural cues necessary to signal differential gene expression
early in embryogenesis.

METHYLATION OF SPERM DNA

In order to detemine whether genes destined to be active in the early
embryo might display a characteristic pattern of methylation of CpG
dinucleotides relative to inactive genes or to tissue-specific genes
activated later in development, we analyzed the methylated state in sperm
DNA of two endogenous avian retroviruses ev-1 and ev-3 (Astrin et al.,
1980; Hayward et al., 1980), the chicken TK gene, and the chicken
β-globin gene cluster. The assay for DNA methylation consists of
digesting DNA with the restriction enzymes Msp I or Hpa II, both of which
cleave at the sequence CCGG. Hpa II will not cut DNA that is methylated
at either C (cytosine) residue, whereas Msp I will cleave this sequence
regardless of the state of methylation of the C in the CpG (guanine)
dinucleotide (Waalwyck and Flavell, 1978). Our former analyses of ev-1

and ev-3 in somatic tissue revealed that, while the inactive ev-1 provirus is highly methylated in all somatic tissues, the active ev-3 provirus, which contains more than 20 Hpa II sites detectable by Southern blotting, displays no CpG methylation at any of these Hpa II sites (Groudine et al., 1981; Conklin et al., 1982). As shown in Figure 5A, Hpa II digestion of DNA obtained from the sperm of an ev-1-containing rooster reveals that this provirus is also highly methylated in sperm DNA. However, analyses of proviral DNA in sperm obtained from a rooster containing both ev-1 and ev-3 reveals a prominent viral related band of 3.7 kbp (Fig. 5, A and B), which is not detected after Hpa II digestion of RBC DNA from the same rooster (Fig. 5B). In addition, no such band is observed upon the digestion of either sperm (Fig. 5, A and B) or RBC DNA (Fig. 5B) from roosters containing only ev-1, indicating that the 3.7 kbp band is ev-3-specific. The 10 kbp fragment (Fig. 5) is also specific to ev-3 sperm DNA, but is variably observed, and represents partial cutting at site D shown in the line drawing in Figure 5. A faint but reproducible 0.25 kbp band is also observed in the ev-3 containing DNA from sperm (Fig. 5A). Mapping of these ev-3 sperm-specific Hpa II bands by redigestion of Hpa II samples with either Eco RI, Sst I, or Bam HI (Fig. 5C) reveals that these sites of undermethylation correspond to sites B, C, and D illustrated on line drawing in Figure 5. The hatched region in the line drawing corresponds to approximately 16 Hpa II sites present in ev-3 (Groudine et al., 1981; Conklin et al., 1982); whereas all of these sites are undermethylated in somatic tissue, only site D is undermethylated in sperm DNA. These results show that the active ev-3 provirus, in contrast to the inactive ev-1 provirus, displays at least three sites that are undermethylated in sperm.

The possible significance of these undermethylated point sites in ev-3 proviral DNA of sperm became apparent when their locations were compared to the location of nuclease hypersensitive sites in these same regions in chromatin of somatic origin. We previously reported that ev-3 contained two hypersensitive sites, one in each of the two long terminal repeats (LTR's) of the provirus (Groudine et al., 1981; Conklin et al., 1982); further mapping of the chromatin structure of this active provirus has revealed the presence of three additional hypersensitive sites. As shown in Figure 6A, Sst I digestion of DNA from cells that contain ev-1 and ev-3 reveals bands of 10 and 6.5 kbp corresponding to ev-1 and ev-3, respectively. Analysis of Sst I digested DNA derived from DNase I or S1 treated nuclei reveals two additional bands of 5.8 nd 4.5 kbp. The DNase I or S1 cleavage sites that generate these subbands map to the locations

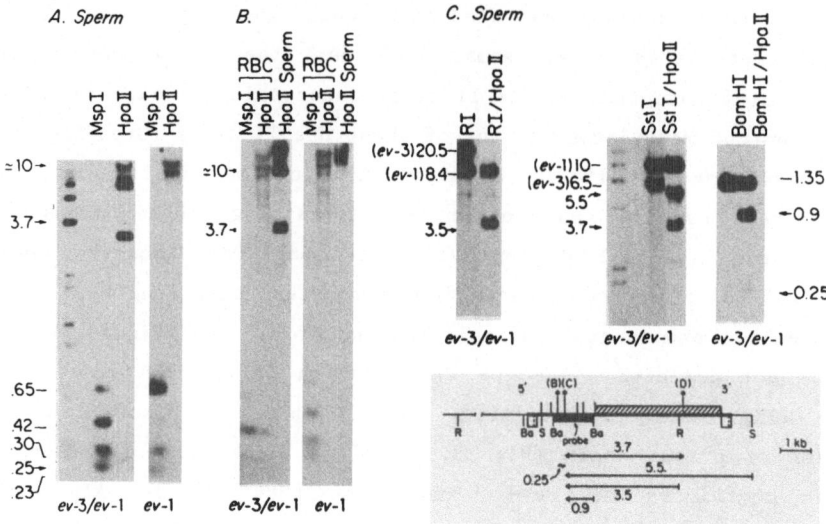

Fig. 5. Undermethylated Hpa II sites in sperm DNA of the
active avian endogenous retrovirus ev-3. (A and B)
DNA from mature sperm and RBC of ev-1 or ev-1 and
ev-3 roosters harboring ev-3 was isolated, digested
with the restriction endonuclease Msp I or Hpa II,
and hybridized to the probe shown in D. The arrows
indicate Hpa II generated fragments specific to
ev-3-containing DNA. (C) The same DNA samples
described in 2A above were digested with Eco RI
(RI), Sst I (S), Bam HI (Ba), Hpa II, or
combinations of the above to map the locations
of the undermethylated Hpa II sites in sperm DNA. The
straight lines on the top of the drawing indicate
known Hpa II sites in ev-3, all of which are
undermethylated in somatic DNA. Derivation of the
locations of undermethylated sites B, C, and D in
sperm DNA is shown below the line drawing.

indicated on the line drawing at the bottom of the figure, and are derived
from ev-3 since digestion of nuclei containing only ev-1 reveals no such
subbands (Fig. 6C) (Groudine et al., 1981; Conklin et al., 1982).
Additional subbands revealed by Eco RI digestion of DNA isolated from
DNase I or S1 treated nuclei correspond to hypersensitive sites A, B, and
C (Fig. 6, line drawing). Although the data in Figure 6 are derived from
an analysis of RBC nuclei, these hypersensitive sites are detectable in
the nuclei of every somatic tissue analyzed thus far, as well as in nuclei
from the testes of 4-week old roosters (unpublished data). Incubation of
intact nuclei with restriction endonucleases has shown that DNA sequences
within S1 or DNase I hypersensitive sites are often preferentially
accessible to digestion with these nucleases as well (McGhee et al.,
1981). Using this assay, we have observed that even though all detectable
Hpa II sites in DNA purified from ev-3-containing cells are

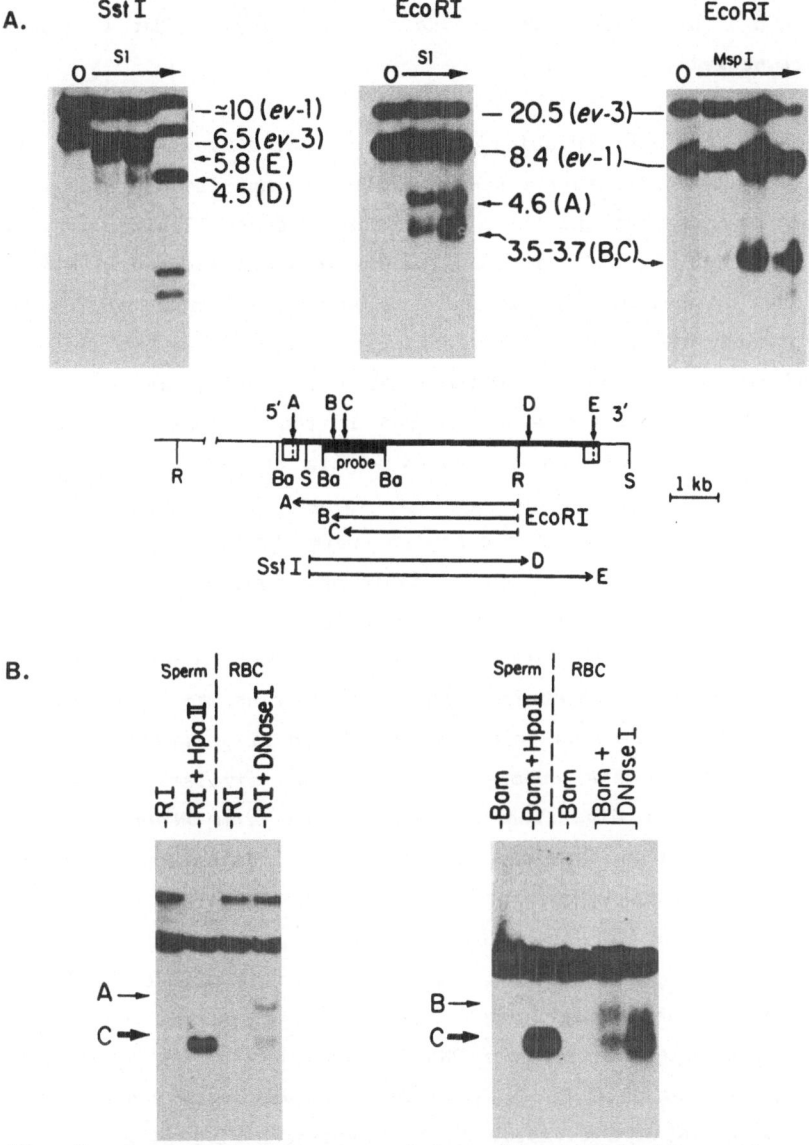

Fig. 6. Location of hypersensitive sites within the active
avian endogenous retrovirus ev-3 correspond to the
location of undermethylated sites in sperm DNA.
(A) The mapping of hypersensitive sites was
performed as in Fig. 4. The sizes of parent bands
are indicated by straight lines, and subbands are
designated by arrows and corresponding sizes. The
locations of hypersensitive sites A through E are
indicated on the line drawing, and examples of the
mapping of these sites are shown under the line
drawing. The open squares indicate the location of
the LTR's of ev-3; R, Eco RI; S, Sst I; Ba, Bam
HI. (B) The identity of the undermethylated point
sites in sperm with hypersensitive regions in
somatic tissue is confirmed by the comigration of
Hpa II generated fragments in sperm DNA with
subbands generated by DNase I in RBC nuclei.

undermethylated in RBC, only those sites associated with sequences in the DNase I and S1 hypersensitive regions are accessible to Msp I on digestion of intact nuclei (Fig. 6A).

Comparison of the location of ev-3-associated hypersensitive sites in somatic tissue and immature testes and sites of Hpa II undermethylation in sperm DNA reveals that the three undermethylated Hpa II sites in sperm DNA (Fig. 6, sites B, C, and D in the line drawing) are located within regions defined by hypersensitivity to DNase I and S1. Because of the lack of CpG's in methylation-sensitive restriction enzyme sites in the LTR's of this endogenous provirus (Hishinuma et al., 1981), we are unable to determine the methyalated state of DNA in the regions corresponding to hypersensitive sites A and E located within the LTR's (Fig. 6A). The coincidence of hypersensitive sites B, C, and D (Fig. 6, line drawing) with the undermethylated point sites in sperm DNA is substantiated by the results in Figure 6B. As shown in Figure 6B, sperm DNA digested with Hpa II has been redigested with either Eco RI or Bam HI and run on the same gel as DNA isolated from DNase I treated nuclei and digested with the same restriction enzymes. Blotting and hybridization to the viral probe (Figs. 5 and 6A) indicates that the Hpa II-Eco RI generated 3.5 kbp band derived from sperm DNA co-migrates with the similarly sized subband generated by Eco RI digestion of DNA isolated from DNase I treated RBC nuclei. The larger subband detected with Eco RI digestion of the DNase I samples is a hypersensitive site located in the LTR (site A); since no Hpa II site is present in the LTR, no Hpa II fragment corresponding to cleavage at site A is observed. Similarly, Bam HI digestion of the same samples reveals the co-migration of the Hpa II generated band from sperm DNA and the smaller of the two DNase I subbands which is generated by DNase I cleavage at site C. Both products are seen with DNase I since the probability of DNase I cleavage at two hypersensitive sites on the same molecule is low (Groudine and Weintraub, 1982; Weintraub et al., 1981). The absence of a Hpa II digestion product corresponding to cleavage at site B is explained by the fact that Hpa II cleaves at sites B and C in the same DNA molecule, resulting in the 0.25 kbp band described in Figure 5.

Similar analyses were carried out on the chicken TK gene. As in the case of ev-3, sites of undermethylation in sperm DNA within the TK gene correspond to the location of hypersensitive sites in cells in which this gene is expressed (not shown).

In order to ascertain whether the correlation between undermethylated point sites and nuclease hypersensitive sites is restricted to genes marked for expression in the early embryo, we also examined the pattern of methylation in the chicken β-globin gene cluster, which does not become transcriptionally active in erythroid cells until 32 hours of embryonic development'(Groudine et al., 1974; Groudine and Weintraub 1981). At that time, the embryonic rho (ρ) and epsilon (ε) β-like globin genes are activated, whereas the β^A- and β^H-globin genes are not transcribed until day 6 to 7 of embryogenesis (Groudine et al., 1974; Groudine and Weintraub, 1981). As summarized in Figure 7, Hpa II sites are present within the 5' and 3' hypersensitive sites of all four β-like genes (Stalder et al., 1980), and are undermethylated in those cells in which the specific gene is transcribed (McGhee and Ginder, 1979; Ginder et al., 1983). No Hpa II cleavage is observed in the 14 kbp of the avian β-globin domain containing sequences 5' and 3' to the ρ gene and 5' to the β^H gene in sperm DNA. Analysis of the β^A and ε gene regions, however, does reveal that one of three Hpa II sites 5' to the adult β^A gene is undermethylated in sperm, as well as every embryonic and adult tissue examined (unpublished data; Stalder et al., 1980). In contrast, all three of these 5' sites are unmethylated in adult erythroid cells (McGhee and Ginder,

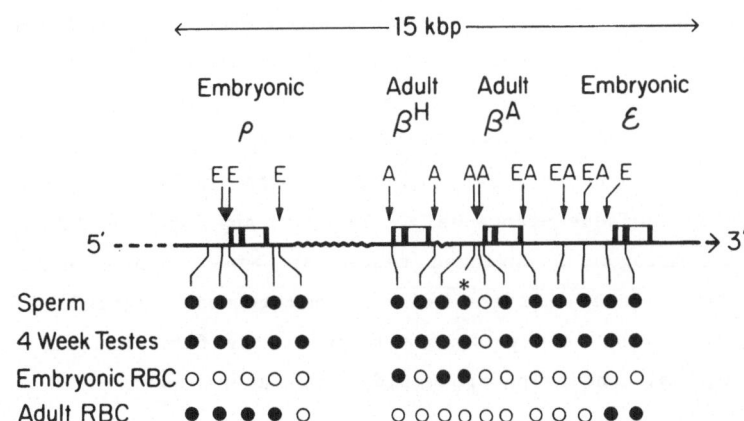

Fig. 7. Summary of HpaII site methylation in sperm, early testicular and red blood cell DNA within erythroid-specific hypersensitive sites in the chicken β-globin locus. Hypersensitive sites found in red blood cells are indicated by arrows on the top of the drawing. E, sites found in embryonic RBC; A, sites in adult RBC; EA, sites in both embryonic and adult RBC. Methylated HpaII sites are indicated by closed circles (o), and unmethylated HpaII sites by open circles (o). The asterick (*) indicates a region composed of three HpaII sites.

1979; Ginder et al., 1983), and are contained within a red cell-specific hypersensitive site (McGhee et al., 1981).

These data reveal that Hpa II sites within hypersensitive sites of the embryonic ρ- and ε-globin genes, the adult β^H gene, as well as all but one Hpa II site with the β^A gene, are fully methylated in 4-week testicular DNA and sperm DNA. Specifically, Hpa II sites within the 5' and 3' hypersensitive sites of the first β-like genes to be activated in the developing embryo are fully methylated in sperm DNA. Thus, in contrast to the two constitutively expressed genes described above, regions of potential regulatory significance within these tissue-specific genes are not marked as different from other regions of the domain in sperm.

TIMING OF METHYLATION DURING SPERMATOGENESIS

The experiments presented above indicate that, for two constitutively expressed genes, but not for globin, point sites of Hpa II undermethylation in mature sperm DNA correspond to sites of altered chromatin structure in somatic tissues and spermatogonia. These results raise the possibility that these regions of undermethylated DNA may be involved in the templating of information in sperm DNA. If this notion is correct, then the time during development when the methylation pattern of sperm DNA is established could be important in determining which genes will become part of this postulated templating. Thus, we assayed the state of methylation of ev-3 and TK in DNA isolated from cells in different stages of spermatogenesis. For this analysis, we compared DNA's from the spermatogonial enriched testes of 4-week old chickens to DNA isolated from later stages of spermatogenesis including primary spermatocytes in the late pachytene stage of the first meiotic prophase, round spermatids, condensing spermatids, and mature sperm.

In the DNA of early testicular cells, the ev-3 and TK genes are as undermethylated as in somatic tissues (Fig. 8), whereas in primary spermatocytes and more mature spermatogenic cells these genes are as methylated as in mature sperm. While we have presented only the analysis of Hpa II site methylation, we have observed similar phenomenon regarding Hha I site methylation of sperm DNA (unpublished data). One possible explanation for these results is that these genes are also hypermethylated in the spermatogonial cells, but that we are failing to detect the

222

contribution of DNA from these cells in our analysis of total testicular DNA. However, since spermatogonia contribute 40% of these early testicular cells and, as previously shown, we are able to detect alternative patterns of DNA methylation and chromatin structure contributed by cells making up less than 10% of the population under investigation (Groudine and Weitnraub, 1981), we think that this is unlikely. Therefore, we conclude that a de novo methylation of these genes occurs at a developmental time between the transition from spermatogonial cell to primary spermatocyte. In contrast to these constitutive genes, we have observed that the pattern of methylation of the ß-globin, ovalbumin, and vitellogenin loci, none of which contain hypersensitive sites in spermatogonial cells, is unchanged between spermatogonial cells and mature sperm (unpublished data). Since the methylation of sequences outside the hypersensitive site regions of these genes occurs between the spermatogonial and primary spermatocyte stages of spermatogenesis, our results suggest that the general increase in methylation of sperm DNA compared to somatic DNA is the result of a de novo methylation process operant during spermatogenesis, a process that excludes those DNA sequences within hypersensitive sites in spermatogonial chromatin. Thus, the observed undermethylation of DNA sequences in the

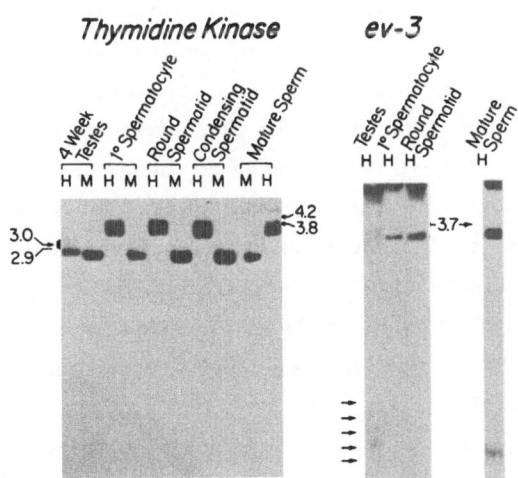

Fig. 8. De novo methylation of TK and ev-3 during spermatogenesis. Total testicular cells from an ev-1/ev-3-containing adult rooster were separated by unit gravity sedimentation (Bellve et al., 1977). DNA from these cells, as well as from mature sperm and cells from 4-week chick testes, were digested with Hpa II (H) or Msp I (M) and blot hybridized to the TK probe or the proviral probe.

hypersensitive sites of constitutively expressed genes appears to reflect a templating of those genes that contain nuclease hypersensitive sites and are expressed in spermatogonial cells.

IMPLICATIONS FOR THE DEVELOPMENTAL ACTIVATION OF THE PATERNAL GENOME

These results raise several questions regarding the mechanism whereby undermethylated point sites are established during spermatogenesis, the generality of the correlation between undermethylated point sites in sperm and regions that include hypersensitive sites in spermatogonial cells, and, finally, the significance of this correlation for regulation of expression of the paternal genome after fertilization. Figure 9 presents a model that combines our results with the idea that the maintenance of hypersensitive sites in an undermethylated state in sperm DNA marks such sites as signals for gene activation early in development.

It is possible that the de novo methylation of DNA during spermatogenesis might play a role in, for example, DNA-protein (Fisher and Caruthers, 1979) (or protamine) interactions important in condensation of the sperm nucleus. The preferential undermethylation at regions of hypersensitive sites could be based on the presence of a protein at such local alterations in chromatin structure (Groudine and Weintraub, 1982; Emerson and Felsenfeld, 1984), thereby inhibiting the procession of the methylase, or based on the lack of specificity of the methylase for regions of altered DNA structure. In the former case, the undermethylated regions would be virtual footprints of such proteins, and if these proteins were lost during the process of spermatogenesis, the undermethylated state of these binding sites might then facilitate either the formation of alternative DNA structures (Behe and Felsenfeld, 1981; Behe et al., 1981; Weintraub, 1983) or the binding of regulatory proteins to these sites early in development. Such a model would thus predict that the maintenance of these hypersensitive sites in an undermethylated state in sperm DNA would mark such sites as signals for gene activation early in development.

If this model is valid, then the correlation between hypersensitive site and undermethylated sites in sperm DNA should be found for other constitutively expressed genes. In fact, reports from other investigators concerning the methylated state of constitutively expressed genes suggest

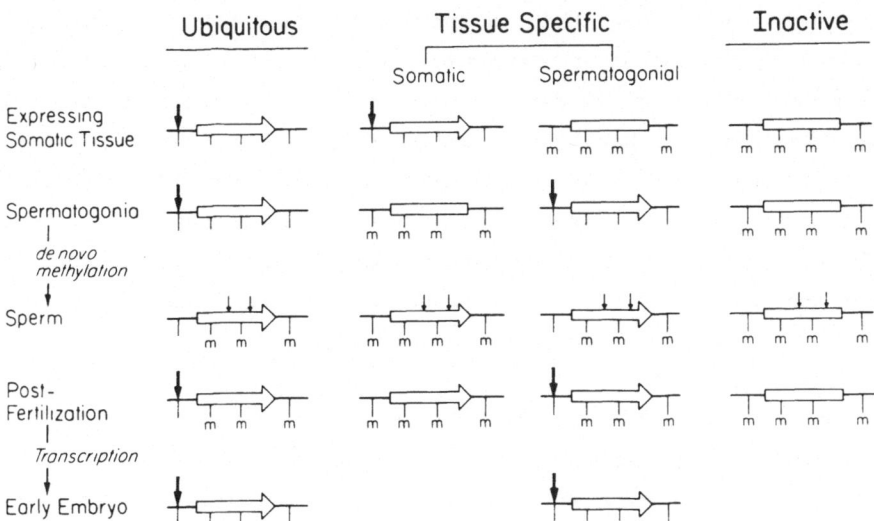

Fig. 9. Model for templating of the paternal genome. In somatic cells, inactive, tissue-specific and constitutive genes display the chromatin structure and pattern of methylation typical of each class; that is, active genes (open boxes with arrowheads) contain hypersensitive sites (thick arrows) and unmethylated CpG dinucleotides (vertical lines), while inactive genes (open boxes without arrowheads) do not contain hypersensitive sites and are methylated (vertical lines marked with an m). These relationships are also evident in the chromatin of spermatogonial cells. During the transition from immature to mature sperm, however, all genes acquire a novel pattern of hypersensitive sites (thin arrows) which more closely resembles that seen in supercoiled plasmid DNA's that contain inserts of these genes. In addition, CpG residues which were included within hypersensitive sites in spermatogonial cells remain specifically undermethylated in mature sperm. These undermethylated point sites may serve to template the paternal genome such that constitutive and spermatogonial-specific genes are activated after fertilization. The methylated sites of constitutive and spermatogonial-specific genes not associated with hypersensitive sites would become progressively unmethylated after continued transcription in the embryo. Activation of tissue-specific genes later in development would require additional events to initiate expression.

that the 5' regions of such genes are undermethylated in sperm DNA (Stein et al., 1983; Tykocinski and Max, 1984). While the relation of these reported undermethylated sites in sperm DNA to hypersensitive sites in spermatogonia and somatic tissues has not yet been addressed directly, the location of such undermethylated sites in sperm DNA is coincident with a region often associated with nuclease hypersensitivity.

Since we have shown that in sperm DNA tissue-specific genes do not exhibit undermethylation at regions of potential hypersensitive sites [and similar conclusions can be derived from the work of others (Stein et al., 1983; Tykocinski and Max, 1984; Sanders-Haigh et al., 1983)], our model regarding the templating of developmental cues in sperm DNA cannot account for the activtion of tissue-specific genes during embryogenesis. In this model, the activation of tissue-specific genes later in development would be the consequence of other developmental cues, which may be present in the egg or activated through subsequent events.

The wave of de novo methylation occurs during a stage when spermatogonial genes are active in transcription, but prior to the reported transcriptional activation of genes specific to the haploid stages of spermatogenesis such as protamines and a testes-specific α-tubulin (Iatrou et al., 1978; Distel et al., 1984). Our model would therefore predict that, in addition to housekeeping genes involved in DNA replication and other nonspecialized functions, germline-specific genes presumed active in spermatogonial cells would also be templated in sperm DNA. Thus, one possibility would be that the observed marking of sperm DNA could be important not only in the early activation of constitutive genes of the paternal genome but also in the initiation of the germline program, due to the maintenance of signals in the hypersensitive site DNA sequences of spermatogonial (that is, germline) specific genes. In contrast, genes whose products are specific to the later stages of spermatogenesis would not be templated in the fashion described above. These possibilities can be tested by isolating spermatogonial- and postmeiotic-specific genes, and by determining the chromatin structure and methylation of these genes during various stages of spermatogenesis, as well as by determining the timing of their activation during embryogenesis.

PROTEIN SYNTHESIS, HYPERSENSITIVE SITES, AND THE ACTIVATION OF TRANSCRIPTION

The above model relies on the notion that hypersensitive sites involved in the templating of information for gene expression are composed, at least in part, of proteins, and that such structures may be essential for the transcriptional activation of subsets of genes. Recently, we have been examining these possibilities, using the avian c-myc gene as a model system. The c-myc gene of mouse, man, and birds appears to be transcribed in all actively dividing cells. The c-myc gene is composed of three exons, two of which encode c-myc protein (Hamlyn and Rabbitts, 1983; Stanton et al., 1983; Shih et al., 1984; Linial and Groudine, 1985). Recent experiments with inhibitors of protein synthesis such as cycloheximide (CH; Kelly et al., 1983; Leder et al., 1983) have suggested that c-myc transcription might be controlled by a labile repressor protein. However, since in some cells CH treatment results in an increase in stability in c-myc RNA (Dani et al., 1984), the CH effect could be at the post-transcriptional level. We have been studying c-myc transcription in cell lines derived from tumors induced by avian leukosis virus (ALV) in which c-myc transcription is augmented by the insertion of the viral LTR in the region of the c-myc gene. Promoter utilization (LTR or cellular), as well as the structure of the c-myc transcripts (Fig. 10), have been determined in some these lines. The chicken system is particularly useful for study of transcription of the c-myc gene, because

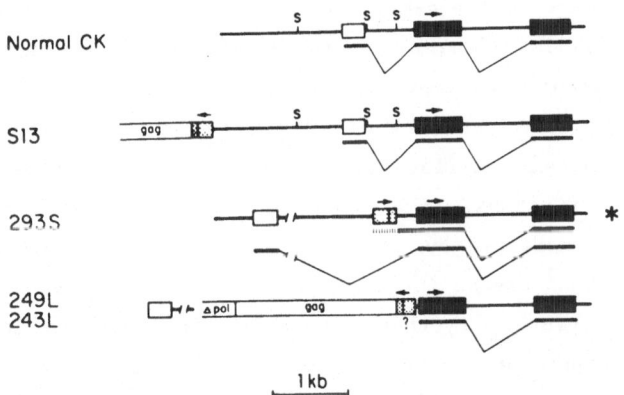

Fig. 10. Structure of the c-myc alleles in normal chicken cells and the bursal lymphoma cell lines used in this study. Solid boxes indicate the c-myc coding exons 2 and 3; the open box indicates the non-coding first exon. Stippled boxes indicate the viral LTR; other known viral sequences are indicated by gene name (gag, viral structural protein gene; pol, RNA dependent DNA polymerase). The broken line indicates that portions of the locus are located further upstream in these transformed cells. S indicates restriction endonuclease Sma I sites. The arrows denote the direction of transcription from both the LTR and c-myc. The major c-myc RNA transcripts are drawn beneath the structure of the c-myc gene. The dotted line denotes uncertainty in the start site for the transcript. Asterisks indicate cell lines which harbor a normal c-myc allele in addition to the rearranged allele depicted.

(unlike the case in mouse and human tumors) gross transloctions have not occurred in the vicinity of c-myc in avian cells exhibiting altered c-myc transcriptional patterns. Furthermore, promoter utilization differs among the cell lines (Fig. 10). It was thus of interest to use these cell lines to attempt to correlate promoter utilization with stabilization of c-myc RNA in the absence of protein synthesis and the response of c-myc RNA levels to protein synthesis inhibition.

Steady state RNA levels were measured by Northern blot analysis in normal cells and transformed cell lines (Fig. 11). In these experiments we utilized glyceraldehyde-3-phosphate dehydrogenase (GAPDH; Dugaiczyk et al., 1983) as a control for the amount of RNA loaded per lane, since all chicken cells examined contained reasonably high steady state levels of GAPDH RNA which did not vary with CH or emetine (another inhibitor of protein synthesis) treatment. We examined the effect of protein synthesis inhibition upon the c-myc RNA in MSB cells (Fig. 11), a chicken T-cell line transformed by Marek's disease virus (Akiyama and Kato, 1974), in which the structure of the c-myc gene and transcript is normal (Linial and Groudine, 1985). In MSB cells, we found about a 12-fold increase in c-myc RNA after treatment with emetine (compare 0' sample with 0' + E sample). Similar results were obtained with CH treatment (not shown). We also examined chicken embryo fibroblasts (CEF) and lymphocytes derived from bursa, spleen, or thymus. All contained low steady state levels of c-myc RNA which were increased by emetine or CH treatment (unpublished data). Thus, there is an increase in steady state level c-myc gene transcripts in the absence of protein synthesis in all cells examined in which c-myc expression is dependant upon normal control signals.

We next examined the amount of c-myc RNA present in the bursal lymphoma cell lines whose c-myc genes are depicted in Figure 10, and other cell lines (not shown). In five cell lines, little or no quantitative change in c-myc RNA was seen after CH or emetine treatment; we found a 1- to 2.7-fold increase in the amount of c-myc RNA relative to that of GAPDH RNA (e.g. S13, Fig. 11, comparing 0' sample to 0' + E sample).

One interpretation of these results is that the insertion of viral sequences in the c-myc domain somehow affects the turnover of c-myc RNA. To examine this, the half-life of the c-myc RNA in MSB and S13 cells was compared, using actinomycin D to prevent RNA synthesis (Dani et al., 1984; Fig. 11). The c-myc RNA in both cell types was found to be equally unstable with a half-life of about 30 minutes. Furthermore, an

228

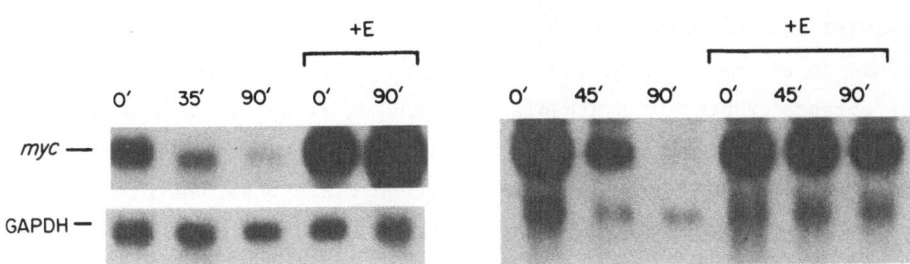

Fig. 11. Determination of the half-life of c-<u>myc</u> RNA in the presence or absence of protein synthesis inhibitors. Exponentially growing MSB, or S13 cells were treated with 0.5 µg/ml of Actinomycin D for the indicated times, rapidly lysed, and RNA extracted. Identical aliquots of MSB or S13 cells were pretreated with 0.1 mM emetine (E) for 20 minutes prior to addition of Actinomycin D (+E).

actinomycin D decay experiment performed in the presence of the protein synthesis inhibitor emetine, shows that the turnover rate of both MSB and S13 c-<u>myc</u> RNA is dramatically reduced in the absence of protein synthesis.

TRANSCRIPTIONAL ACTIVITY OF C-<u>MYC</u> IS INHIBITED BY CYCLOHEXIMIDE IN ALV TRANSFORMED CELLS

In order to examine whether differences in the observed steady state levels of c-<u>myc</u> RNA in normal and bursal lymphoma cells, in the presence or absence of protein synthesis, are a result of differences in transcriptional rates or post-transcriptional events, nuclear runoff experiments were performed. In all experiments, the rate of c-<u>myc</u> transcription was normalized to that of either <u>fos</u> (Curran et al., 1982) or GAPDH (Piechaczyk et al., 1984). Transcription from all three genes is sensitive to concentrations of α-amanitin that specifically inhibit polymerase II transcription (not shown). In initial analyses (Fig. 12A), the rate of transcription was measured in MSB and S13 cells, either untreated or treated with cycloheximide or emetine for three hours. In the untreated cells, we found that the transcription of c-<u>myc</u> in S13 cells was about 9-fold greater than that of MSB cells, consistent with the observed differences in steady state levels of c-<u>myc</u> RNA in the two cell types (Fig. 11). The use of inhibitors had little or no effect on the c-<u>myc</u> transcription observed in MSB cells. Thus, the increase in steady state c-myc RNA in these cells in the absence of protein synthesis can be accounted for by stabilization of RNA. In contrast, protein synthesis inhibition decreased the relative level of c-myc transcription about 7-fold in S13 cells (Fig. 12). The rate of transcription in S13 cells in the absence of protein synthesis was found to be approximately the same as that in treated or untreated MSB cells. Thus, we conclude that the differences in steady state levels of c-myc RNA observed in MSB and S13 cells is a result of different rates of transcription of a rapidly degraded mRNA.

To determine if the results with S13 and MSB cells were generally true for bursal lymphoma cell lines and nonestablished normal cells, runoff transcription assays were performed using nuclei from the five other cell lines (those depicted in Fig. 10 and others), as well as CEF and lymphocytes from normal chicken bursa (B). We found no decrease in c-myc transcriptional activity after protein synthesis inhibition in any of the normal cells examined (Fig. 12B). In contrast, five bursal lymphoma cell lines showed decreases (ranging from 5- to 10-fold) in the level of c-myc transcript measured in the runoff assay (S13 and others not shown). All of the bursal lymphoma cell lines contain numerous copies of integrated ALV and produce infectious virus. Thus, we could also measure the transcriptional activity of the LTR promoted viral gag gene, and found that changes in transcription of viral gag sequences after CH treatment paralleled that of c-myc sequences in the cell lines (Fig. 12C and data not shown).

The cycloheximide sensitive factor, presumably a protein, which interacts with the viral LTR to regulate transcription, could be either of viral or cellular origin. Since we found the level of viral genes in most bursal lines to be CH sensitive, we analyzed the transcription of the viral gag gene in CEF infected with the avian leukosis virus RAV-1, or the ALV released from S13 cells (S13 ALV; Fig. 12D). As in the bursal lymphoma experiments, greater than 90% of protein synthesis in the infected CEF was inhibited by CH (unpublished observations). However, no difference in the transcription of gag was noted between control and CH treated infected CEF.

QUALITATIVE CHANGES IN C-MYC TRANSCRIPTS AFTER CYCLOHEXIMIDE TREATMENT

293S cells contain two c-myc alleles, the rearranged allele with viral LTR sequences inserted within intron 1 depicted in Figure 10, as a

230

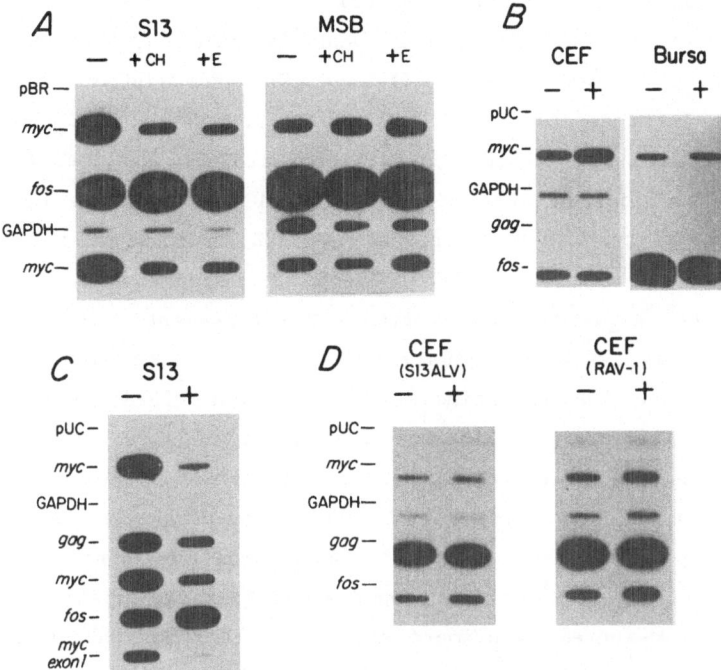

Fig. 12. Runoff transcription analysis of normal and transformed cells. Cells were treated with 10 μg/ml cycloheximide (CH) or 0.1mM emetine (E) for 3 hours. Isolation of nuclei, nuclear runoff reactions, isolation of ^{32}P RNA, hybridization conditions, and analyses of hybrid were performed as detailed (Groudine et al., 1981). Nitrocellulose filters containing 5 μg of the indicated plasmids were hybridized to the runoff products for 36 hours. Equal numbers of cpm (1-2 x 10^7) of runoff products were added per filter for nuclei that were being directly compared. Plasmids used for these experiments were: myc (a 2.4 kb SstI-EcoRI fragment of chicken c-myc containing exons 1 and 2 and intron 2 cloned in SP65); U5 LTR (a 0.6 kb EcoRI-BamHI fragment from the U3 region to gag of RSV cloned in SP63); gag (a 1.3 kb Bam HI fragment cloned in pBR); GAPD (a 1.3 kb cDNA insert in pBR, Dugaiczyk et al., 1983); fos (a 1.1 kb Pst-BglII fragment of v-fos, Curran et al., 1982); env (a 2.1 kb BamHI-XbaI fragment in pBR obtained from E. Hunter; control plasmid pUC19 (Yanisch-Perron et al., 1985).

normal allele. We have detected at least two distinct c-myc RNAs in 293S cells. To ascertain which sequences comprise the different 293S c-myc RNAs, primer extension experiments were performed with a synthetic oligonucleotide Primer (primer A) specific for the normal c-myc promoter at the end of exon 1, P1 (Linial and Groudine, 1985), and another oligonucleotide primer (Primer B) specific for intron 1 seqences (Fig. 13). In untreated 293S cells we found a small amount of transcription from P1, yielding an extension product identical in size to that of the bonafide P1 start in CEF (Linial and Groudine, 1985). (In this experiment, the CEF were treated with CH to increase the level of c-myc RNA present in the sample.) The 293S c-myc transcripts could be from the normal c-myc allele, or from LTR enhancement of exon 1 in the rearranged allele, as we previously found for cell line BK25 (Linial and Groudine, 1985). Using primer B, we found 293S cell specific transcription initiating in intron 1, yielding an extension product of about 250 bases, which represents either transcripts originating in the viral LTR, or at a cryptic promoter in the intron as has been seen in some mouse plasmacytomas (Prehn et al., 1984). Upon treatment of the 293S cells with CH for three hours, we found a shift in the relative abundance of transcripts detected by primers A and B. There was about a 10-fold increase in the circa 50 base P1 extension products and an approximately 5-fold decrease in the 250 base product from the intronic region. These results, in conjunction with the nuclear transcription analyses (Fig. 12), demonstrate that in the absence of protein synthesis there is a shift in c-myc promoter utilization in 293S cells.

CHANGES IN CHROMATIN STRUCTURE AFTER CYCLOHEXIMIDE TREATMENT

Since protein synthesis inhibition led to both quantitative and qualitative changes in c-myc transcription in bursal lymphoma cells, we examined the structure of chromatin surrounding the c-myc alleles. We had previously reported that after integration of the ALV LTR into the c-myc region of bursal lymphoma cells, the major hypersensitive site within this chromosomal domain is within the proviral LTR (Schubach and Groudine, 1984). If a labile protein were involved in the enhanced transcription directed by the ALV LTR, this protein might be involved in the formation of the hypersensitive site within the LTR, and diminution of this protein by CH treatment might result in the loss of this altered chromatin structure. To test this, we chose 243L cells, which contain only one c-myc allele, and in which the LTR integration has occurred in the first

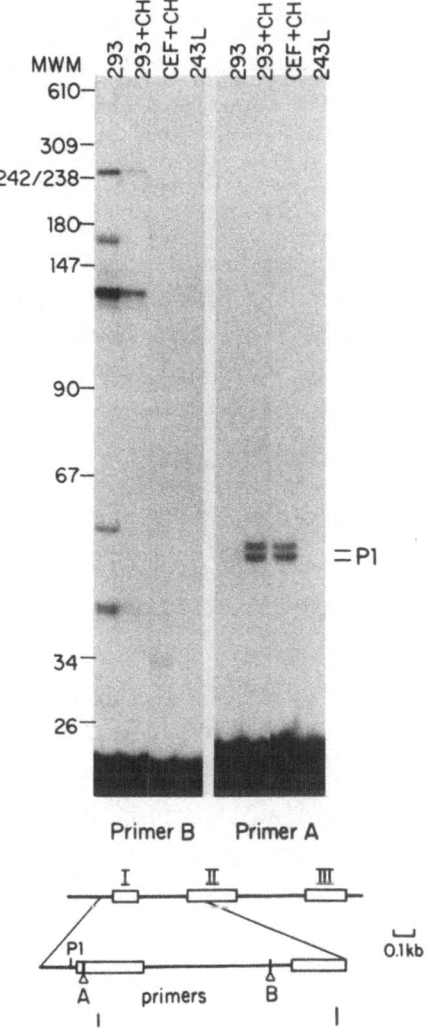

Fig. 13.   Primer extension analysis of c-<u>myc</u> transcripts
encoded by 293S cells. Synthetic oligonucleotides
based upon the chicken c-<u>myc</u> sequence were
synthesized commercially.   The location of the
primers relative to the c-<u>myc</u> locus is shown on
the bottom of the figure.   The oligonucleotides
were   end-labeled   using ^{32}P-γ-ATP   and   T4
polynucleotide kinase.   RNAs used were from 293S
cells (+) or (-) CH treatment for 3 hours, chick
embryo fibroblasts (CEF) treated with CH, or
untreated 243L cells. Hybridization to total cell
RNA was at 65°C for 3 hours and elongation of
primers utilized AMV reverse transcriptase as
previously described (McKnight et al., 1981).
Products are analyzed on an 8% polyacrylamide
denaturing gel.   Molecular weight markers (MWM)
used were MspI digested lambda DNA end-labeled
with ^{32}P-TTP and -CTP.

intron (Fig. 10), sufficiently distant from the 5' end of the first c-myc exon to permit us to distinguish the major c-myc hypersensitive sites from that of the LTR. As shown in Figure 14, when DNA from DNase I digested 243L nuclei is redigested with Cla I and Bam HI and assayed for the presence of subbands by Southern blotting, a prominant subband (indicated by the arrow) is found within the LTR region. When the same assay is performed on 243L cells that have been treated with CH for three hours, this major site is barely detectable. As a control, when the same DNA samples are digested with Bgl II and blot hybridized to a thymidine kinase (TK) probe (Fig. 14B), the major TK hypersensitive site in the 3' region of the TK locus (Groudine and Conklin, 1985) is detectable in both treated and untreated nuclei, indicating that the loss of the LTR hypersensitive site is not due to the inability to detect such sites in DNA from the CH-treated 243L cells.

IMPLICATIONS FOR THE REGULATION OF GENE EXPRESSION

In summary, we have found that treatment of all normal avian cells by inhibitors of protein synthesis dramatically increases the steady state level of c-myc RNA, as had been previously found for some mammalian cells and regenerating rat liver (Dani et al., 1984; Kelly et al., 1983; Makino et al., 1984). This is caused by stabilization of c-myc RNA that continues to be synthesized at the normal level. Five bursal lymphoma

Fig. 14. DNase I digestion of 243L nuclei. Nuclei from cell line 243L either untreated or treated with CH (30 µg/ml) for 3 hours, were digested with increasing concentrations of DNase I (as designated by the direction of the arrow: 0, 0.3, 0.6, 1.2, 2.4, 4.8, 7.2 µg/ml), DNA was isolated, digested with the enzymes denoted, and subjected to blot hybridization after electrophoresis in 1% agarose gels. The blots were probed with the nick-translated probes designated below the autoradiograms. (A) DNA samples were digested with Cla I and Bam HI and blots were hybridized to a 1 kb ClAI-SalI fragment that essentially spans the second c-myc intron as indicated in the line drawing. A 2.4 kb parent band and 1.7 kb subband mapping to the LTR sequences are evident in the control samples. (B) Duplicate of samples used in (A) above were digested with BglII and blots were hybridized to a 675 bp fragment from the 3' region of the chicken TK (cTK) gene (Groudine and Conklin, 1985).

cell lines which contain viral LTR sequences integrated in the vicinity of c-myc, behaved quite differently compared to normal cells upon protein synthesis inhibition. Cycloheximide or emetine teatment had little or no effect upon the amount of steady state c-myc RNA, but a 5- to 10-fold decrease was observed in c-myc transcription as measured in nuclear runoff

A c-myc

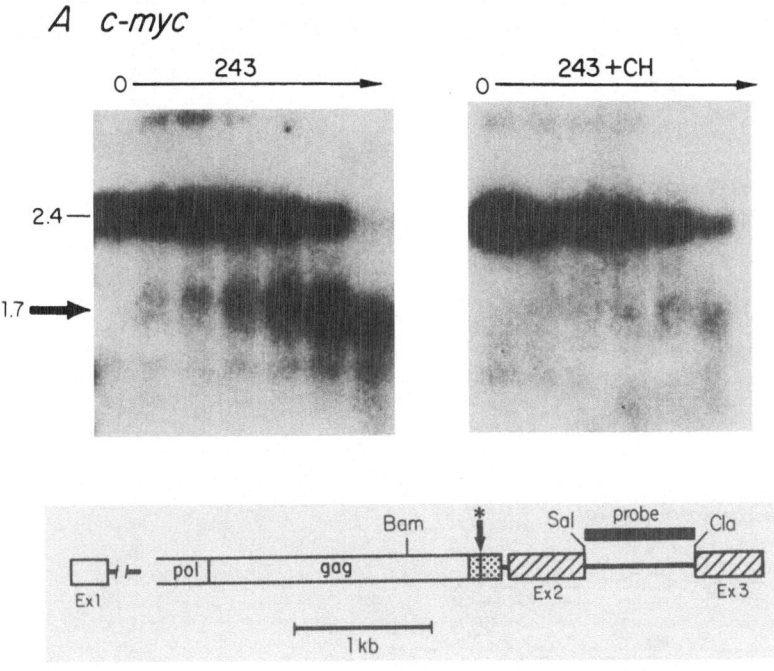

B cTK

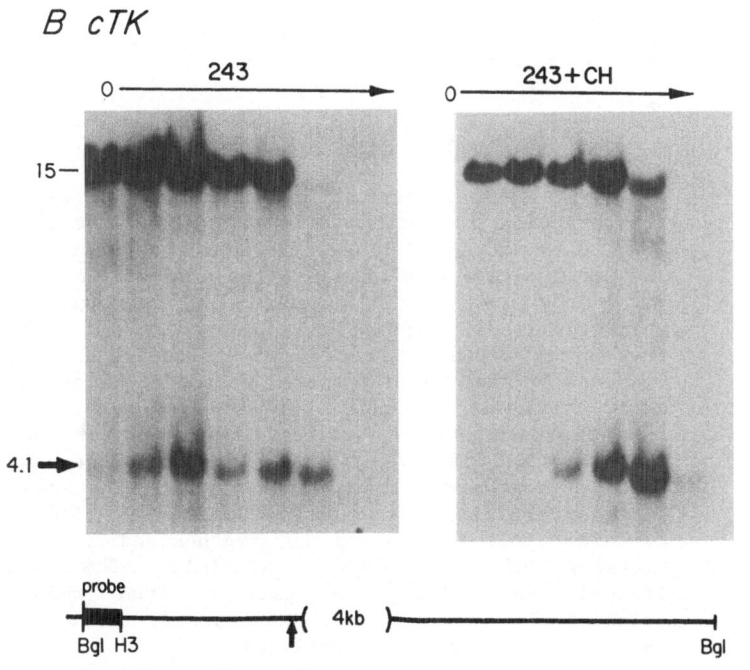

experiments.   In these bursal lymphoma cell lines, c-myc RNA is unstable,
and the RNA turnover can be reduced by CH treatment as in normal cells.
However, in these cell lines, transcription of c-myc is additionally
controlled by proximal viral LTR sequences acting as enhancers on the
normal c-myc promoter, or as promoters and enhancers.   We suggest that in
the bursal lymphoma cell lines, a rapidly turning over protein interacts
with the viral LTR to increase the transcription from the c-myc gene.   In
the absence of protein synthesis, the putative regulatory protein
disappears, and the transcription level reverts to that seen in normal
cells.   Thus, the lack of change in the steady state level of c-myc RNA in
CH treated bursal lymphoma cell lines can be explained by a decrease in
the rate of c-myc transcription and a concommitant increase in the half
life of c-myc RNA (Fig. 15).

Normal Cells :

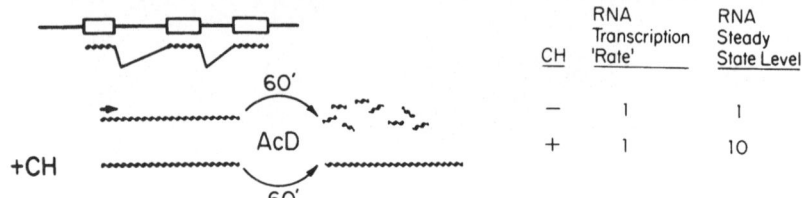

Bursal Lymphoma Cells (S13) :

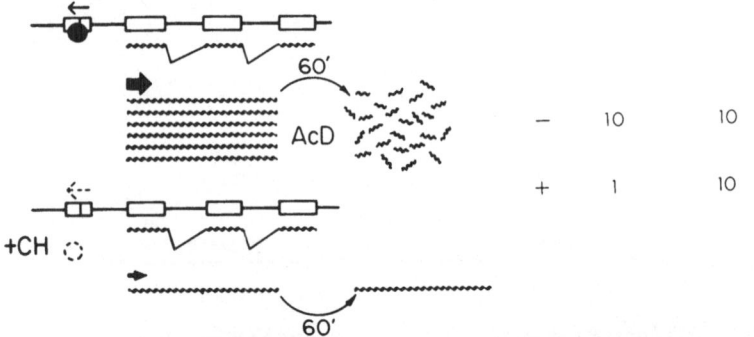

Fig. 15. Model for the effect of cycloheximide on c-myc
transcription innormal and bursal lymphoma cells.
Open boxes and solid lines represent the c-myc gene
and, in the case of S13 cells, the placement of the
viral LTR upstream of exon 1 in the opposite
transcriptional orientation of c-myc (<--).   The
wavy lines represent the c-myc transcript in the
presence or absence of cycloheximide (CH) before or
after treatment of cells with Actinomycin D (AcD)
for 60 minutes to inhibit synthesis of new
transcripts.   The size of the arrows above the wavy
lines indicates the transcription rate.   The solid
circle represents the putative labile LTR binding
protein which ceases to be synthesized in the
presence of CH (dotted circle).   The RNA
transcription "rate" is derived from runoff
transcription analyses and the steady state RNA
level from Northern blot analyses.

This putative regulatory protein could interact with either LTR enhancer sequences alone (S13) or with promoter and enhancer sequences (293S) to increase transcription of genes under LTR control, including c-myc and gag. One possibility is that the control protein is a cellular protein found in the bursal target cells for ALV-induced transformation. This normal cellular protein might then fortuitously interact with regions of the ALV LTR containing sequences similar to cellular DNA sequences which are normal binding sites for the protein.

The loss of the LTR-associated hypersensitive site in cell line 243L is reminiscent of similar changes in the MMTV LTR upon withdrawal of hormone (Zaret and Yamamoto, 1984), and is compatible with the notion that a labile protein involved in the enhanced transcription directed by the LTR might also be involved in the generation of the hypersensitive site. The notion that protein may be involved in the formation of hypersensitive sites has been suggested by reconstitution experiments with the chicken β^A globin gene and nuclear extracts from avian red blood cells (Emerson and Felsenfeld, 1984). In addition, a HeLa cell factor which interacts with the sequences in the SV40 enhancer has been described (Sassone-Corsi et al., 1985), as has a cellular protein which binds to the MMTV LTR and the BK virus enhancer, as well as to sequences 5' to the chicken lysozyme gene (Nowock et al., 1985).

Our results measuring runoff transcription levels in ALV-infected CEF show that the gag gene is actively transcribed in these cells, but that transcription of gag and c-myc genes is not cycloheximide sensitive. The level of gag transcription in infected CEF is at least as high as that in the bursal lymphoma cell lines. This would tend to suggest that the CH sensitive protein is tissue or cell type, rather than viral, specific. However, this leaves us with a paradox, since the viral LTR functions in CEFs and other normal cells to drive gag gene transcription. There are at least three different hypotheses to explain this: 1) Transcription from the LTR does not normally require any unique regulatory proteins, but in bursal cells a tissue specific unstable protein fortuitously interacts with the LTR. This model seems unlikely, since it would predict that in the absence of the protein (as after CH treatment), the level of transcription from the LTR would remain high. 2) The same regulatory protein functions in all cell types but the protein is modified in the target cells for ALV induced bursal lymphomas so that it becomes unstable, or regulation of the gene coding for this protein differs in bursal and other cells. 3) Different cellular regulatory proteins which recognize

237

LTR sequences are present in different cells or tissues. Although at present we have no data to support this last idea, it is the most attractive of the three, and may account for the tissue-specific activation of subsets of genes during embryogenesis.

ACKNOWLEDGEMENTS

We thank many of our colleagues, especially Hal Weintraub, for discussion and suggestions during the course of this work and Helen Devitt for putting together this manuscript. This work was supported by NIH grants CA18282 (ML) and CA28151 (ML and MG), and NSF grant PCM 82-04696 (MG). MG is a Scholar of the Leukemia Society of America.

REFERENCES

Akiyama, Y., and Kato, S., 1974, Two cell lines from lymphomas of Marek's disease, Biken. J., 17:105-116.

Alberts, B. M., Bray, D., Lewis, J., Raff, M., Roberts, K., and Watson, J., 1983, "Molecular Biology of the Cell," Garland Publishing, Inc., New York, pp. 689-690, 834-890.

Astrin, S. M., Robinson, H. L., Crittenden, L. B., Buss, E. G., Wyban, J., and Hayward, W. S., 1980, Ten genetic loci in the chicken that contain structural genes for endogenous avian leukosis viruses, Cold Spring Harbor Symp. Quant. Biol., 44:1105-1109.

Behe, M., and Felsenfeld, G., 1981, Effects of methylation on a synthetic polynucleotide: The B-Z transition in poly(dG m^5-dC)-poly(dG m^{5}dC), Proc. Natl. Acad. Sci. USA, 78:1619-1623.

Behe, M., Zimmerman, S., and Felsenfeld, G., 1981, Changes in the helical repeat of poly(dG m^{5}dC)-poly(dG m^{5}dC) and poly(dGdC)-polyGdc associated with the B-Z transition, Nature, 293:233-235.

Bellve, A. R., 1979, "Oxford Review of Reproductive Biology", C. Finn, ed., Oxford University Press, London, 1:159-261.

Bellve, A. R., and O'Brian, D. A., 1983, "Mechanisms and Control of Animal Fertilization," J. F. Hartmann, ed., Academic Press, New York, pp. 56-137.

Bellve, A. R., Cavicchia, J. C., Millette, C. F., O'Brian, D. A., Bhatnagar, Y. M., and Dym, M., 1977, Spermatogenic cells of the prepuberal mouse: Isolation and morphological characterization, J. Cell Biol., 74:68-85.

Conklin, K. F., and Groudine, M., 1984, Chromatin structure and gene expression, in: "DNA Methylation and Biological Significances," A. Razin, H. Cedar, and A. Riggs, eds., Springer-Verlag, New York, pp. 293-351.

Conklin, K. F., Coffin, J. M., Robinson, H. J., Groudine, M., and Eisenman, R., 1982, Role of methylation in the induced and spontaneous expression of the avian endogenous virus ev-1: DNA structure and gene products, Mol. Cell Biol., 2:638-652.

Curran, T., Peters, G., Van Beveren, C., Teich, N. M., and Verma, I. M., 1982, FBJ murine osteosarcoma virus: Identification and molecular cloning of biologically active proviral DNA, J. Virol., 44:674-682.

Dani, C., Blanchard, J. M., Piechaczyk, M., El Sabouty, S., Marty, L., Jeanteur, P., 1984, Extreme instability of myc mRNA in normal and transformed human cells, Proc. Natl. Acad. Sci. USA, 81:7046-7050.

Distel, R. J., Kleene, K. C., and Hecht, N. B., 1984, Haploid expression of a mouse testis alpha-tubulin gene, Science, 224:68-70.

Dugaiczyk, A., Haron, J. A., Stone, E. M., Dennison, O. E., Rothblum, K. N., and Schwartz, R. J., 1983, Cloning and sequencing of a deoxyribonucleic acid copy of glyceraldehyde-3-phosphate dehydrogenase messenger ribonucleic acid isolated from chicken muscle, Biochemistry, 22:1605-1613.

Elgin, S. C., 1981, DNase I-hypersensitive sites of chromatin, Cell, 27:413-415.

Emerson, B. M. and Felsenfeld, G., 1984, Specific factor conferring nuclease hypersensitivity at the 5' end of the chicken adult β-globin gene, Proc. Natl. Acad. Sci. USA, 81:95-99.

Fisher, E. F., and Caruthers, M. H., 1979, Studies on gene control regions. XII: The functional significance of a lac operator constitutive mutation, Nucl. Acids Res., 7:401-416.

Garry, R., Moyer, M., Bishop, J., Moyer, R., and Waite, M., 1981, Transformation parameters induced in chick cells by incubation in media of altered NaCl concentration, Virology, 111:427-439.

Ginder, G., Whitters, M., Kelley, K., and Chase, R., 1983, "Hemoglobin Switching," G. Stamatoyannopoulos, A. Neinhuis, eds., Liss, New York, pp. 463-474.

Groudine, M., and Casimir, C., 1984, Post-transcriptional regulation of the chicken thymidine kinase gene, Nucl. Acids Res., 12:1427-1446.

Groudine, M., and Conklin, K., 1985, Chromatin structure and de novo methylation of sperm DNA: Implications for activation of the paternal genome, Science, 228:1061-1068.

Groudine, M., and Weintraub, H., 1975, Rous sarcoma virus activates embryonic globin genes in chicken fibroblasts, Proc. Natl. Acad. Sci. USA, 72:4464-4468.

Groudine, M., and Weintraub, H., 1980, Activation of cellular genes by avian RNA tumor viruses, Proc. Natl. Acad. Sci. USA, 77:5351-5354.

Groudine, M., and Weintraub, H., 1981, Activation of globin genes during chicken development, Cell, 24:393-401.

Groudine, M., and Weintraub, H., 1982, Propagation of globin DNase I hypersensitive sites in absence of factors required for induction: A possible mechanism for determination, Cell, 30:131-139.

Groudine, M., Eisenman, R., and Weintraub, H., 1981, Chromatin structure of endogenous retroviral genes and activation by an inhibitor of DNA methylation, Nature, 292:311-317.

Groudine, M., Holtzer, H., Scherrer, K., and Therwath, A., 1974, Lineage-dependent transcription of globin genes, Cell, 3:243-247.

Groudine, M., Peretz, M., and Weintraub, H., 1981, Transcriptional regulation of hemoglobin switching in chicken embryos, Mol. Cell. Biol., 1:281-301.

Hagopian, H. K., Lippke, J. A., and Ingram, V., 1972, Erythropoietic cell cultures from chick embryos, J. Cell Biol., 54:98-106.

Hamlyn, P. H., and Rabbitts, T. H., 1983, Translocation joins c-myc and immunoglobulin gamma 1 genes in a Burkitt lymphoma revealing a third exon in the c-myc oncogene, Nature, 304:135-139.

Harper, M. I, Fosten, M., and Monk, M., 1982, Preferential paternal X inactivation in extraembryonic tissues of early mouse embryos, J. Embryol. Exp. Morphol., 67:127-135.

Hayward, W. S., Baverman, S. B., and Astrin, S. M., 1980, Transcriptional products and DNA structure of endogenous avian proviruses, Cold Spring Harbor Symp. Quant. Biol., 44:1111-1121.

Hishinuma, F., DeBona, P. J., Astrin, S., and Skalka, A. M., 1981, Nucleotide sequence of the acceptor site and termini of integrated avian endogenous provirus ev-1: Integration creates a 6 bp repeat of host DNA, Cell, 23:155-164.

Iatrou, K., Spira, A. W., and Dixon, G. H., 1978, Protamine messenger RNA: Evidence for early synthesis and accumulation during spermatogenesis in rainbow trout, Dev. Biol., 64:82-98.

Kelly, K., Cochran, B. H., Stiles, C. D., and Leder, P., 1983, Cell-specific regulation of the c-myc gene by lymphocyte mitogens and platelet-derived growth factor, Cell, 35:603-610.

Larsen, A., and Weintraub, H., 1982, An altered DNA conformation detected by S1 nuclease occurs at specific regions in active chick globin chromatin, Cell, 29:609-622.

Leder, P., Battey, J., Lenoir, G., Moulding, C., Murphy, W., Potter, H., Stewart, T., and Taub, R., 1983, Translocations among antibody genes in human cancer, Science, 222:765-771.

Lezzi, M., and Robert, M., 1976, Chromosomes isolated from unfixed salivary glands of Chironomous, in: "Developmental Studies on Giant Chromosomes: Results and Problems in Cell Differentiation," W. Beerman, ed., Springer-Verlag, New York, vol. 4:35-57.

Lilley, D. M. J., 1980, The inverted repeat as a recognizable structural feature in supercoiled DNA molecules, Proc. Natl. Acad. Sci. USA, 77:6468-6472.

Lilley, D. M. J., 1981, Hairpin-loop formation by inverted repeats in supercoiled DNA is a local and transmissible property, Nucl. Acids Res., 9:1271-1289.

Lillie, F., 1952, "Development of the Chick," 3rd Edition, Holt, Rinehart and Winston, New York.

Linial, M., and Groudine, M., 1985, Transcription of three c-myc exons is enhanced in chicken bursal lymphoma cell lines, Proc. Natl. Acad. Sci. USA, 82:53-57.

Makino, R., Hayashi, K., and Sugimura, T., 1984, C-myc transcript is induced in rat liver at a very early stage of regeneration or by cycloheximide treatment, Nature, 310:697-698.

McGhee, J., and Ginder, G. D., 1979, Specific DNA methylation sites in the vicinity of chicken β-globin genes, Nature, 280:419-420.

McGhee, J. D., Wood, W. I., Dolan, M., Engel, J. D., and Felsenfeld, G., 1981, A 200 base pair region at the 5' end of the chicken adult β-globin gene is accessible to nuclease digestion, Cell, 27:45-55.

McGrath, J., and Solter, D., 1984, Completion of mouse embryogenesis requires both the maternal and paternal genome, Cell, 37:179-183.

McKnight, S. L., Gavis, E. R., Kingsbury, R., and Axel, R., 1981, Analysis of transcriptional regulatory signals of the HSV thymidine kinase gene: Identification of an upstream control region, Cell, 25:385-398.

Nowock, J., Borgmeyer, U., Puschel, A. W., Rupp, R. A. W., and Sippel, A. E., 1985, The TGGCA protein binds to the MMTV-LTR, the adenovirus origin of replication, and the BK virus enhancer, Nucl. Acids Res., 13:2045-2061.

Panayotatos, N., and Wells, R. D., 1981, Cruciform structures in supercoiled DNA, Nature, 289:466-470.

Piechaczyk, M., Blanchard, J. M., Marty, L., Dani, C., Panabieres, F., El Sabouty, S., Fort, P., and Jeanteur, P., 1984, Post-transcriptional regulation of glyceraldehyde-3-phosphate-dehydrogenase gene expression in rat tissues, <u>Nucl. Acids Res.</u>, 12:6951–6963.

Prehn, J., Mercola, M., and Calame, K., 1984, Translocation affects normal c–<u>myc</u> promoter usage and activates fifteen cryptic c–<u>myc</u> transcription starts in plasmacytoma M603, <u>Nucl. Acids Res.</u>, 12:8987–9007.

Razin, A., and Riggs, A. D., 1980, DNA methylation and gene function, <u>Science</u>, 210:604–610.

Riggs, A. D., 1984, "DNA Methylation and Biological Significances," A. Razin, H. Cedar, and A. Riggs, eds., Springer-Verlag, New York, pp 269–278.

Samarut, J., Jurdic, P., and Nigon, V., 1979, Production of erythropoietic colony-forming units and erythrocytes during chick embryo development: An attempt at modelization of chick embryo erythropoiesis, <u>J. Embryol. Exp. Morphol.</u>, 50:1–20.

Sanders-Haigh, L., Blanchard-Owens, B., and Ingram, V., 1983, "Hemoglobin Switching", G. Stamatoyannopoulos and A. Neinhuis, eds., Liss, New York, pp. 39–52.

Sassone-Corsi, P., Wildeman, A., and Chambon, P., 1985, A <u>trans</u>-acting factor is responsible for the simian virus 40 enhancer <u>in vitro</u>, <u>Nature</u>, 313:458–463.

Schubach, W., and Groudine, M., 1984, Alteration of c–<u>myc</u> chromatin structure by avian leukosis virus integration, <u>Nature</u>, 307:702–708.

Shih, C.-K., Linial, M., Goodenow, M. M., and Hayward, W. S., 1984, Nucleotide sequence 5' of the chicken c–<u>myc</u> coding region: Localization of a noncoding exon that is absent from <u>myc</u> transcripts in most avian leukosis virus-induced lymphomas, <u>Proc. Natl. Acad. Sci. USA</u>, 81:4697–4701.

Stalder, J., Groudine, M, Dodgson, J. B., Engel, J. D., and Weintraub, H., 1980, Hb switching in chicken, <u>Cell</u>, 19:973–980.

Stalder, J., Larsen, A., Engel, J. D., Dolan, M., Groudine, M., and Weintraub, H., 1980, Tissue-specific DNA cleavages in the globin chromatin domain introduced by DNase I, <u>Cell</u>, 20:451–460.

Stanton, L. W., Watt, R., and Marcu, K. B., 1983, Translocation, breakage and truncated transcripts of c–<u>myc</u> oncogene in murine plasmacytomas, <u>Nature</u>, 303:401–406.

Stein, R., Sciaky-Gallili, N., Razin, A., and Cedar, H., 1983, Patterns of methylation of two genes coding for housekeeping functions, <u>Proc. Natl. Acad. Sci. USA</u>, 80:2422–2426.

Surani, M. A., Barton, S. C., and Norris, M. L., 1984, Development of reconstituted mouse eggs suggests imprinting of the genome during gametogenesis, Nature, 308:548–550.

Takagi, N., Wake, N., and Sasaki, M., 1978, Cytologic evidence for preferential inactivation of the paternally derived X chromosome in XX mouse blastocytes, Cytogenet. Cell Genet., 20:240–248.

Tykocinski, M. L., and Max, E. E., 1984, CG dinucleotide clusters in MHC genes and in 5' demethylated genes, Nucl. Acids Res., 12:4385–4396.

Waalwyck, C., and Flavell, R., 1978, DNA methylation of CCGG sequence in the large intron of the rabbit B-globin gene: Tissue-specific variations, Nucl. Acids Res., 5:4631–4641.

Weintraub, H., 1983, A dominant role for DNA secondary structure in forming hypersensitive structures in chromatin, Cell, 32:1191–1203.

Weintraub, H., and Groudine, M., 1976, Chromosomal subunits in active genes have an altered conformation, Science, 193:848–856.

Weintraub, H., Campbell, G., and Holtzer, H., 1971, Primitive erythropoiesis in early chick embryogenesis I. Cell cycle kinetics and the control of cell division, J. Cell Biol., 50:652–668.

Weintraub, H., Larsen, A., and Groudine, M., 1981, α-Globin gene switching during the development of chicken embryos: Expression and chromatin structure, Cell, 24:333–344.

West, J. D., Frels, W. I., Chapman, V. M., and Papaioannou, V. E., 1977, Preferential expression of the maternally derived X chromosome in the mouse yolk sac, Cell, 12:873–882.

Yanisch-Perron, C., Viera, J., and Messing, J., 1985, Improved M13 phage cloning vectors and host strains: Nucleotide sequences of the M13mp18 and pUC19 vectors, Gene, 33:103–119.

Zaret, K. S., and Yamamoto, K. R., 1984, Reversible and persistent changes in chromatin structure accompany activation of a glucocorticoid-dependent enhancer element, Cell, 38:29–38.

ORGANIZATION AND EXPRESSION OF GONADOTROPIN GENES

William W. Chin and Soheyla D. Gharib

Department of Medicine, Brigham and Women's Hospital
and Howard Hughes Medical Institute Laboratories
Harvard Medical School
Boston, Massachusetts   02115

INTRODUCTION

Lutropin (LH) and follitropin (FSH) are gonadotropins produced and secreted by the mammalian anterior pituitary gland. They serve to regulate gonadal function by stimulating the production of sex steroid hormones (estrogen, progesterone, and testosterone) and gametogenesis. The gonadotropins are the effectors of well-regulated gonadal development and function orchestrated in a finely tuned manner by factors from the hypothalamus. Hence, LH and FSH play critical roles in sexual development and maturation as well as normal reproductive function (Kalra and Kalra, 1985).

LH and FSH are members of a family of structurally-related protein hormones known as the glycoprotein hormones. In addition to the gonadotropins, the other members include pituitary thyroid-stimulating hormone (TSH) and placental chorionic gonadotropin (CG). Each hormone contains two different subunits called α and β (Fig. 1). Within a species, the protein portion of the α-subunit is identical and hence the α-subunit is often referred to as the common subunit. In contrast, the subunits are different and confer specific biologic activities to the dimeric proteins. The molecular weight of the α-subunit is approximately 22,000 daltons and the β-subunit is 18,000 to 28,000 daltons. Each subunit is glycosylated and contains multiple disulfide linkages. In general, the carbohydrate moieties are attached to the polypeptide backbone via N-glycosidic bonds. The exceptions are the presence of O-linked oligosaccharides present in

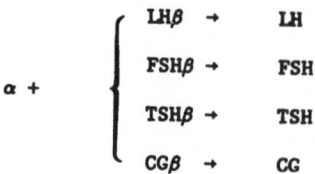

Fig. 1. The glycoprotein hormone family. The glycoprotein hormone family consists of 3 to 4 members in most mammalian species. The hormones are lutropin (LH), follitropin (FSH), thyroid-stimulating hormone (TSH), and chorionic gonadotropin (CG). Each glycoprotein hormone is made up of two different subunits called α and β that are associated in a non-covalent fashion to form the biologically active dimer.

the free α-subunit and C-terminal extension peptide of the hCGβ subunit (Pierce and Parsons, 1981; Chin, 1985). The β-subunits display limited protein sequence homology which varies from 80% (LH-CG) to 30-40% (others). In addition, there is sufficient homology among the α- and β-subunits to suggest their derivation from a common ancestor gene (Acher, 1980; Stewart and Stewart, 1977). Recent studies have indicated that the carbohydrate moieties are critical for both the proper synthesis of bioactive dimers and the achievement of the ultimate biologic activity of the mature hormone. Specifically, the inhibition of glycosylation early in the biosynthesis of the glycoprotein hormones results in inappropriate subunit folding and diminished dimerization (Weintraub et al., 1983). Also, alterations of the subunit carbohydrate moieties may result in dimers which have diminished biologic activity without diminution of the binding of the hormones to their respective receptors (Sairam and Bhargavi, 1985; Keutmann et al., 1983). In addition, all cells that produce the glycoprotein hormones produce the α-subunit in excess. Interestingly, this free α-subunit is biochemically distinguishable from the α-subunit which is found in association with the β-subunit in the bioactive dimer. The difference is the localization of an O-linked carbohydrate moiety at position 43 (position 39 in human free α-subunit). This region corresponds to the well-established α-β contact point described by several workers. The removal of the O-linked sugar from free or unassociated α-subunits results in a subunit which now can combine with any β-subunit to obtain biologically active dimers (Parsons and Pierce, 1984). Thus, the addition of an O-linked sugar onto the α-subunit may play an important biologic role.

The glycoprotein hormones are expressed in a tissue-specific fashion; LH and FSH, for instance, are produced only in a few specialized cells in the anterior pituitary gland known as gonadotropes. Similarly, TSH is produced in another group of cells known as thyrotropes. Finally, CG is produced in the syncytiotrophoblast cell of the placenta, in non-gestational trophoblastic tissues, and in non-trophoblastic tissues such as tumors of the lung, breast, brain and gastrointestinal tract (Chin, 1985).

The regulation of production and synthesis of the gonadotropins is complex (Chin, 1985). As an example, shown in Figure 2, LH stimulates the gonadal tissues to produce gonadal sex steroid hormones which enter the general circulation to influence cells in the hypothalamus and other regions of the central nervous system, and the pituitary gonadotrope. The major stimulator of LH synthesis and secretion is hypothalamic gonadotropin-releasing-hormone (GnRH). The gonadal sex steroid hormones

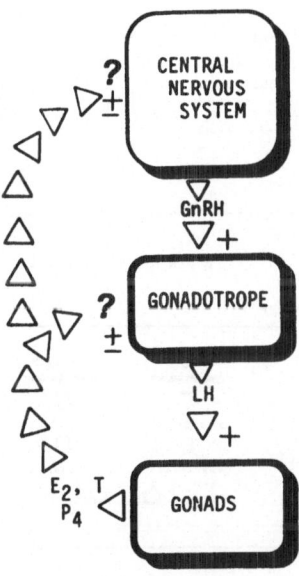

Fig. 2. Regulation of LH synthesis and release. The gonadotrope, a cell found in the anterior pituitary gland, synthesizes and releases LH. LH, in turn, stimulates the gonads to produce vital sex steroid hormones including estrogen, testosterone, and progesterone. These gonadal products then act in a negative feed-back manner to regulate central nervous system outputs and possibly gonadotropes directly. They probably affect the GnRH pulse generator to alter amplitude and frequency of GnRH pulsations and other factors in the hypothalamus. In addition, these gonadal sex-steroid hormones may act directly at the gonadotrope to alter the sensitivity of the gonadotrope to endogenous GnRH and/or to exert effects on the subunit genes.

affect the GnRH pulse generator in the hypothalamus by decreasing pulse frequency and pulse amplitude. On the other hand, the effect of these hormones on the gonadotrope is less certain although it is likely that they alter the sensitivity of the GnRH receptor to GnRH. However, it is unclear whether these steroid hormones have a direct effect on the expression of the α and LHβ genes in these cells. Hence, the major feature of this regulatory circuit is a negative feedback loop involving the gonadal sex steroid hormones with their effects exerted at both the hypothalamic and gonadotrope levels. In contrast, the peptide hormones produced by gonadal tissues are less well characterized. The inhibins are ovarian peptide factors which appear to decrease the release of FSH from the gonadotrope and whose protein sequences and genes in the pig have recently been described (Mason et al., 1985).

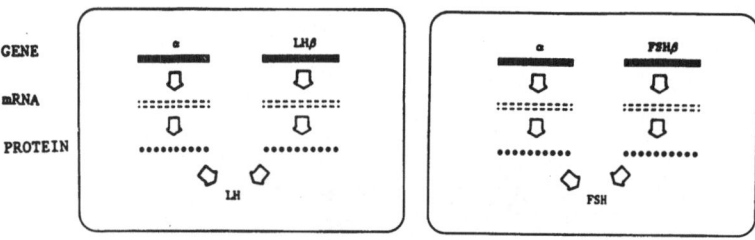

GONADOTROPE

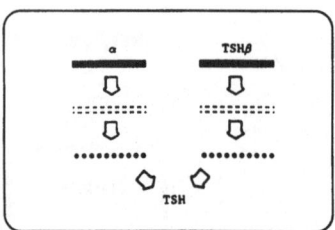

THYROTROPE

Fig. 3. Tissue-specific expression of the glycoprotein hormones. The α- and β-subunits are encoded by separate genes. They are expressed in a tissue-specific fashion so that LH and FSH are produced only in gonadotropes, and TSH only in thyrotropes in the anterior pituitary gland. The single α gene is evidently activated in each one of these cells. However, the subunit β genes are activated only in specific cells while other glycoprotein hormone subunit β genes are suppressed in order to achieve this tissue-specific expression of hormones. Although the gonadotropes are depicted here as producing either LH or FSH, they generally synthesize both.

Recent data indicate that the subunits are encoded by separate genes which are localized on separate chromosomes and are coordinately expressed in a tissue-specific manner (Fig. 3; Chin, 1985). As shown in Figure 4, our present understanding of the biosynthesis of a protein, in general, requires that a specific gene be transcribed initially into a precursor

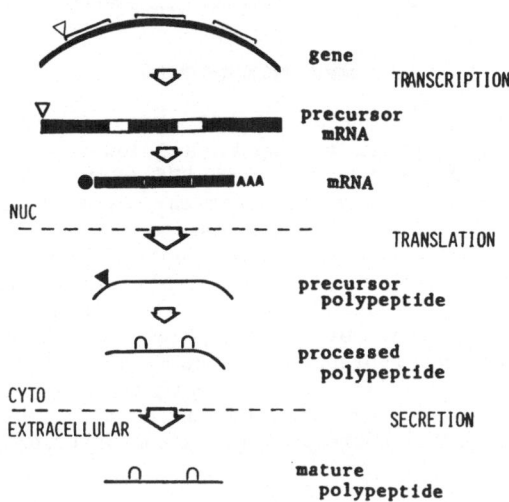

Fig. 4. Biosynthesis of a protein. A gene encoding a protein is transcribed (the ∇ indicates the start of transcription) to yield a precursor mRNA. This precursor mRNA contains both coding (black boxes; exons) and non-coding (white boxes; introns) regions. Then very rapid events occur in the nucleus including 5'-capping (•) and 3'-poly-adenylation (AAA) as well as intron splicing to yield the mature mRNA. This mRNA is transported from the nucleus to the cytoplasm where it interacts with the protein synthetic machinery including the ribosome and the endoplasmic reticu-lum (ER). There, the mRNA is translated (the ∇ depicts the start of translation) into a precursor polypeptide. This direct translation product is rapidly processed co-translationally to remove the leader or signal peptide and to add N-linked sugar cores (∩). The partially processed polypeptides obtain the appropriate conformations to interact with one another to yield a biologically active dimer and are then transferred through the remainder of the ER and Golgi apparatus. During this period of transport further modifications occur, including maturation of carbohydrate moieties, phosphoryla-tion, acetylation and sulfation. Finally, these mature proteins are stored in secretory granules to await the appropriate regulatory signals for their release from the cytoplasm into the extracellular space in the process of secretion.

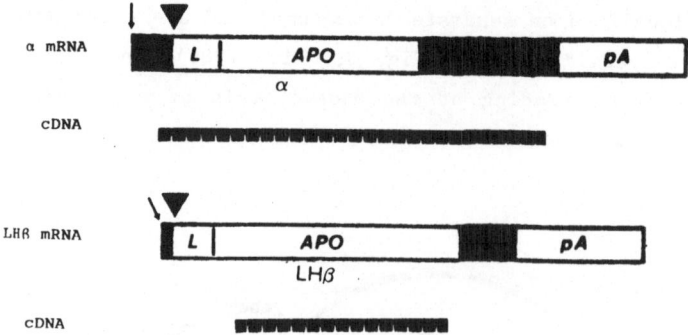

Fig. 5. cDNAs encoding α and LHβ mRNAs in the rat. The
large bars in the upper and lower panels of this
diagram represent α and LHβ mRNAs, respectively.
The black boxes represent the untranslated regions.
The first open box represents the precursor of the
subunits (L = leader peptide, APO = subunit
apoprotein) and the second box represents the poly
A tail (pA). The small arrow indicates the cap
site or start of transcription, and the large black
arrowhead represents the start of translation. The
serrated solid bar underneath the mRNA represents a
characteristic cDNA (Godine et al., 1982; Chin et
al., 1983) corresponding to the particular mRNA.

mRNA, which is then processed rapidly in the nucleus to yield a mature
mRNA (Darnell, 1982). This is then transported from the nucleus to the
cytoplasm, where it is translated into protein. The translation product
is a precursor polypeptide that is rapidly processed. During translation,
the leader or signal hydrophobic peptide found in the N-terminus of most
secretory and membrane proteins is cleaved and N-linked carbohydrate
moieties may be added. At this point, a partially digested glycoprotein
hormone subunit is folded and may combine non-covalently to its
complementary subunit. They are then transported through the Golgi stack,
during which the carbohydrate moieties undergo a maturation process. In
addition, a proportion of α-subunits do not associate and are
O-glycosylated rendering them incapable of dimerization. These free
α-subunits are produced in most cells that produce these hormones but lack
defined function. The mature hormone dimers and free subunits are then
sequestered in secretory granules, where they await signals for release
into the extracellular space. Thus, the processes of transcription,
translation, and secretion are integrated within the thyrotrope and the
gonadotrope to produce the α- and β-subunits, and the biologically active
dimers in the synthesis of glycoprotein hormones (Chin, 1985). A large
body of information has been gathered about the effects of various
hormones on the secretion of LH and FSH from the pituitary gonadotrope.

In addition, recent data have been gathered concerning the events occurring at the translational and post-translational levels. However, we wish to understand how these hormones control the synthesis of gonadotropins at the pre-translational level. In particular we would like to know whether these hormones control the synthesis of LH and FSH by altering the levels of their subunit mRNAs. However, these studies required the isolation of molecular probes for mRNAs and genes encoding the subunits of the gonadotropins.

MOLECULAR CLONING OF cDNAs AND GENES ENCODING THE SUBUNITS OF LH

Using the techniques of recombinant DNA technology and molecular biology we have been successful, over the last several years, in the molecular cloning of cDNAs and genes encoding the α and LHβ subunits in the rat. Typical cDNAs are depicted in Figure 5. Such cDNAs are useful for hybridization studies in order to measure subunit mRNA levels, and to isolate and localize the subunit genes to specific chromosomes in various species.

Figure 6 shows in schematic form the structure of the human α gene. The α gene is 9.4 Kb in size with three intervening sequences or introns (Fiddes and Goodman, 1981; Boothby et al., 1982); a similar structure for the bovine α-subunit has been demonstrated (Goodwin et al., 1983). Figure 7 illustrates the structures of the human LHβ and CGβ genes. Of note, two groups have determined that there are one LHβ subunit and six to seven CGβ genes in man which are located in clusters arranged in tandem inverted pairs (Talmadge et al., 1983; Talmadge et al., 1984; Policastro et al., 1983). It is unclear at this moment whether they are located on a contiguous segment of DNA. However, studies described below indicate that they are located on the same chromosome. In addition, data show that

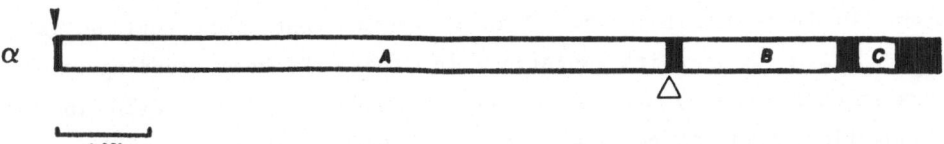

Fig. 6. Structure of the human α-subunit gene. The human α-subunit gene is 9.4 Kb in size and contains three introns labeled a, b, and c. The small solid arrowhead indicates the start of transcription or cap site and the large open arrowhead signifies the start of translation.

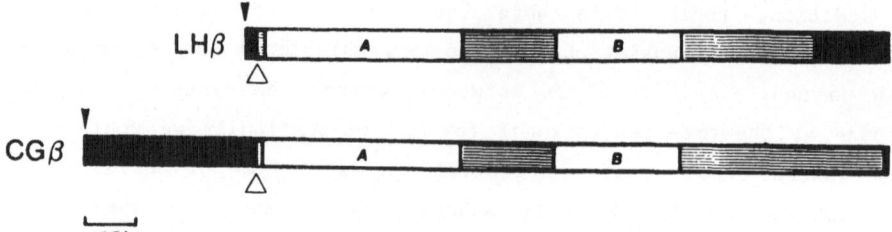

Fig. 7.   Structures of the human LHβ and CGβ subunit genes.
The upper panel displays the human LHβ subunit gene
in schematic form.   The gene is 1.2 Kb in size and
contains two introns A and B.   The black areas
represent the untranslated regions and the shaded
areas represent the coding regions.   The human CGβ
gene is represented in the lower panel.   The gene
is approximately 1.5 Kb in size and contains two
introns located in the same sites as in the LHβ
gene.   The introns are also approximately the same
sizes.   The small solid arrowhead represents the
transcriptional cap site and open triangle the
start of translation.   Note the larger 5'-untrans-
lated region in the CGβ compared to LHβ subunit
gene 5'-untranslated region as well as the virtual
absence of a 3'-untranslated region in the CGβ gene
as result of a frame-shift and read-through in the
3'-end of the coding region or third exon.

primates and perhaps the horse may be the only animals that contain the
CGβ gene (Tepper and Roberts, 1984; Moore et al., 1980).   In support of
this view, there are some convincing data that suggest the rat does not
contain a CGβ separate from the single LHβ gene (Tepper and Roberts, 1984;
Carr and Chin, 1985).

The human LHβ-CGβ genes are smaller and approximately 1.3-1.5 KB in
size and contain two introns in similar positions within the coding
regions (Fiddes and Talmadge, 1984).   However, Figure 7 also indicates
that there are two major differences between these genes in man.   Although
the two genes are highly homologous, the extreme 3'-end of the CGβ gene
encodes an additional 24 amino acid carboxy-terminal peptide which is
absent in human LHβ subunit.   This C-terminal extension peptide is a
result of a frame-shift mutation in the LHβ gene leading to a
translational read-through until the termination codon located in the
polyadenylation signal consensus sequence, AATAAA, is encountered at the
very end of the CGβ gene (Fiddes and Goodman, 1980; Fiddes and Talmadge,
1984).   In addition, there is a major difference in the 5'-untranslated
regions of these two mRNAs.   Interestingly, the LHβ 5'-untranslated region
consists of only seven bases, a finding which has been confirmed in the

rat and cow LHβ subunit genes (Jameson et al., 1984; Virgin et al., 1985). In contrast, however, the 5' flanking region of the CGβ subunit gene is larger, with a size of approximately 350 bases. A comparison of these regions between LHβ and CGβ genes shows extensive sequence homology. This high level of homology includes the retention of a TATAAA, a classic promoter element, located approximately 35 to 37 bases upstream from the start of translation. Thus the subtle differences between the two genes in the 5' flanking regions likely result in an alteration of gene expression which is manifest in different transcription start sites and ultimate formation of different 5'-untranslated regions (Fiddes and Talmadge, 1984). Whether this variation accounts for differential hormonal regulation is yet unknown.

In addition, work from a number of groups has provided the structures of the TSHβ and FSHβ genes in several species. The structures of these genes are similar to that of the LHβ and CGβ genes, in that the introns are located in similar if not identical positions relative to the respective precursor mRNAs (Carr et al., 1985; Hayashizaki et al., 1985; Whitfield et al., 1985; Beck et al., 1985).

CHROMOSOMAL LOCALIZATION OF THE SUBUNIT GENES

It is possible using cloned subunit cDNAs to determine the precise chromosomal localization of these genes within the human and mouse genomes. In collaboration with Dr. Susan L. Naylor and her colleagues at the University of Texas at San Antonio, we have determined the chromosomal localization of the subunit genes in man and mouse (Naylor et al., 1983; Naylor et al., 1986). Figure 8 depicts the general approach to determining the localization of genes to specific chromosomes. Briefly, human-mouse somatic cell hybrids are isolated and characterized. These somatic cell hybrids contain the entire complement of mouse chromosomes while possessing only a small number of the human complement. The exact human chromosomes that remain in these hybrids may be determined by karyotype and enzyme marker analysis. These somatic cell hybrids are then analyzed for the presence of a particular subunit gene by blot hybridization techniques using genomic DNAs from specific somatic cell hybrids. As indicated in Figure 8, the results of blot hybridization studies along with the presence or absence of a specific chromosome in a hybrid cell allow a concordance analysis and provide the localization of particular genes to chromosomes. As shown in Figure 9, human chromosome 6 contains

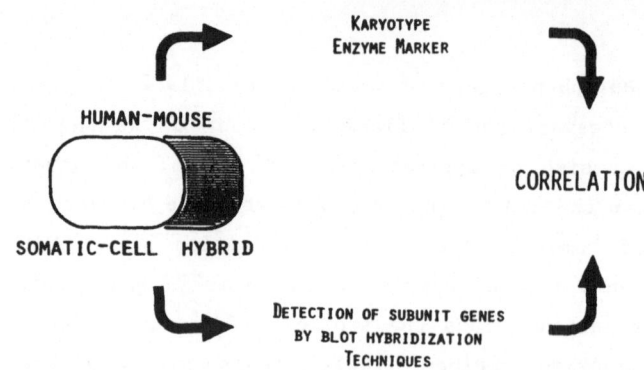

KARYOTYPE
ENZYME MARKER

HUMAN-MOUSE

CORRELATION

SOMATIC-CELL HYBRID

DETECTION OF SUBUNIT GENES
BY BLOT HYBRIDIZATION
TECHNIQUES

Fig. 8. Chromosomal localization of specific genes. Human-
         mouse somatic hybrids are characterized in terms of
         the human chromosome content by karyotype and enzyme
         marker analysis. The cells are also analyzed by
         blot hybridization techniques to determine the
         presence of the subunit genes in its DNA. Correla-
         tion of the presence or absence of specific subunit
         genes with a specific chromosome is performed to
         determine the chromosomal localization of the
         subunit genes.

HUMAN

Subunit Gene	Chromosome #
α	6
LHβ CGβ	19
FSHβ	11
TSHβ	1

MOUSE

α	4
LHβ	7

Fig. 9. Chromosomal localization of subunit genes in man
         and mouse. The genes encoding the subunits of the
         glycoprotein hormones have been localized to
         specific human and mouse chromosomes as shown in
         Figure 8 (Naylor et al., 1983; Naylor et al., 1986;
         Julier et al., 1984; Kourides et al., 1984; Watkins
         et al., 1985). The LHβ-CGβ subunit gene cluster in
         man is located in a nearly contiguous segment on
         the 19th chromosome (Naylor et al., 1983; Julier et
         al., 1984).

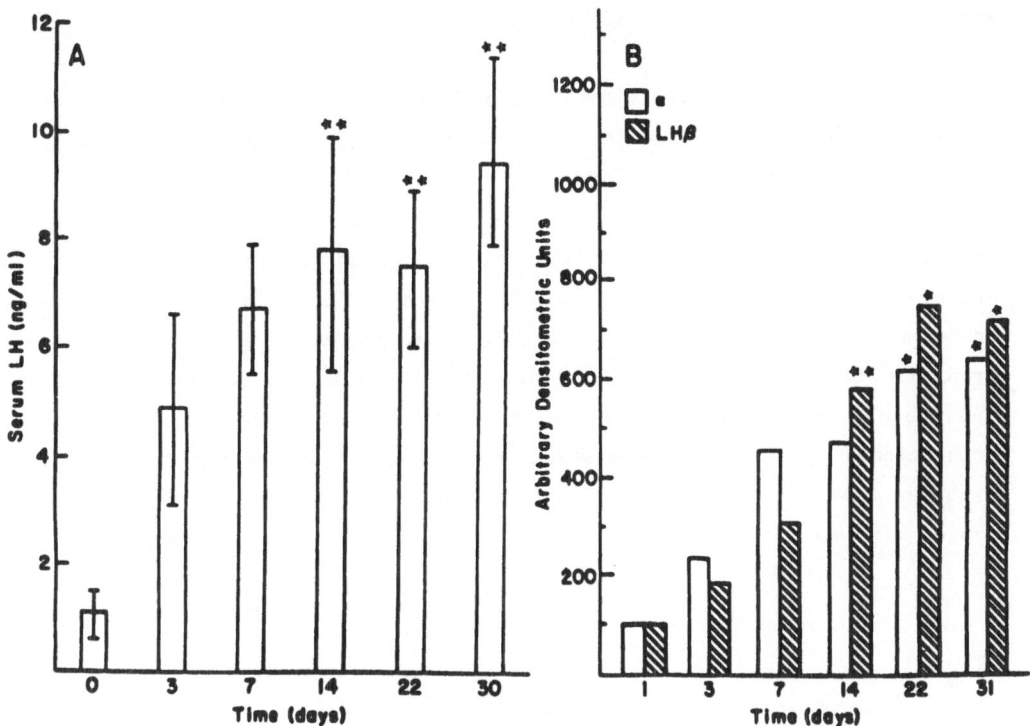

Fig. 10. The effect of castration on serum LH and subunit
mRNAs. Castration was performed in rats and
animals sacrificed at various times after opera-
tion. Panel A shows serum LH levels determined by
RIA. Panel B shows α and LHβ mRNA levels in
pituitary glands of castrated animals after
various times. mRNA levels are expressed in
arbitrary densitometric units and are normalized
to 100 for convenience. Stars indicate statistical significance of data points compared to
control levels: * = $p < 0.05$; and ** = $p < 0.01$.
(Reprinted from Gharib et al., 1986.)

the α-subunit gene (Naylor et al., 1983). Chromosome 19 contains the
LHβ-CGβ gene cluster (Naylor et al., 1983; Julier et al., 1984), chromosome 11 contains the FSHβ gene (Watkins et al., 1985) and chromosome 1
contains the TSHβ gene (Naylor et al., 1986). Studies have also been
performed examining the mouse genome with similar results. Hence, the
separate genes which encode the glycoprotein hormone subunits are located
on separate chromosomes in both man and mouse. It is interesting to note
that the subunit genes are associated with certain genes in a given human
chromosome with which the subunit genes are also found in the mouse
chromosomes. Thus, the subunit genes appear to have evolved in gene
clusters and are syntenic with these other marker genes in specific
chromosomes (Naylor et al., 1983).

In order to determine whether gonadal sex steroid hormones influence the synthesis of LH at the pre-translational level, we utilized two well known experimental paradigms involving the castrate animal (Gharib et al., 1986). We studied gonadectomized male and female rats and examined the pituitary content of α and LHβ mRNAs as a function of time after surgery. These results were correlated with the serum LH level as determined by radioimmunoassay (RIA). Figure 10 shows that the α and β mRNAs increase gradually from day 1 to day 7 and are significantly increased at day 14. The subunit mRNAs levels then plateau by the third to fourth week after gonadectomy. Of interest, the serum LH level increases markedly at day 3, a change that precedes the increase in mRNAs in the pituitary gland. The rapid increase in serum LH levels in castrate rats is well known,

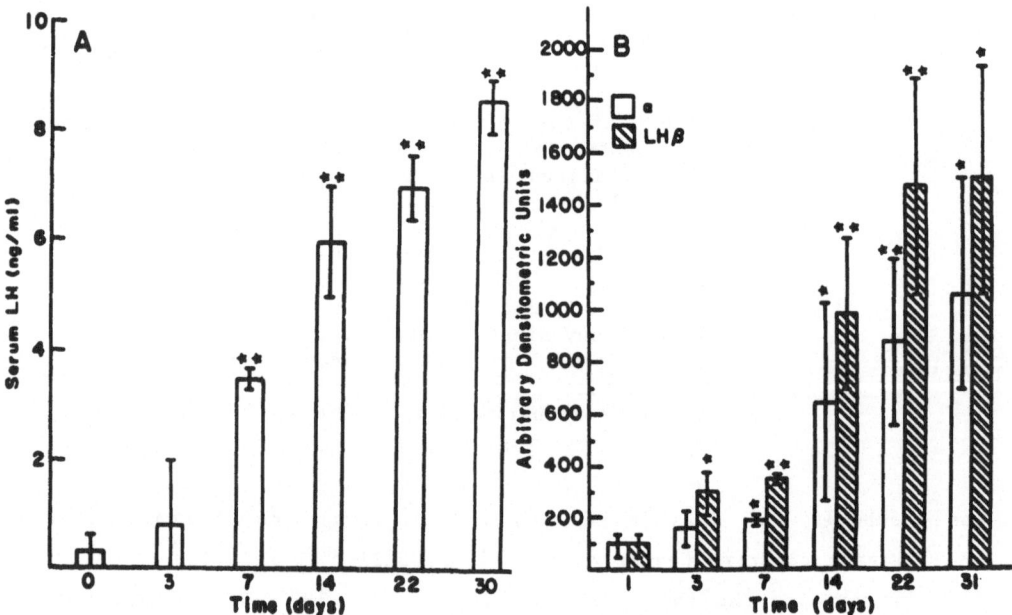

Fig. 11. The effect of ovariectomy on serum LH and subunit mRNA levels. Rats were ovariectomized and sacrificed at various times after surgery. Panel A shows the serum LH levels at various times after ovariectomy. Panel B shows the levels of α and LHβ mRNAs at the same times. They are expressed and normalized as discussed in the legend to Figure 10. Bars represent means ± S.D. (n = 3-5). Stars indicate statistical significance of data points compared to control levels: * = p<0.05; and ** = p<0.01. (Reprinted from Gharib et al., 1986.)

reflecting the effect of decreasing testosterone on increasing the sensitivity of the gonadotrope to endogenous GnRH which results in increased LH release. However, as seen in this experiment, this sensitivity is shown only with regard to secretion but not synthesis of LH. Similar studies performed in the ovariectomized female rat are shown in Figure 11. Again, gradual increases from day 1 to day 14 of the α and LHβ mRNAs are observed between day 7 and day 14 with a plateau at 3 weeks. These increases appear to parallel the increases in serum LH.

The second experimental paradigm involves the treatment of gonadectomized animals with either testosterone or estradiol. In these experiments castrate rats, 3 weeks after surgery, were treated with testosterone propionate at 500μg/100g bw/d for varying times up to 7 days.

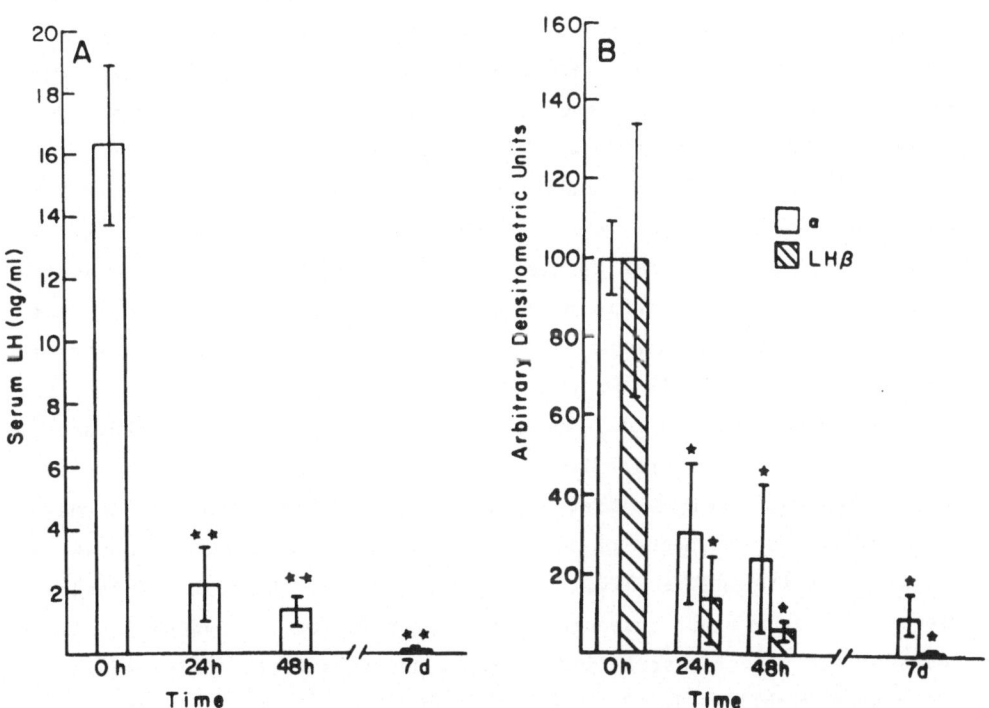

Fig. 12. Effect of testosterone replacement in castrate rats. Castrate rats, 3 weeks after surgery, were treated with testosterone propionate as indicated in the text. Panel A indicates the serum LH values and panel B the α and LHβ mRNA levels at various times after hormonal replacement. For other details, see legend to Figure 10. Bars represent means ± S.D. (n = 3-5). Stars indicate statistical significance of data points compared to control levels: * = p<0.05; and ** = p<0.01. (From Gharib et al., 1986.)

Figure 12 shows that the LHβ mRNA level decreases markedly even after 24 hours of treatment. However, the α-subunit mRNA level, though diminishing in time, responds more slowly. An additional point is that the LHβ mRNA declines to 15% of control levels, whereas the α mRNA decreases to 30% of control levels at 24 hours. This disparity in relative rates of decline of the subunit mRNAs persists even after 7 days of hormonal replacement. Thus, the effect of testosterone to decrease the subunit mRNAs is a striking and rapid one with an effect which is greater on the LHβ than on the α-subunit mRNA. Similar results were obtained in the studies of the ovariectomized treated with 17β-estradiol-3-benzoate at 10μg/100g bw/d. These results are shown in Figure 13. Again, at 24 hours the α-subunit mRNA has declined to 30% of control and LHβ mRNA to 20% of control levels. Similarly, the LHβ mRNA is virtually undetectable at 7 days whereas the α-subunit mRNA is still approximately 15% of control levels. Hence, as seen in males, the female data indicate that the LHβ subunit mRNA is more rapidly regulated that the α-subunit. All these changes occur in parallel with the decline of the LH secreted into the plasma. Thus, these data demonstrate that gonadal sex steroid hormones regulate the synthesis of LH in vivo at the pre-translational level. Studies carried out in the sheep by Nilson et al. (1983) and Landefeld et al. (1984) and Landefeld and Kepa (1984) indicate similar results. Counis et al. (1983) and Corbani et al. (1984), using translational assays, have found analogous results for the regulation of LH biosynthesis at the pre-translational level in the rat. It is unclear at this moment whether the regulation by gonadal steroid hormones on LH synthesis at the pre-translational level occurs at the level of gene transcription. However, recent data from our laboratory, in collaboration with Drs. Margaret Shupnik, E. Chester Ridgway, and Joel F. Habener, indicate that in a related system, there is a major effect of thyroid hormones on the inhibition of transcription of the TSH subunit genes (Shupnik et al., 1985, 1986). Hence, the negative-feedback effect of these various hormones on glycoprotein hormone synthesis likely occurs at the gene transcription level in addition to possible regulation at the mRNA stability level. Table I reviews the data on the regulation of LH and FSH biosynthesis by steroid hormones in various animal models. It is clear from these results that sex steroid hormones regulate the synthesis of both LH and FSH at the pre-translational levels in a negative fashion. In addition, there is some evidence in sheep that α and LHβ mRNAs change significantly during the estrous cycle. Landefeld et al. (1985) have shown that α-subunit mRNA increases 24 hours before the surge in the ewe and persists in its elevation until the end of the estrous

258

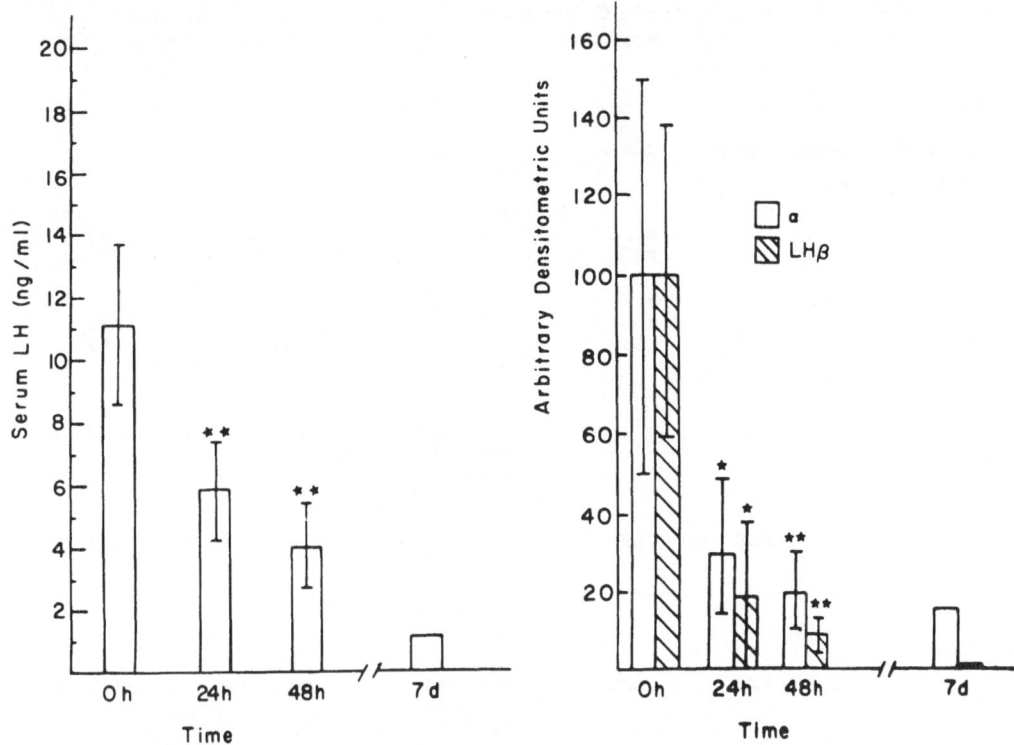

Fig. 13. Effect of estrogen replacement in ovariectomized rats. Rats, ovariectomized 3 weeks prior to experimentation, were treated with 17-β-estradiol-3-benzoate and animals sacrificed at various times as described in the text. LH levels were determined (panel A) and pituitary levels of α and LHβ mRNAs (panel B) were assayed. See details of legend to Figure 10. Values are expressed as means ± S.D. (n = 3-5). Stars indicate statistical significance of data points compared to control levels: * = p<0.05; and ** = p<0.01. (From Gharib et al., 1986.)

cycle. In contrast, Tepper et al. (1984) have demonstrated, in a preliminary report, that there is no change in the steady-state levels of LHβ mRNA in the rat during the estrous cycle. However, they have shown that there may be as great as a 4-fold increase in the transcriptional rate of the LHβ gene at metestrus. Finally, Jonassen and Roberts (1984), in another preliminary report, have shown that a single dose of GnRH in a cell culture system can increase LH gene transcription by 5- to 20-fold. These data taken in total indicate that many factors including sex steroid hormones and GnRH influence the activity of the genes encoding α and LHβ subunits. It is likely that the careful integration of these events results in the ultimate levels of α and LHβ subunit mRNAS.

Table I. Regulation of gonadotropin subunit mRNAs by sex steroid hormones and GnRH and in the estrous cycle.

hormone/physio	animal	mRNAs		effects	
SEX STEROID					
	sheep	*α*	*(T)*	*decreased*	*Landefeld et al, 1983*
	"	*α, LH β*		*"*	*Nilson et al, 1983*
	"	*FSH β*	*(T)*	*"*	*Alexander and Miller, 1982*
	rat	*α, LH β*		*decreased*	*Gharib et al, 1986*
	"	*α, LH β*	*(T)*	*"*	*Counis et al, 1983*
	"	*LH β*		*"*	*Tepper et al, 1984*
ESTRUS CYCLE					
	sheep	*α*		*increased @ 24h pre-surge; persistent elevation at end*	*Landefeld et al, 1985*
	rat	*LH β*		*no change in mRNAs; 4X increase in TX @ metestrus*	*Tepper et al, 1984*
GnRH					
	rat	*LH β*		*5--20X increase in TX*	*Jonassen and Roberts, 1984*

Summary data from several groups in two species indicate that various hormones regulate the subunit mRNA levels at the pre-translational level. T = translation assay; TX = transcription level.

SUMMARY

We have examined the question of how hormones regulate the biosynthesis of gonadotropins. In particular we have studied the effects of gonadal sex steroid hormones on the regulation of LH subunit gene expression to determine whether they may occur at the pre-translational level. Success in the molecular cloning of cDNAs in genes encoding the subunits of LH has allowed such studies. It is now known that the subunit genes are located in separate genes on separate chromosomes in man and mouse. These genes must be coordinately expressed in a tissue-specific fashion in gonadotropes to yield subunit mRNAs which are ultimately translated to form the protein backbones of the subunits. It is clear that the gonadal sex steroid hormones in both castration and castration-replacement experimental paradigms negatively regulate the subunit mRNAs in vivo in a rapid and effective manner. Also, it is interesting to note that α-subunit RNA is regulated to a lesser extent than the LHβ. This observation is reminiscent of those previously observed in the studies of the biosynthesis of TSH in which the subunit mRNA is less well controlled than the TSHβ mRNA. These studies were performed initially using thyrotropic tumor which lacked confounding gonadotropes. However, studies have also been performed in the pituitary gland of the hypothyroid mouse with similar results. Hence, it appears that the α subunit gene is also under regulation by hormones but to a lesser extent than the LHβ. These findings provide hope that future studies will allow us to understand further the molecular mechanisms involved in the regulation of these genes by various hormonal influences.

ACKNOWLEDGEMENTS

We thank Nancy Patterson for expert assistance in typing this manuscript. We also acknowledge the major contributions of Drs. John E. Godine and J. Larry Jameson to this work. This effort is supported in part by NIH grant HD19938.

REFERENCES

Acher, R., 1980, Molecular evolution of biologically active polypeptides, Proc. R. Soc. London (Biol.), 200:21-43.

Alexander, D. C., and Miller, W. L., 1982, Regulation of ovine follicle-stimulating hormone β chain mRNA by 17β-estradiol in vivo and in vitro, J. Biol. Chem., 257:2282-2286.

Beck, A., Vellucci, V., and Curry, K., 1985, Cloning and expression of DNAs coding for follicle-stimulating hormone, DNA, 4:76.

Boothby, M., Ruddon, R. W., Anderson, C., McWilliams, D., and Boime, I., 1982, A single gonadotropin α subunit gene in normal tissue and tumor derived cell-lines, J. Biol. Chem., 256:5121-5127.

Carr, F. E., Need, L. R., Shin, L., and Chin, W. W., 1985, Isolation and characterization of the gene encoding the β-subunit of rat thyrotropin, 9th Intl. Thyroid Cong., Sao Paulo, Brazil, Abstract #13.

Carr, F. E., and Chin, W. W., 1985, Absence of detectable chorionic gonadotropin mRNAs in rat placenta during gestation, Endocrinology, 116:1151-1157.

Chin, W. W., 1985, Organization and expression of glycoprotein hormones gene, in: "The Pituitary Gland," H. Imura, ed., Raven Press, New York, pp. 103-125.

Chin, W. W., Godine, J. E., Klein, D. R., Chang, A. S., Tan, L. K., and Habener J. F., 1983, Nucleotide sequence of the cDNA encoding the precursor of the β subunit of rat lutropin, Proc. Natl. Acad. Sci. USA, 80:4649-4653.

Corbani, M., Counis, R., Starzec, A., and Jutisz, M., 1984, Effect of gonadectomy on pituitary levels of mRNA encoding gonadotropin subunits and secretion of luteinizing hormone, Mol. Cell. Endo., 35:83-87.

Counis, R., Corbani, M., and Jutisz, M., 1983, Estradiol regulates mRNAs encoding precursors to rat lutropin (LH) and follitropin (FSH) subunits, Biochem. Biophys. Res. Comm., 114:65-72.

Darnell, J., 1982, Variety in the level of gene control in eucaryotic cells, Nature, 297:365-371.

Fiddes, J. C., and Goodman, H. M., 1980, The cDNA for the β-subunit of human chorionic gonadotropin suggests evolution by gene by read-through into the 3'-untranslated region, Nature, 286:684-687.

Fiddes, J. C., and Goodman, H. M., 1981, The gene encoding the common alpha subunit of the four human glycoprotein hormones, J. Molec. App. Gen., 1:3-18.

Fiddes, J. C., and Talmadge, K., 1984, Structure, expression, and evolution of the genes for the human glycoprotein genes, Recent Prog. Horm. Res., 40:43-78.

Gharib, S. D., Bowers, S. M., Need, L. R., and Chin, W. W., 1986, Regulation of rat luteinizing hormone (LH) subunit mRNAs by gonadal-steroid hormones, J. Clin. Invest., 77:582-589.

Godine, J. E., Chin, W. W., and Habener, J. F., 1982, α subunit of rat pituitary glycoprotein hormones. Primary structure of the precursor determined from the nucleotide sequence of cloned DNAs, J. Biol. Chem., 257:8368-8371.

Goodwin, R. G., Moncman, C. L., Rottman, F. M., and Nilson, J. H., 1983, Characterization and nucleotide sequence of the gene for the common α-subunit of the bovine pituitary glycoprotein hormones, Nucleic Acids Res., 11:6873-6882.

Hayashizaki, Y., Miyai, K., Kato, K., and Matsubara, K., 1985, Molecular cloning of the human thyrotropin-β subunit gene, FEBS Lett., 188:394-400.

Jameson, J. L., Chin, W. W., Hollenberg, A. N., Chang, A. S., and Habener,

J. F., 1984, The gene encoding the β-subunit of rat luteinizing hormone: Analysis of gene structure and evolution of nucleotide sequence, J. Biol. Chem., 259:15474-15480.

Jonassen, J. A., and Roberts, J. L., 1984, Transcription of the rat beta LH gene is stimulated by GnRH, 14th Annual Meeting of the Society for Neuroscience, Abstract #308.4.

Julier, C., Weil, D., Couillin, P., Cote, J., Nguyen, V. C., Foubert, C., Boue, A., and Junien, C., 1984, The beta chorionic gonadotropin-beta luteinizing hormone gene cluster maps to human chromosome 19, Hum. Genet., 67:174-177.

Kalra, P. S., and Kalra, S. P., 1985, Control of gonadotropin secretion, in: "The Pituitary Gland", H. Imura, ed., Raven Press, New York, pp. 189-220.

Keutmann, H. T., McIlroy, P. J., Bergert, E. R., and Ryan, R. J., 1983, Chemically deglycosylated human chorionic gonadotropin subunits: Characterization and biological properties, Biochemistry, 22:3067-3072.

Kourides, I. A., Barker, P. E., Gurr, J. A., Pravtcheva, D. D., and Ruddle, F. H., 1984, Assignment of the genes for the α and β subunits of thyrotropin to different mouse chromosomes, Proc. Natl. Acad. Sci. USA, 81:517-519.

Landefeld, T. D., Kepa, J., and Karsch. F. J., 1983, Regulation of α subunit synthesis by gonadal steroid feedback in the sheep anterior pituitary, J. Biol. Chem., 258:2390-2393.

Landefeld, T. D., and Kepa, J., 1984, Regulation of LH beta subunit mRNA in the sheep pituitary gland during different feedback states of estradiol, Biochem. Biophys. Res. Comm., 122:1307-1313.

Landefeld, T.D., Kepa, J., and Karsch, F., 1984, Estradiol feedback effects on the α-subunit mRNA in the sheep pituitary gland: Correlation with serum and pituitary luteinizing hormone concentrations, Proc. Natl. Acad. Sci., 81:1322-1326.

Landefeld, T., Kaynard, A., and Kepa, J., 1985, Pituitary α-subunit messenger ribonucleic acid remains elevated during the latter stages of the preovulatory luteinizing hormone surge, Endocrinology, 117:934-938.

Mason, A. J., Hayflick, J. S., Long, N., Esch, F., Ueno, N., Ying, S.-Y., Guillemin, R., Niall, H., and Seeburg, P., 1985, Complementary cDNA sequences of ovarian follicular fluid inhibin show precursor structure and homology with transforming growth factor-β, Nature, 318:659-663.

Moore, W. T., Jr., Burleigh, B. D., and Ward, D. N., 1980, Chorionic gonadotropin; Comparative studies and comments on relationships to other glycoprotein hormones, in: "Chorionic Gonadotropins," S. J. Segal, ed., Plenum Press, New York, pp. 89-126.

Naylor, S. L., Chin, W. W., Goodman, H. M., Lalley, P. A., Grzeschik, K. H., and Sakaguchi, A. Y., 1983, Chromosome assignment of genes encoding the α and β subunits of glycoprotein hormone in man and mouse, Som. Cell. Gen., 9:757-770.

Naylor, S. L., Sakaguchi, A. Y., McDonald, L., Todd, S., and Lalley, P. A., 1986, Mapping the thyrotropin β subunit gene in man and mouse, Som. Cell. Genet., in press.

Nilson, J. H., Nejedlik, M. T., Virgin, J. B., Crowder, M. E., and Nett, T. M., 1983, Expression of α subunit and luteinizing hormone β genes in the ovine anterior pituitary. Estradiol suppresses accumulation of mRNAs for both α subunit and luteinizing hormone β, J. Biol. Chem., 258:12087-12090.

Parsons, T. F., and Pierce, J. G., 1984, Free alpha-like material from bovine pituitaries: Removal of its O-linked oligosaccharide permits combination with lutropin-beta, J. Biol. Chem., 259:2662-2666.

Pierce, J. G., and Parsons, T. F., 1981, Glycoprotein hormones: Structure and function, Annu. Rev. Biochem., 50:465-495.

Policastro, P., Ovitt, C. D., Hoshina, M., Fukuoka, H., Boothby, M. R., and Boime, I., 1983, The subunit of human chorionic gonadotropin is encoded by multiple genes, J. Biol. Chem., 258:11492-11499.

Sairam, M. R., and Bhargavi, G. N., 1985, A role for glycosylation of the alpha subunit in transduction of biological signal in glycoprotein hormones, Science, 229:65-67.

Shupnik, M. A., Chin, W. W., and Ridgway, E. C., 1986, T3 regulation of thyrotropin subunit genes, in: "Mechanism of Action of Thyroid Hormone", R. Tolman and L. DeGroot, eds., in press.

Shupnik, M. A., Chin, W. W., Habener, J. F., and Ridgway, E. C., 1985, Transcriptional regulation of the thyrotropin subunit genes by thyroid hormones, J. Biol. Chem., 260:2900-2903.

Stewart, M., and Stewart, F., 1977, Constant and variable regions in glycoprotein beta subunit sequences: Implications for receptor binding specificity, J. Mol. Biol., 116:175-179.

Talmadge, K., Boorstein, W. R., and Fiddes, J. C., 1983, The human genome contains seven genes for the β-subunit of chorionic gonadotropin but only one gene for the β-subunit of luteinizing hormone, DNA, 2:279-287.

Talmadge, K., Vamvakopoulos, N. C., and Fiddes, J. C., 1984, Evolution of the genes for the β subunits of human chorionic gonadotropin and luteinizing hormone, Nature, 307:37–40.

Tepper, M. A., Dionne, F. R., Eberwine, J. H., Wilcox, J. N., and Roberts, J. L., 1984, Regulation of beta LH gene expression during the estrous cycle and castration in the rat, 7th International Congress of Endocrinology, Quebec City, Canada, Abstract #2293.

Tepper, M. A., and Roberts, J. L., 1984, Evidence for only one β luteinizing and no β chorionic gonadotropin gene in the rat, Endocrinology, 115:385–391.

Virgin, J. B., Silver, B. J., Thomason, A. R., and Nilson, J. H., 1985, The gene for the β subunit of bovine luteinizing hormone encodes a gonadotropin mRNA with an unusually short 5'–untranslated region, J. Biol. Chem., 260:7072–7077.

Watkins, P., Eddy, R., Beck, A., Vellucci, V., Gusella, J., and Shows, T., 1985, Assignment of the human gene for the β subunit of follicle-stimulating hormone (FSHB) to chromosome 11, Cytogenet. Cell Genet., 40:773.

Weintraub, B. D., Stannard, B. S., and Meyers, L., 1983, Glycosylation of thyroid-stimulating hormone in pituitary tumor cells: Influence of high mannose oligosaccharide units on subunit aggregation, combination and intracellular degradation, Endocrinology, 112:1331–1345.

Whitfield, G. K., Powers, R. E., Gurr, J. A., Wolf, O., and Kourides, I. A., 1985, Isolation of a gene encoding human thyrotropin beta subunit, 9th Intl. Thyroid Cong., Sao Paulo, Brazil, Abstract #14.

STRUCTURE AND EXPRESSION OF HUMAN PLACENTAL HORMONE GENES

Irving Boime, Mark Boothby, Robert B. Darnell, and
Paul Policastro

Departments of Pharmacology and Obstetrics and Gynecology
Washington University School of Medicine
St. Louis, Missouri 63110

INTRODUCTION

One of the important functions of the human placenta is to produce peptide hormones during pregnancy; e.g. human chorionic gonadotropin (hCG) and human placental lactogen (hPL). The appearance of these hormones in maternal serum during pregnancy is quite different. Whereas hCG peaks in the first trimester, hPL reaches maximal levels at term. Since the levels of these hormones differ during the course of gestation, it is apparent that the factors controlling their synthesis are not the same. Thus the human placenta represents a convenient and unique tissue for studying expression of human hormonal genes during development.

Placental lactogen is a single non-glycosylated polypeptide chain which shares greater than 90% homology with human growth hormone. Chorionic gonadotropin consists of two non-identical glycosylated subunits (alpha and beta) linked non-covalently. The amino acid sequence of hCGα is identical to that of the α-subunits contained in human pituitary gonadotropins, and in thyrotropin (Bahl, 1977). The β-subunit confers the unique biologic specificity on each hormone. Although there is also significant homology between the β-subunits (Bahl, 1977), hCGβ contains an extra 29 amino acid peptide at the carboxyl end of the molecule, which is not found in the other β-subunits (Morgan et al., 1975). Although only the dimer is biologically active (Morgan et al., 1974), the free α-subunit is observed at low levels in the serum during the first trimester and progressively increases during gestation. As gestation progresses and absolute levels of hCG decline, the ratio of free to combined α-subunit increases; no free β-subunit is detectable during pregnancy. Thus the accumulation of a net excess of α-subunit seems characteristic of hCG in the placenta (Vaitukaitis, 1974).

Expression of chorionic gonadotropin is evoked in tumorigenesis. A significant number of non-trophoblastic tumors secrete hCG subunits. This is referred to as ectopic hCG production. Cell lines from some of these cancers and from choriocarcinomas (the malignant tumors of trophoblastic cells) produce hCG in vitro (Ruddon et al., 1979) and change production levels in response to physiologic effectors (Hussa et al., 1978). In addition, most cell lines produce more α- than β-subunit (Ruddon et al., 1980), reminiscent of the subunit imbalance of placenta.

A unique feature of the human placenta is its continued differentiation during gestation. Progeny of mitotically active mononucleated cytotrophoblasts fuse to form mitotically inactive syncytiotrophoblast (Pierce and Midgley, 1963; Enders, 1965; Wynn, 1972). Gestational abnormalities of trophoblast development such as choriocarcinoma and hydatidiform mole occur at different stages of trophoblast differentiation. Such changes in patterns of differentiation lead to alterations in the expression of hCG and hPL (Hoshina et al., 1982, 1983).

Despite the extensive information available regarding hCG and hPL, the factors regulating their synthesis is obscure. Here we will attempt to address this point by dealing with the following issues:

1. Structure and organization of the hPL and hCGα- and hCGβ-subunit genes in normal placenta and in trophoblast and non-trophoblast derived tumors. Attempts will be made to determine how genetic organization of the genes might be related to control.

2. What are the mechanisms by which net hCG subunit production is regulated during pregnancy? Are their translational and/or transcriptional controls operative during subunit synthesis?

3. Expression of chorionic gonadotropin during tumorigenesis.

4. The linkage between trophoblast differentiation and placental hormone production.

STRUCTURE AND ORGANIZATION OF HPL AND HCG SUBUNIT GENES

HPL together with human growth hormone (hGH) comprise a multigene family coding for distinct proteins with similar biologic activities. At

268

least two hGH and three hPL genes have been identified, and all are localized to chromosome 17 (Selvanayagam et al., 1984; Kidd and Saunders, 1982; Owerbach et al., 1980; Moore et al., 1982). These genes share 95% sequence homology in the mRNA coding sequence, but yet are expressed in a tissue-specific manner: hGH in anterior pituitary and hPL in the syncytiotrophoblast of the placenta. Expression of two of the hPL genes in vivo resulting in authentic hPL has been confirmed. A third gene has a splice site mutation in the second intron and deviates from a consensus sequence in the promotor regions (Selvanayagam et al., 1984). Such changes suggest that this gene may be a pseudogene.

The hPL gene contains multiple sites from which transcription can initiate. It has been proposed that these sites may play a role in tissue-specific expression of the placental lactogen gene in placenta, reminiscent to that seen with the different mRNA species encoding α-amylase transcribed in salivary and pancreatic tissue from an otherwise identical gene (Young et al., 1981).

Several studies have shown that certain sequences, e.g. enhancers, responsible for tissue-specific expression reside in the 5' flanking sequences of genes. However, hGH and hPL genes share a homologous stretch of over 400 bp in the 5' flanking region. Thus candidate regions for tissue-specific expression of hPL (and hGH) may lie between the hGH and hPL clusters or within intragenic sequences per se. Recently an enhancer sequence has been identified in the 3' flanking region of at least one of the expressed hPL genes (B. Rogers, M. Sobnosky, and G. Saunders, personal communication); this may account for the specificity for hPL activation in the placenta.

STRUCTURE OF HCG SUBUNIT GENES

hCGα

That the amino acid sequence of the hCGα-subunit is identical to that of the α-subunits contained in the pituitary hormones LH, FSH, and TSH raises the question whether or not the α-subunits are encoded by more than one gene. In humans, only one gene bearing hCGα sequences could be detected per haploid complement (Boothby et al., 1981; Fiddes and Goodman, 1981). The structure of this gene is the same in DNA from both first trimester and term placenta (Fig. 1). Polymorphisms were observed for the

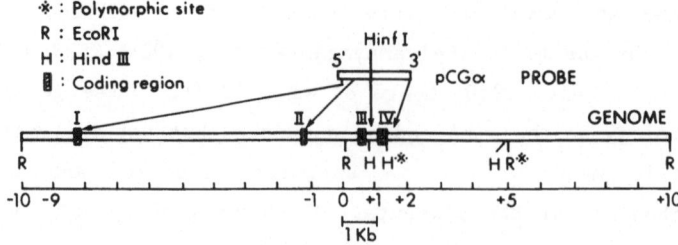

Fig. 1. Restriction map of the hCGα gene including the polymorphic* HindIII (H) and EcoRI (R) sites. The cDNA probe shown, pCGα, corresponds to the entire mRNA encoding the α-subunit.

presence of HindIII in exon 4 and an EcoRI site at position +5 in the 3' flanking sequence (Fig. 1) of the gene. These polymorphic sites were used as markers of tissue genotype in normal placenta and trophoblastic disease (Hoshina et al., 1984).

hCGβ

The hCGβ-subunit is encoded of a multigene family (Boorstein et al., 1982; Policastro et al., 1983), comprised of six genes (or pseudogenes) and a single LHβ gene (Fig. 2; Policastro et al., in press). Regarding the LHβ-subunit, gene sequence analysis indicates that a one-base pair deletion and a two-base pair insertion in hCGβ relative to LHβ coding sequences account for the extended reading frame of hCGβ (Talmadge et al., 1984). To date, there is evidence for the in vivo expression of only two hCGβ genes: CGβ3 and CGβ5 (Talmadge et al., 1984). CGβ1, CGβ2, and possibly CGβ7 are nonfunctional or poorly expressed genes (Policastro et al., 1983; Policastro et al., in press). The reason for the multiplicity of β-subunits is not clear. Though LHβ and hCGβ apparently evolved from the same ancestral coding sequence, there are undoubtedly differences in the regulatory sequences that are associated with differential hCGβ and

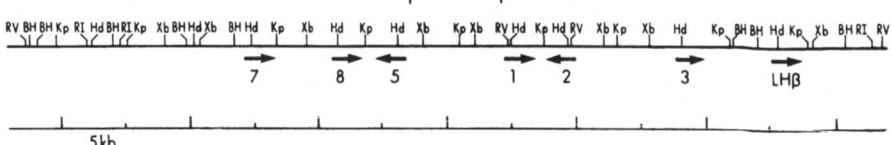

Fig. 2. Complete map of CGβ-LHβ gene cluster. Bracket above map defines a ∼9 Kb region not represented in cloned genomic DNA. Arrows below map show 5'-3' transcription orientation proposed for the six hCGβ genes (1-3, 5, 7, and 8) and single LHβ gene.

270

LHβ expression in placenta and anterior pituitary. Analogous to globin gene switching, the hCGβ gene family is a developmentally regulated system that is sequentially expressed from embryogenesis to a fully developed placenta at term. Alternatively, a different β-subunit gene may be activated in tumorigenesis.

EXPRESSION OF CHORIONIC GONADOTROPIN DURING TUMORIGENESIS

As discussed above, synthesis of hCG subunits, particularly the α-subunit genes, is evoked in a variety of tumors. Of particular interest is ectopic production, since such tumors are not trophoblast derivatives. There is evidence that differences exist in the control of the α gene in trophoblast and non-trophoblast derived cell lines. The induction of α-subunit mRNA expression during the differentiation of the human placenta (Hoshina et al., 1982) is correlated with the exit of those cells from the cell cycle (Midgley et al., 1963), while α transcription in an ectopic secretory tumor cell line (HeLa) is independent of the cell cycle (Darnell, 1984). In both types of cell lines, α-subunit expression responds differently to cAMP and sodium butyrate. Cyclic AMP induces α protein synthesis in choriocarcinoma cell lines (Chou, 1978; Hussa, 1980), but is without effect in HeLa cells (Chou, 1978); butyrate stimulates α-subunit transcription in HeLa cells, but has no effect on α-subunit protein levels in choriocarcinoma cells (Chou, 1978; Hussa, 1980). The morphology of eutopic and ectopic hCG secreting cells is quite different. While mononucleated cells are evident, a truly multinucleated syncytial layer is not observed (Hoshina et al., 1983). Instead, a population of "syncytial-like" or multinucleated cells are seen. Non-trophoblast-derived ectopically hCG producing tumors are devoid of multinucleated cells, and appear to be a homogeneous population of mononucleated cells. This implies that the pathways of cellular differentiation are dissimilar in the two classes of tumors. Regarding hormonal production, choriocarcinomas are always associated with synthesis of both hCG subunits, while hCGα synthesis is the predominant feature of ectopic production. Further evidence linking hCG production with tumorigenesis was reported by Stanbridge et al., (1982). Somatic cell fusions of HeLa cells with normal human fibroblasts result in cessation of hCGα-subunit production and tumorigenicity. When these stable fusions lose a defined pair of chromosomes, tumorigenicity and hCGα production are concomitantly restored.

271

CONTROL OF HCGα EXPRESSION IN EUTOPIC AND ECTOPIC TUMOR CELLS

In choriocarcinoma cells hCGα production is induced by cyclic AMP, whereas in HeLa cells it is induced by sodium butyrate. Using plasmid expression-vector systems we examined the role of 5' flanking hCGα sequences in the differential expression of this gene in eutopic- and ectopic-hCG-producing cells, since sequences governing tissue-specific expression and regulation of some genes are in this region (Chandler et al., 1983; Karim et al., 1984; Majors and Varmus, 1983; Walker et al., 1983).

Plasmid vectors containing hCGα 5'-flanking genomic sequences were fused to the bacterial gene coding for chloramphenicol acetyl transferase (CAT; Howard, 1983; Gorman et al., 1982a, 1982b). These convenient vectors permit the assay of exogenous promotors, and since mammalian cells lack this enzyme, exogenously inserted α-subunit sequences can be distinguished from the endogenous α gene in host cells.

The construct pαCAT, which contained 1.5 Kb of 5'-flanking sequence of the hCGα gene, was transfected into HeLa (ectopic) and JAr (eutopic) cells, and CAT activity determined as a measure of hCGα-dependent transcription. The amount of hCGα expressed from pαCAT paralleled the levels of expression seen with the endogenous α-subunit gene. CAT expression was stimulated in JAr cells by cAMP but not by butyrate, whereas in HeLa cells, synthesis was stimulated by butyrate. These results further imply that there are cell-specific differences in the regulation of α-subunit gene expression.

A series of deletion mutants of pαCAT were constructed that had truncated sequences of 5'-flanking α-subunit DNA. Transfection studies with these constructs indicated that expression and regulation observed with pαCAT was dependent on the presence of 140 bp of 5'-flanking α-subunit DNA. Sequencing of this region revealed the presence of canonical sequences associated with promotor regions such as TATA and CAAT boxes. In addition, the 140 bp flanking sequence contains a 9 bp DNA sequence homologous with the concensus viral enhancer sequence. This may be significant, since tissue-specific expression of a variety of genes, including immunoglobulins (Bannerji et al., 1983; Queen and Baltimore, 1983; Gillies et al., 1983), and chymotrypsin and insulin (Walker et al., 1983), are thought to be dependent on the presence of enhancer-like sequences. Thus an hCGα-subunit enhancer-like sequence may also be associated with cell-specific expression or induction of the hCGα gene in cells synthesizing hCG eutopically and ectopically. Such features of hCGα expression common to both cell types may be involved in the coordinate expression of the α gene and the tumorigenic phenotype observed in each cell type.

MECHANISMS REGULATING THE NET HCG SUBUNIT PRODUCTION DURING PREGNANCY

As discussed above, previous studies have demonstrated an imbalance in placental levels of hCG subunits. Free α-subunit was present in first trimester placenta, and the imbalance was accentuated as gestation approached parturition. Two sets of experiments were performed to assess the control on the synthetic levels of each subunit. Synthesis of the α- and β-subunits was determined by labeling the nascent chains of polysomes derived from first trimester and term placenta, since production from polysomes reflects the biosynthetic rates. The products of these reactions were immunoprecipitated with subunit-specific antisera, and the labeled subunits were quantitated; the ratio of α- to β-subunit synthesized was 1.7 (Boothby et al., 1983). To examine whether this imbalanced synthesis reflected differences in the amount of subunit mRNAs, or differing mRNA translational efficiencies, the ratio of the steady-state levels of these mRNAs was determined (Boothby et al., 1983). These experiments demonstrated the presence of twice as much hCGα mRNA as hCGβ mRNA. In term placenta, the amounts of excess α-subunit are greater than at first trimester; the ratio of α to β mRNAs in term RNA was 12:1. These data show that the progressive rise in production of free hCGα-subunit during gestation must for the most part be attributed to independent regulation of the accumulation of mRNAs for the two subunits. Translational control is not the major determinant for subunit production in vivo; rather the ratio of peptide production is comparable to the ratio of the corresponding mRNAs in first trimester tissue and thus the subunit mRNA levels are the major determinant of the imbalance of hCG subunits observed during gestation.

The data also establish that the observed excess in hCGα-subunit in placental tissue at first trimester is due to a difference in synthetic rate, rather than only from subsequent degradation of the β-subunit after equimolar synthesis of the subunits. Ruddon et al. (1981) studied explants of first trimester placentae that had been maintained in culture for 3 days before pulse-label experiments, and found a somewhat greater imbalance in the subunit secretion than the 2:1 ratio determined here. Although these explant studies may not exactly reflect the in vivo situation, we cannot exclude that some selective degradation of the β-subunit occurs in placentae.

The imbalance of peptide production by placentae contrasts with the findings of Chin et al. (1981) on production of murine thyrotropin subunits by a tumor cell line. They performed pulse-chase experiments and observed that equal labeling of the α- and β-subunits progressed to α-subunit excess as label was chased. Studies on another pituitary hormone, bovine lutropin, by Fetherston and Boime (1982) showed a more complex behavior. When physiologic stimulation of hormone release was greatest, production of β-subunit was comparable to that of α. At lower levels of stimulation, production in both subunits was diminished and the α-subunit level was observed to exceed that of the β substantially. When the production of hCG was stimulated in choriocarcinoma cells by cAMP, the β-subunit was affected more than the α-subunit (Ruddon et al., 1980). Thus, it appears that for several gonadotropins, gross synthesis of both subunits is amenable to stimulation, and that the subunit responds more to such stimuli. Similar conclusions were made by Gurr and Kourides (1984) regarding the biosynthesis of thyrotropin. They suggested that synthesis of TSH parallels the levels of TSHβ mRNA.

Production of hCG by placenta may be promoted by some factor present in lesser concentration at term than at first trimester. Because its population dynamics parallel production of hCG, the cytotrophoblasts could be a source of a factor regulating the expression of hCG subunit genes (as discussed below).

RELATIONSHIP OF TROPHOBLAST DIFFERENTIATION TO HORMONE PRODUCTION

As discussed above, the main populations of placental trophoblasts are a set of mitotically active cells, cytotrophoblasts, that form the syncytial layer. Using in situ hybridization to detect specific mRNAs in individual cells, hCGα mRNA was shown not only to be present in syncytial regions but also in some differentiating cytotrophoblasts in placental villi (Hoshina et al., 1982). The signal for hCGβ mRNA in first trimester placenta was primarily localized to syncytiotrophoblast regions and in a few cytotrophoblasts (Hoshina et al., 1983). These data are in agreement with those of Gaspard and colleagues who demonstrated, using immunofluorescent methods, the presence of hCG in cytotrophoblastic elements (Gaspard et al., 1980). HPL mRNA, by contrast, was exclusively localized to the syncytium (Hoshina et al., 1982). These data imply that the hCG subunit and hPL genes are activated at different stages of placental differentiation.

During pregnancy there is a gradual decrease of cytotrophoblasts until, at term, the syncytiotrophoblast is the major cellular component of placental villi. Thus expression of hCGα- and β-subunit genes may be dependent on the presence of cytotrophoblasts, and the decreasing levels of the subunits at term discussed above are associated with the depletion of these cells. While hCGα and β mRNAs were identified in the syncytial regions, this does not necessarily indicate that hCG is expressed in mature syncytium. HCGα and β mRNAs may be localized to a multinucleated intermediate that cannot be resolved from mature syncytium by the methods used here. Thus during early gestation, when the placenta contains numerous cytotrophoblasts, they would serve to maintain a level of such a transient intermediate. Apparently, the presence of differentiating cytotrophoblasts is not necessary for maintaining the production of hPL in syncytiotrophoblast. It is also clear that hPL is fully expressed only in advanced stages of trophoblast differentiation, whereas, at least in the case of hCGα, expression is initiated prior to syncytial formation.

Data obtained from sections of neoplastic trophoblast tissue are consistent with this model. Choriocarcinoma is composed of trophoblasts that are not fully differentiated. There are clusters of cytotrophoblastic-like and multinucleated cells that are apparently immature syncytial elements. In these sections, hCGα and β mRNA signals were seen in the syncytial-like cells, while the signal for hPL mRNA was almost at background level. Nevertheless, it appears that certain cytotrophoblast elements also contain the α and β mRNAs. These mononucleated trophoblasts may represent the equivalent of an intermediate cell formed during the development of syncytiotrophoblast from progenitor cytotrophoblasts.

In the case of hydatidiform mole, the hPL gene and the hCG subunit genes are expressed. It is significant that one of the distinguishing features of molar tissue, compared with choriocarcinoma tissue, is that the former retains a significant placental villous morphology. Thus it seems reasonable that a certain level of differentiation, such as that indicated by villus formation, is essential for maximal hPL gene expression.

275

Recently it has been demonstrated that oncogenes are expressed in trophoblast (Muller et al., 1983; Pfeifer-Ohlsson, 1984). Muller et al. (1983) demonstrated that c-fms, the oncogene of the McDonough strain of feline sarcoma virus, is expressed in human first trimester and term placentae and cultured choriocarcinoma cells (Muller et al., 1983). These data suggest that expression of fms is associated with advanced stages of trophoblast differentiation. That fms was reported in cultured choriocarcinoma cells implies it is not exclusively localized to mature syncytium. Pfeifer-Ohlsson et al. (1984) showed that the c-myc oncogene was localized to undifferentiated cytotrophoblasts in very early placenta; no signals were seen in the syncytium. Recently Goustin et al., (1985) presented evidence that the placenta expresses the c-cis oncogene.

Taken together, we proposed the following scheme for the coupled events of hormone expression and trophoblast differentiation (Hoshina et al., 1985): The first stage of differentiation (i.e., the proliferation of undifferentiated cytotrophoblasts) is associated with production of myc; the hormonal genes remain unexpressed. The next stage involves formation of committed cytotrophoblasts which is accompanied with activation of the hCGα gene; at this point myc expression is attenuated. Later, these cells further differentiate into an intermediate which begins to express the β-subunit. This cell-type thus elaborates both subunits and is similar to the multinucleated structure expressing the hCG subunits in choriocarcinoma. Coincident with the formation of this intermediate is synthesis of other oncogenes (e.g., fms). The population of these intermediate cells exists transiently, differentiating further with subsequent attenuation of the hCGβ gene. Expression of the hPL gene may begin in a multinucleated intermediate trophoblast cell, but it is not fully activated until a true syncytium is formed. This hypothesis predicts that formation of the syncytial layer is not a prerequisite for hCG synthesis. To test this and other predictions derived from this model, a cell culture system employing normal placental cytotrophoblasts, which would allow us to examine their differentiation in vitro, is essential. Such in vitro models could elucidate the factors governing differentiation and regulation of placental hormones during gestation and in tumorigenesis.

ACKNOWLEDGEMENTS

This research was supported by NIH grant HD13481 and a grant from the Monsanto Corporation.

REFERENCES

Bahl, O., 1977, Human chorionic gonadotropin, its receptor and mechanism of Action, Fed. Proc., 36:2119-2129.

Bannerji, J., Olson, L., and Schaffner, W., 1983, A lymphocyte-specific cellular enhancer is located downstream of the joining region in immunoglobulin heavy chain genes, Cell, 33:729-740.

Boorstein, W., Vamvakopoulos, N., and Fiddes, J., 1982, Human chorionic gonadotropin is encoded by at least light genes arranged in tandem and inverted pairs, Nature, 300:419-422.

Boothby, M., Kukowska, J., and Boime, I., 1983, Imbalanced synthesis of human choriogonadotropin and subunits reflects the steady-state levels of the corresponding mRNAs, J. Biol. Chem., 258:9250-9253.

Boothby, M., Ruddon, R., Anderson, C., McWilliams, D., and Boime, I., 1981, A single gonadotropin α-subunit gene in normal tissue and tumor-derived cell lines, J. Biol. Chem., 256:5121-5127.

Chandler, V. L., Maler, B. A., and Yamamato, K. R., 1983, DNA sequences bound specifically by glucocorticoid receptor in vitro render a heterologous promotor hormone responsive in vivo, Cell, 33:489-499.

Chin, W., Maloof, F., Habener, J., 1981, Thyroid-stimulating hormone biosynthesis, J. Biol. Chem., 256:3059-3066.

Chou, J. Y., 1978, Establishment of clonal human placental cells synthesizing human choriogonadotropin, Proc. Natl. Acad. Sci. USA, 75:1854-1858.

Darnell, R. B., 1984, Independent regulation by sodium butyrate of gonadotropin alpha gene expression and cell cycle progression in HeLa cells, Mol. Cell. Biol., 4:829-839.

Enders, A., 1965, Formation of syncytium from cytotrophoblast in the human placenta, Obstet. Gynecol., 25:378-386.

Fetherston, J., and Boime, I., 1982, Synthesis of bovine lutropin in cell-free lysates containing pituitary microsomes, J. Biol. Chem., 257:8143-8147.

Fiddes J., and Goodman, H., 1981, The gene encoding the common α-subunit of the four human glycoprotein hormones, J. Molec. Appl. Genet., 1:3-18.

Gaspard, V., Hustin, J., and Rentes, A., 1980, Immunofluorescent localization of placental lactogen, chorionic gonadotropin and its α- and β-subunits in organ cultures of human placenta, Placenta, 1:135-148.

Gillies, S. D., Morrison, S. L., Oi, V. T., and Tonegawa, S., 1983, A tissue-specific transcription enhancer element is located in the major intron of a rearranged immunoglobulin heavy chain gene, Cell, 33:717-728.

Gorman, C. M., Merlino, G. T., Willingham, M. C., Pastan, I., and Howard, B. H., 1982b, The rous sarcoma virus long terminal repeat is a strong promoter when introduced into a variety of eukaryotic cells by DNA-mediated transfection, Proc. Natl. Acad. Sci. USA, 79:6777-6781.

Gorman, C. M., Moffat, L. F., and Howard, B. H., 1982a, Recombinant genomes which express chloramphenicol acetyl transferase in mammalian cells, Mol. Cell. Biol., 2:1044-1051.

Goustin, A., Betsholtz, C., Ohlsson, S., Persson, H., Rydnect, J., Bywater, M., Holmgren, G., Heldin, C., Westermak, B., and Ohlsson, R., 1985, Co-expression of the sis and myc proto-oncogenes in developing human placenta suggests autocrine control of trophoblast growth, Cell, 41:301-312.

Gurr, J., and Kourides, I., 1983, Regulation of thyrotropin biosynthesis, J. Biol. Chem., 258:10208-10211.

Hoshina, M., Boothby, M. and Boime, I., 1982, Cytological localization of chorionic gonadotropin α and placental lactogen mRNAs during development of the human placenta, J. Cell Biol., 93:190-198.

Hoshina, M., Boothby, M., Hussa, R., Pattillo, R., Camel, H., and Boime, I., 1985, Linkage of human chorionic gonadotropin and placental lactogen biosynthesis to trophoblast differentiation and tumorigenesis, Placenta, 6:163-172.

Hoshina, M., Hussa, R., Pattillo, R., and Boime, I., 1983, Cytological distribution of chorionic gonadotropin subunit and placental lactogen mRNA in neoplasms derived from human placenta, J. Cell Biol., 97:1200-1206.

Hoshina, M., Boothby, M., Hussa, R., Pattillo, R., Camel, M., and Boime, I., 1984, Segregation patterns of polymorphic restriction sites of the gene encoding the α-subunit of human chorionic gonadotropin in trophoblastic disease, Proc. Natl. Acad. Sci. USA, 81:2504-2507.

Howard, B. H., 1983, Vectors for introducing genes into cells of higher eukaryotes, Trends in Biochem. Sci., 8:209-212.

Hussa, R. O., 1980, Biosynthesis of human chorionic gonadotropin, Endocrine Rev., 1:268-294.

Hussa, R., Pattillo, R. Rueckert, A., Scheuerman, K., 1978, Effects of butyrate and dibutyryl cyclic AMP on hCG-secreting trophoblastic and non-trophoblastic cells, J. Clin. Endocrinol. Metab., 46:69-76.

Karin, M., Haslinger, A., Holtgreve, H., Richards, R., Krauter, P., Westphal, H. M., Beato, M., 1984, Characterization of DNA sequences through which cadmium and glucocorticoid hormones induce human metallothionein-IIa gene, Nature, 308:513-519.

Kidd, V., and Saunders, G., 1982, Linkage arrangement of human placental lactogen and growth hormone genes, J. Biol. Chem., 257:10673-10680.

Majors, J., and Varmus, H., 1983, A small region of the mouse mammary tumor virus long terminal repeat confers glucocorticoid hormone regulation on a linked heterologous gene, Proc. Natl. Acad. Sci. USA, 80:5866-5870.

Midgely, A. R., Jr., Pierce, G. B., Jr., Deneau, G. A., and Gosling, J. R. G., 1963, Morphogenesis of syncytiotrophoblast in vivo: An autoradiographic demonstration, Science, 141:349-350.

Moore, D., Conkling, M., and Goodman, H., 1982, Human growth hormone: A multigene family, Cell, 29:285-288.

Morgan, F., Birken, S., and Canfield, R., 1975, The amino acid sequence of human chorionic gonadotropin, J. Biol. Chem., 250:5247-5258.

Morgan F., Canfield, R., Vaitukaitis, J., and Ross, G., 1974, Properties of the subunits of human chorionic gonadotropin, Endocrinology 94:1601-1606.

Muller, R., Tremblsy, J., Adamson, E., and Verma, I., 1983, Tissue and cell type expression of two human c-onc genes, Nature, 304:454-456.

Owerbach, D., Rutter, W., Martial, J., Baxter, J., and Shows, T., 1980, Genes for growth hormone, chorionic somatomammotropin and growth hormone-like gene on chromosome 17 in humans, Science, 209:289-292.

Pfeifer-Ohlsson, S., Goustin, A., Rydnect, J., Bjersing, L., Wahlstrom, T., Stehlin, P., and Ohlsson, R., 1984, Spatial and temporal pattern of cellular myc oncogene expression in developing human placenta: Implications for embryonic cell proliferation, Cell, 38:585-596.

Pierce, G., and Midgley, A., 1963, The origin and function of human syncytiotrophoblastic giant cells, Amer. J. Pathol., 43:153-173.

Policastro, P., Daniels-McQueen, S., Carle, G., Boime, I., A map of hCGβ-LHβ gene cluster, J. Biol. Chem., in press.

Policastro, P., Ovitt, C., Hoshina, M., Fukuoka, H., Boothby, M., and Boime, I., 1983, The β-subunit of human chorionic gonadotropin is encoded by multiple genes, J. Biol. Chem., 258:11492-11499.

Queen, C., and Baltimore, D., 1983, Immunoglobulin gene transcription is activated by downstream sequence elements, Cell, 33:741-748.

Ruddon, R., Anderson, C., and Meade-Coburn, K., 1980, Stimulation of synthesis and secretion of chorionic gonadotropin subunits by eutopic and ectopic hormone-producing human cell lines, Cancer Res., 40:4519.

Ruddon, R., Anderson, C., Meade, K., Aldenferer, P., and Neuwald, P., 1979, Content of gonadotropins in cultured human malignant cells and effects of sodium butyrate treatment on gonadotropin secretion of HeLa cells, Cancer Res., 39:3885-3892.

Ruddon, R., Hortel, R., Peters, B., Anderson, C., Huot, R., and Stromberg, K., 1981, Biosynthesis and secretion of chorionic gonadotropin subunits by organ cultures of first trimester placenta, J. Biol. Chem., 256:11389-11392.

Selvanayagam, C., Tsai, S., Tsai, M., Selvanayagam, P., and Saunders, G., 1984, Multiple origins of transcription for the human placental lactogen genes, J. Biol. Chem., 259:14642-14646.

Stanbridge, E. J., Channing, J. D., Doerson, C., Nishimi, R. Y., Pechl, D. M., Weissman, B. E., and Wilkinson, J. E., 1982, Human cell hybrids: Analysis of transformation and tumorigenicity, Science, 215:252-259.

Talmadge, K., Boorstein, W., Vamvakopoulos, N., Gething, M., and Fiddes, J., 1984, Only 3 of the 7 human chorionic gonadotropin beta subunit genes can be expressed in the placenta, Nucleic Acids Res., 12:8415-8436.

Talmadge, K., Vamvakopoulos, M., and Fiddes, J., 1984, Evolution of human chorionic gonadotropin beta-subunit: Gene sequence comparison with human luteinizing hormone beta-subunit, Nature, 307:37-40.

Vaitukaitis, J., 1974, Changing placental concentrations of human chorionic gonadotropin and its subunits during gestation, J. Clin. Endocrinol. Metabolism, 38:755-760.

Walker, M. D., Edlund, T., Boulet, A. M., and Rutter, W. J., 1983, Cell-specific expression controlled by the 5'-flanking region of insulin and chymotrypsin genes, Nature, 306:557-561.

Wynn, R., 1972., Cytotrophoblastic specialization: An ultrastructural study of the human placenta, Am. J. Obstet. Gynecol., 114:339-355.

Young, R., Hagenbuchle, and Schibler, V., 1981, A single mouse α-amylase gene specifies two different tissue-specific mRNAs, Cell, 23:451-460.

STRUCTURE AND REGULATED EXPRESSION OF BOVINE PROLACTIN

AND BOVINE GROWTH HORMONE GENES

Fritz Rottman, Sally Camper, Edward Goodwin, Robert Hampson,
Robert Lyons, Dennis Sakai, Richard Woychik, and Yvonne Yao

Department of Molecular Biology and Microbiology
Case Western Reserve University
Cleveland, Ohio 44106

INTRODUCTION

Prolactin and growth hormone are members of a gene family which has been extensively studied during the past several years (Miller and Eberhardt, 1983). Along with the gonadotropins, the members of this gene family are primarily expressed in the anterior pituitary. The expression of these genes and their corresponding regulatory factors has been of interest from the perspective of differential regulation during developmental stages, hormonal effects on gene expression, and the possibility of expression at ectopic sites other than the pituitary (Gluckman et al., 1981). Studies described in this manuscript are restricted to the growth hormone and prolactin genes and more specifically to aspects of gene structure which influence the level of their respective mRNAs.

Ultimately, the spectrum of proteins made in any differentiated tissue, such as the anterior pituitary, is a direct reflection of the steady-state levels of mRNA contained in that tissue. Collectively, bovine prolactin (bPRL) and bovine growth hormone (bGH) mRNAs account for well over 50% of the polyA-containing RNA in the bovine anterior pituitary (Nilson et al., 1983). This immediately raises questions as to how these two mRNAs attain such high levels of expression in this tissue. Three possible levels of regulation which influence steady-state levels of mRNA include transcriptional regulatory signals, processing signals and stability of mature mRNA (Nevins, 1983). These regulatory events could also play a role in the differential expression of these two genes, as observed in embryogenesis (Nilson, et al., 1983). Aspects of each of

281

these potential regulatory events for the bPRL and bGH genes are considered in more detail in the sections that follow.

Although transcriptional rates, as indicated by nuclear runoff experiments, are relatively low for both bPRL and bGH genes (Maurer, 1982; Evans et al., 1982; Barinaga et al., 1983), the prolactin promoter is still sufficiently active to be used in transfection studies using chimeric genes to assess tissue-specific and hormonal regulatory signals. Processing events include the reactions involved in the conversion of nuclear mRNA precursors to mature forms of the cytoplasmic mRNA. Polyadenylation of the mRNA precursor is one essential step in mRNA processing, and the features of the polyadenylation signal also can be defined by transfection of chimeric genes. The polyadenylation signal contained in the bGH gene is particularly well suited for such studies. Splicing of mRNA precursors is a frequent but not universal event in mRNA maturation. Where intervening sequences occur, splicing is usually necessary to yield stable cytoplasmic messengers. Both GH and PRL mRNAs undergo major as well as minor alternative splicing events. Finally, mRNA stability plays a major role in determining the steady-state level of a given message and estimated half-lives range from several days for vitellogenin mRNA to minutes for histone and myc mRNAs. Growth hormone mRNA appears to lie at one end of this spectrum, being a very stable mRNA; this likely plays a major role in the accumulation of high steady-state levels of this message. Bovine GH mRNA is a good candidate for defining some of the features of a stable mRNA.

Collectively, these processes represent several examples of potential transcriptional and post-transcriptional mechanisms for the regulation of bGH and bPRL expression, but should definitely not be viewed as exhaustive. For example, modified nuclear transport and stabilization via interaction with specific proteins, to name just a few, represent other mechanisms for potential regulation. What follows is a description of several studies which utilize the transfection of cloned chimeric genes in an attempt to analyze the regulatory signals found in the bPRL and bGH genes.

COMPARISON OF GROWTH HORMONE AND PROLACTIN GENE SEQUENCES BETWEEN SPECIES

Potential regulatory sequences in 5' flanking, 3' flanking and internal regions of cloned genes are frequently deduced by preliminary comparisons of nucleotide sequences between species. Perhaps as a result of the inherent interest in the physiological properties of growth hormone, this gene was an early subject for cloning and expression experiments and consequently has been characterized in several species. In the human, rat, and bovine, the growth hormone gene is contained in a small genomic fragment of approximately 2 Kb and includes 5 exons and 4 introns, (DeNoto et al., 1981; Barta et al., 1981; Woychik et al., 1982). The human GH gene family contains at least one nonexpressed GH-related gene in addition to three closely related chorionic somatomammotropin (CS sequences; Seeburg, 1982), whereas there is a single GH gene sequence in both rat and bovine. The major difference in overall genomic structure of the GH gene in these three species is the existence of a longer intervening sequence (IVS 2) in the rat gene, apparently containing a pol III transcription initiation site (Barta et al., 1981).

Nucleotide sequence comparisons between the flanking and intervening sequence regions of homologous genes in different species have resulted in the description of several conserved sequences. It has been argued that these conserved sequences are involved in regulation of gene expression and that their intrinsic biological importance is reflected in the retention of the sequence during evolution. Comparison of GH genes between these three species (human, rat, and bovine) showed a remarkably conserved sequence of 37 nucleotides in the 5' flanking region of these genes beginning at position -105 relative to the transcription initiation site of the bGH gene (Woychik et al., 1982). There is a maximum variation of only four nucleotides in this sequence when any two of the three members are compared. Assuming a constraint in retention of this sequence for reasons mentioned above, one might expect the sequence from -105 to -141 to be a particularly attractive target for site directed mutagenesis. Unfortunately, as will be discussed below, the 5' flanking region of the bGH gene appears to be inactive in transfection experiments and therefore this hypothesis has been difficult to test directly.

Examination of the 5' flanking region of PRL genes reveals a high degree of sequence homology between the bovine, human, and rat species (Cooke and Baxter, 1982; Camper et al., 1984; Truong et al., 1984). This homology, which extends almost 500 nucleotides upstream from the start site of transcription, may reflect the conservation of short, but overlapping sequence elements, each mediating the actions of one of many effectors known to modulate prolactin transcription. Of note is a 15 nucleotide sequence which is homologous to a subfragment of the 37 nucleotide conserved GH sequence. No other significant homologies are apparent throughout the remainder of the flanking regions of these two pituitary expressed genes. As will be discussed below, the bPRL promoter is active during transfection experiments and is hormonally responsive.

```
              -140        -130        -120        -110
               .           .           .           .
  rGH:       ...T...G......C...............C......
  hCS:       ...T...G......C...............C.......
  hGH:       ...T...G......C....................
  bGH:       ATGAGTGAGAGGAGGTTCTAAATTATCCATTAGCACA
  bPRL:      GAATACCATTCAATGTTTGAAATTATGGGGGTAATCT
  hPRL:      A.........TG.........................
  rPRL:      A.........TG......A.......T....CT....
               .           .           .
              -130        -120        -110
                         GG   CT
15-mer Consensus:    GAG  TT  AAATTAT
                         TC   TG
```

Comparison of nucleotide sequences in the 3' flanking region of GH genes is of interest from the perspective of polyadenylation signals known to be present in this region (Birnstiel et al., 1985). Aside from the existence of GC rich sequences and short nucleotide sequences which are possibly complementary to U4 snRNA, there is no obvious retention of sequence in this region of the GH genes.

Analysis of codon usage of GH transcripts indicates a marked preference for G or C in the redundant third position (Miller and Eberhardt, 1983). This reflects the overall G + C richness of GH mRNAs and possibly correlates with the high degree of secondary structure of these messages. Among the bovine, human, and rat GH genes, the bGH gene shows the strongest preference (82%) for G or C in the third codon position.

ANALYSIS OF 5' FLANKING REGULATORY SEQUENCES BY TRANSFECTION INTO HOMOLOGOUS AND HETEROLOGOUS CELLS

In order to assess the existence of possible regulatory sequences in a more direct manner, we transfected homologous and heterologous cells

284

with chimeric gene constructs containing possible regulatory sequences derived from both the bPRL and bGH genes (Camper et al., 1985; Lyons, 1985). The transcriptional activity of 5' flanking portions of the bPRL and bGH genes contained in chimeric constructs was studied as a first step in defining the tissue specific and hormonal regulation of these genes. To perform these experiments, portions of the 5' flanking regions of bPRL and bGH were ligated to a recorder structural gene coding for the bacterial enzyme chloramphenicol acetyltransferase (CAT; Fig. 1).

The bPRL-CAT constructs were transfected into three cell types: HeLa, COS, and GH_3, a rat pituitary cell line. Promoter activity was determined by examining the level of CAT activity in transfected cells in the presence and absence of added hormones. Details of the experimental procedures are presented in table and figure legends. When the bPRL-CAT construct, containing approximately 1 kb of 5' flanking sequence derived from bPRL, was transfected into HeLa, COS, or GH_3 cells, only the GH_3 cells provided expression of CAT (Table I).

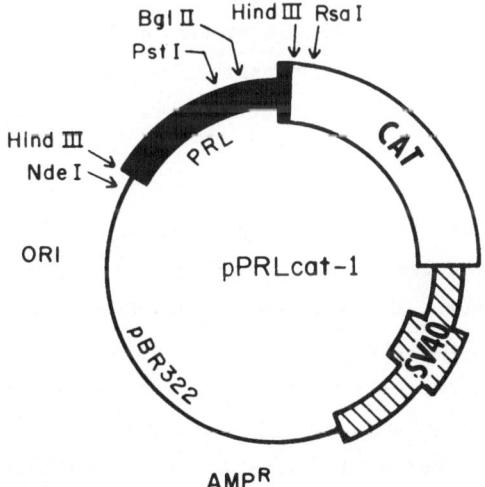

Fig. 1. Bovine prolactin chloramphenicol acetyltransferase fusion clones. pPRL-CAT-1 is depicted with 1 kb of bovine prolactin DNA (darkened box), including the prolactin promoter and 16 nucleotides of exon 1, fused to the bacterial chloramphenicol acetyltransferase gene (open box) with HindIII linkers. Also shown are SV40 sequences necessary for splicing and polyadenylation (hatched area) and pBR322 sequences (thin line), including the bacterial origin of replication and the ampicillin resistance gene.

Table I. Relative expression of pPRL-CAT-1 in three cell lines.

Plasmid	COS 1	HeLa	GH_3
pRSV-CAT	2.8	4.8	83
pSV2-CAT	75	9.4	0.031
pPRL-CAT-1	0.028	0.023	0.20
pPRL-CAT-3	0.021	0.023	0.022
None	0.021	0.023	0.024

Chloramphenicol acetyltransferase activity of the indicated plasmids transfected into each cell line is expressed as per cent [14C]chloramphenicol converted to acetylated forms/1 h/ 25 µg of protein.   pRSV-CAT contains the RSV promoter sequence fused to the bacterial chloramphenicol acetyl-transferase gene while pSV2-CAT contains the SV40 early promoter in a similar construction (Gorman et al., 1982). pPRL-CAT-3 is a control containing the prolactin promoter sequence in opposite transcription orientation relative to the chloramphenicol acetyltransferase.   Chloramphenicol acetyltransferase assays were performed essentially as described by Gorman et al., (1982).

This result suggests that signals may be present in this bPRL flanking region which define tissue specificity and that these signals are only recognized in homologous tissue represented by the rat pituitary cell line.   Furthermore, GH_3 specific expression was retained when this 5' flanking region was shortened to contain only 250 nucleotides of bPRL sequence (data not shown).

In order to assay for the presence of hormonal responsive elements within the 5' flanking region of the bPRL gene, transfected GH_3 cells were treated with 2 hormones, EGF and the glucocorticoid, dexamethasone (DEX), both known to regulate the expression of the PRL gene in vivo.   EGF stimulates prolactin expression in vivo while DEX is inhibitory (Murdoch et al., 1982; Carillo et al., 1985).   Both hormones had an effect on CAT expression in transfected cells, with up to a 12-fold stimulation by EGF and a 50-fold inhibition by DEX (Table II).

The degree of stimulation or inhibition by added hormone was dependent upon the amount of transfected DNA, as might be expected if regulatory sequences contained in the added DNA bind hormone specific factors which are limiting.

When analogous experiments were attempted with constructs containing up to 2,000 bp of 5' flanking sequence derived from the bGH gene, we were

Table II. Relative chloramphenicol acetyltransferase activity
of clones after hormone treatment.

Plasmid	Experiment A 1 µg DNA	Experiment B 2 µg DNA	Experiment C 5 µg DNA	Experiment D 10 µg DNA
pPRL-CAT 1				
No treatment	1.0(0.08)	1.0(0.50)	1.0(4.9)	1.0(6.6)
Dexamethasone- treated	0	0.05	0.07	0.02
EGF-treated	12	6.6	3.0	2.2
pRSV-CAT				
No treatment	1.0(16)	1.0(26)	1.0(82)	1.0(105)
Dexamethasone- treated	1.4	0.89	0.84	n.d.
EGF-treated	0.94	1.8	1.4	0.91

Chloramphenicol acetyltransferase activity for each hormone
treatment is expressed relative to the basal level of
chloramphenicol acetyltransferase activity from pPRL-CAT-1
and pRSV-CAT plasmids transfected with no hormones added.
In parentheses, the conversion of [^{14}C] chloramphenicol to
acetylated forms is presented as percent converted in one
hour, with background subtracted, per 25 µg of protein
extract. The amount of DNA transfected varied as indicated.

unable to obtain any expression of CAT in all transiently transfected
cells tested, including the GH$_3$ cells (Fig. 2). This inactivity of the
bGH flanking region may be a reflection of the inherent weakness of this
promoter, the absence of appropriate enhancers in adjacent sequences or
lack of recognition of the bGH promoter in these heterologous rat cells.
In any event, the 5' flanking region of the bGH gene affixed to a recorder
CAT gene is not expressed in the rat pituitary cells under conditions in
which the comparable bPRL gene is active.

ANALYSIS OF THE POLYADENYLATION SIGNAL CONTAINED IN THE bGH 3' FLANKING
SEQUENCE

In earlier studies we used the enzyme reverse transcriptase to extend
specific primers which spanned the polyA junction to determine precisely
the site of polyA addition in several mRNA species (Sasavage et al., 1980,
1982). Whereas prolactin mRNA and ovalbumin mRNA exhibited microheterogeneity around a single AAUAAA consensus polyadenylation signal, bGH mRNA
was polyadenylated at a single, homogeneous site. The homogeneity of this
polyadenylation event suggested the presence of a strong polyadenylation

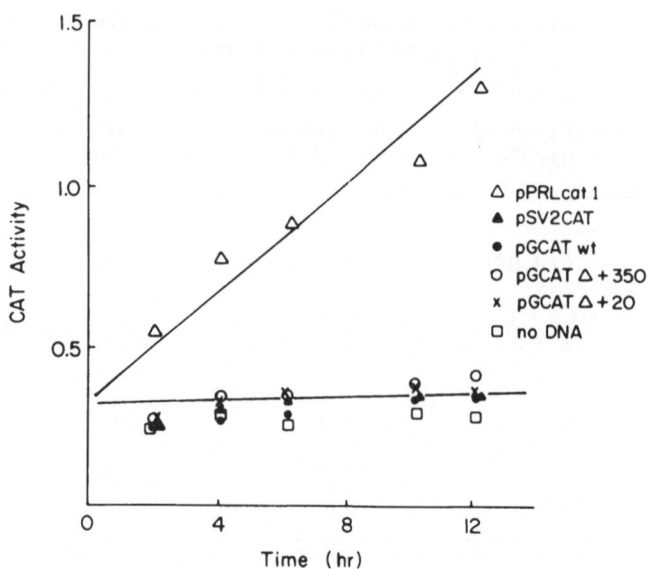

Fig. 2.   Inactivity of bGH promoter in GH$_3$ cells.   Rat
pituitary   GH$_3$ cells   were   transfected   with
GH-promoter/CAT   gene   constructs   containing   2
Kb(pG-CAT wt),   0.35 Kb(pG-CATΔ + 350), and 20
nucleotides (pG-CATΔ + 20) of bGH 5' flanking DNA.
The conversion of chloramphenicol to the acetylated
forms was measured over 12 hours for extracts of
the   transfected   cells.   Values   plotted   are
expressed   as   percent   of   the   ^{14}C-chloramphenicol
converted   to   the   acetylated   form   per 25 µg of
protein assayed.

signal in the bGH gene.   Consequently, the bGH gene appeared to be a good
candidate for characterizing the essential elements of a polyadenylation
signal in addition to the previously defined AAUAAA sequence located 20 to
30 nucleotides upstream from the polyadenylation site.

We were particularly interested in assessing the contribution of 3'
flanking sequences to the complete bGH mRNA polyadenylation signal.   As
this region is presumably retained in primary transcripts derived from RNA
polymerase II templates (Nevins, 1983), the 3' flanking sequence may be
involved in directing the processing by endonucleolytic cleavage and
subsequent polyadenylation (Birnstiel et al., 1985).

To identify sequences located in the 3' flanking region of the bGH
gene which may possibly participate in defining the polyadenylation event,
we generated a series of Bal 31 deletion mutants containing progressively
shorter fragments of the 3' flanking portion of the bGH gene (Fig. 3).
Due to the intrinsic inactivity of the bGH promoter, transcription was

driven from the SV40 late promoter in a chimeric construct designed for efficient expression in COS-1 cells. Following the transfection and transient expression of constructs containing various lengths of the 3' flanking sequence, the site of polyadenylation in the RNA transcripts was defined by S1 nuclease mapping (Woychik et al., 1984). The results indicated that a deletion mutant containing only 18 nucleotides of bGH sequence downstream from the polyA site was still effective in generating the homogeneous, wild-type site of polyadenylation. However, a further deletion leaving only 14 nucleotides of 3' flanking bGH sequence yielded transcripts which were primarily polyadenylated 5 nucleotides upstream of the normal site as well as some residual polyadenylation at the wild-type site. The presence of only a single nucleotide in the 3' flanking sequence resulted in heterogeneous polyadenylation at 3 positions near the wild-type site as well as the utilization of a cryptic polyadenylation site in the adjacent pBR322 plasmid sequence. These data indicate that a signal sequence which is required for accurate polyadenylation begins 15 to 18 nucleotides downstream of the bGH polyadenylation site (Table III).

Two consensus sequences, derived by sequence comparisons, have been proposed as components of the polyadenylation signal in addition to the common hexanucleotide AAUAAA. In a survey of 61 vertebrate genes, the sequence CAYUG (Y=U or C) was often observed within 10 nucleotides of the

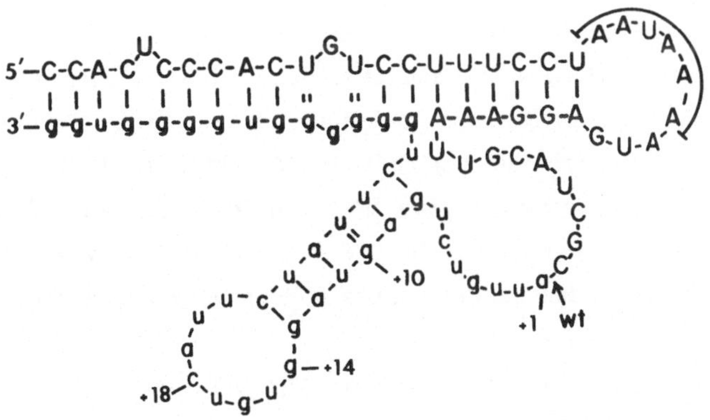

Fig.3. Computer-generated secondary structure of the sequence near the 3' end of the primary RNA transcript produced from the wild-type bGH gene. The positions of various deletions are denoted by the distance from the deletion to the wild-type (wt) polyadenylation site. The conserved hexanucleotide AAUAAA is outlined, the 3' flanking region is in capital letters, and the 3' flanking sequence is in lower case letters.

Table III. Polyadenylation sites observed with various 3'
deletions of the bGH gene.

Nucleotides from deletion to wild-type polyadenylation site	Polyadenylation sites relative to wild-type
+18	WT
+14	-5 and WT
+10	-5 and WT
+1	-5, +3, +35, and crytic pBR322

COS-1 cells were transfected with 25 µg of the indicated
deletion clone and poly(A)+ RNA was prepared after 48 hours.
The position of the polyadenylation site was determined by
hybridization of poly(A)+ RNA with a 3' end-labeled DNA
probe followed by digestion with S1 nuclease and sizing on a
denaturing gel (Woychik et al., 1984). Positive numbers
represent 3' flanking sequences and negative numbers
represent sequence upstream of the wild-type (WT) polyadenylation site.

polyadenylation site (Berget, 1984). The bGH gene contains two such
sequences, a near-consensus CAUCG which is immediately followed by a
consensus CAUUG (Fig. 3). This second "CAYUG" sequence contains the
wild-type polyadenylation site. It is striking that the bGH deletion
leaving only 14 nucleotides of 3' flanking sequence leads to a shift in
the preferred site of polyadenylation from the A in the second consensus
sequence to the A of the first. Birnstiel et al. (1985) have observed the
frequent presence of the trinucleotide TGT in a region about 30 nucleotides downstream of the hexanucleotide. This sequence is found 15
nucleotides downstream of the polyadenylation site in bGH, precisely
mapping in the 4 nucleotide region whose deletion leads to utilization of
the alternate polyadenylation site 5 nucleotides upstream of the wild-type
site. This data suggests that the TGT sequence may play an important role
in the accurate polyadenylation of the bGH primary transcript.

When the nucleotide sequence for the bGH precursor transcript was
examined for potential secondary structure by computer analysis, several
intramolecular structures of reasonably high stability were projected in
which the common hexanucleotide and the wild-type site of polyadenylation
are contained in open loops attached to adjacent stem structures (Woychik
et al., 1984). The wild-type 3' end is also projected to contain such
structures (Fig. 3), but confirmation of these putative secondary
structures awaits characterization by ribonuclease mapping.

290

Both growth hormone and prolactin have been reported to undergo alternative mRNA splicing events. Use of an alternative 3' splice junction 45 nucleotides downstream from the normal splice junction in exon 3 results in deletion of 45 nucleotides from the coding region of human growth hormone (hGH) mRNA (DeNoto et al., 1981). This alternative species of hGH mRNA, which would code for a GH polypeptide with 15 amino acids deleted in the middle, accounts for approximately 10% of the total hGH mRNA species in human pituitary and may account for the 20,000 MW variant of growth hormone reported by several laboratories (Lewis, 1984). Examination of the bGH genomic sequence (Woychik et al., 1982) indicates that a homologous alternative splice site might also exist in exon 3 of bGH mRNA 42 nucleotides downstream from the normal splice site. However,

```
                     32                                 47
         Intron B    Glu Glu              Leu Gln    Asn Pro Gln
Human      cctag     GAA GAA.............CTC CAG  /  AAC CCC CAG

Bovine     cccag     GAG CGC.............ATC CAG     AAC ACC CAG
                     Glu Arg              Ile Gln    Asn Thr Gln
                     33                                 47
```

our preliminary results using S1 mapping have not detected evidence for use of this alternative splice site.

Comparison of rat prolactin (rPRL) cDNA sequences (Cooke et al., 1980; Gubbins et al., 1980) and direct sequencing of rPRL mRNA using an oligonucleotide primer (Taylor et al., 1981) indicated the variable presence of an alanine codon in mRNA at position -20 of the signal sequence. Subsequent publication of the genomic sequence (Maurer et al., 1981) indicated that this codon occurred precisely at the Intron A/Exon 2 boundary with the sequence tag–CAG. Thus there are two adjacent potential splice sites situated such that splicing after the first ag would result in an RNA containing a GCA triplet coding for alanine, while splicing at the second ag would result in an mRNA lacking the triplet coding for alanine. The amino acid sequence of translation products synthesized from rPRL mRNA indicated that the predominant preprolactin polypeptide contained alanine at position -20 (McKean and Maurer, 1978).

The bPRL gene also contains these potential alternative splice junctions (Camper et al., 1984), and preliminary evidence indicates that both of these splice sites are utilized. However, unlike rPRL mRNA, in

```
                      G-                                    -ly        -19
            Arg Lys                                    Thr Leu
                      A-                          -la Gly
rat         CGG AAA G gt...........tttctttag CA GGG ACA CTC

bovine      CAG AAA G gt...........ttatttaag ca gGG TCC CGC
                      G-                          -ly
            Gln Lys                                    Ser Arg
                      A-                          -la Gly        -19
```

which the 5' most splice junction appears to be used predominantly, in
bPRL, the 3' most splice junction is preferentially used.

STUDIES ON THE STABILITY OF GROWTH HORMONE mRNA

Several studies have demonstrated GH synthesis in transfected
non-pituitary cell lines such as mouse fibroblasts (Robins et al., 1981;
Doehmer et al., 1982; Karin et al., 1984) and monkey kidney cells
(Pavlakis et al., 1981; Woychik et al., 1984), indicating that GH mRNA is
fairly stable, even in a heterologous environment. Measurements of GH
mRNA half-life, however, have been attempted only for pituitary cells,
yielding estimates of 20-50 h under normal serum or hormone-supplemented
culture conditions and an even larger range of values following serum or
hormone withdrawal (Yaffe and Samuels, 1984; Nyborg et al., 1984; Diamond
and Goodman, 1985; Gick and Bancroft, 1985). Thus, GH mRNA appears to be
quite stable in comparison to most cellular messages (half-life = 8-15 h).
Some studies have also suggested a hormonal affect of glucocorticoids and
thyroid hormones upon GH mRNA stability although alterations in pre-mRNA
processing or cytoplasmic transport could equally well account for the
data (Karin et al., 1984; Nyborg et al., 1984; Diamond and Goodman, 1985).

The remarkably high stability of bGH mRNA could be the result of
either intrinsic properties of the mRNA molecule or, alternatively, due to
the cellular environment in which this mRNA is expressed (i.e., GH mRNA
might be stable in pituitary cells but not cells of non-pituitary origin).
To test this hypothesis, the bGH mRNA was expressed in HeLa and CHO cells
by transfection. As in the case of analysis of polyadenylation signals in
the bGH gene, these constructs also required heterologous promoters to
drive expression of the structural bGH gene and thereby accummulate
sufficient numbers of transcripts over short time periods. Transfected
cells were treated with the polymerase II transcription inhibitor
2,5-dichloro-1-β-D-ribofuranosylbenzimidazole such that incorporation of
[3]H-uridine into total and GH-specific RNA was inhibited by 95%. By

292

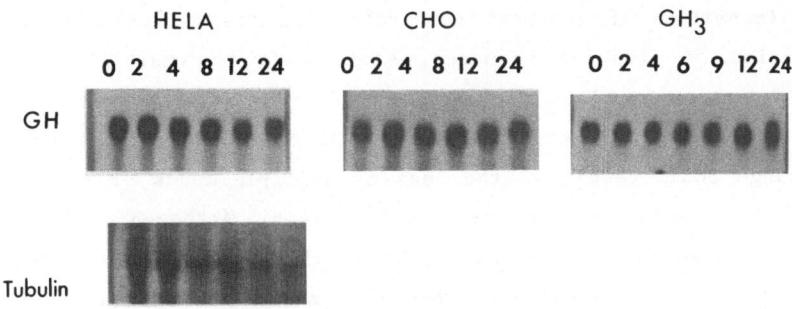

Fig. 4. Stability of growth hormone mRNA in DRB-treated cells. Total RNA was isolated from HeLa or CHO cell lines which express a transfected bGH gene or from rat GH_3 cells at 0 to 24 h. following treatment with 50-100 μM DRB. RNA (10 μg) was fractionated on a denaturing agarose gel, transferred to a nylon membrane, and probed with ^{32}P-labeled cDNAs. Upper panels show bGH mRNA, while the lower panel shows the turnover of α-tubulin mRNA in transfected HeLa cells.

monitoring the decay of mRNA following drug treatment (Khalili and Weinmann, 1984), we found GH mRNA to be stable regardless of the cell in which it is expressed (Fig. 4). Conversely, endogenous β-actin or α-tubulin mRNAs turned over with short half-lives. Only the determination of α-tubulin mRNA in HeLa cells is shown in Figure 4. These results suggest that a major factor governing mRNA stability is RNA structure, although cell type specific factors are also likely to influence the actual half-life.

DISCUSSION

Sequence comparisons between GH and PRL genes from several species indicate a number of retained regions. Among these sequences, perhaps the most interesting is the stretch of 37 nucleotides found in the 5' flanking region of each of the GH genes. Unfortunately, the inactivity of the bGH promoter region containing this sequence precluded further analyses of this conserved region by transfection into currently available recipient cells. The marked retention of such sequences will be of interest in future studies, however, and might possibly reflect the presence of multiple overlapping signals rather than a single regulatory sequence. This is presumably the case with the PRL promoter region. The net effect of such overlapping signals on sequence retention within gene sets would be the imposition of inherent constraints on the evolutionary drift of this region of flanking sequence, similar to the multiple structural coding sequences found in the virus ϕX 174. If regulatory sequences are indeed arranged in an overlapping array, the introduction of a single base pair change could result in the loss of multiple modes of regulation of the gene in question and thus be a lethal event. Although there currently is no data supporting such a model, it should be possible to determine in future transfection experiments whether a molecular rationale exists for the maintenance of multiple regulatory sequences within a confined region of 5' flanking sequence.

The availability of a transfection system utilizing promoter regions of the bPRL gene affixed to the recorder CAT sequence offers several advantages. First, relatively short DNA sequences derived from the 5' flanking region can be studied for promoter activity independent of other sequences contained in this gene which might possibly include additional regulatory elements such as enhancers. Furthermore, the small size of the 5' flanking sequence which is active in this system will permit site-directed mutagenesis and the analysis of specific nucleotides within this region. Rapid turn-around time for analyses will facilitate the study of a number of mutated sequences. Finally, such transfection experiments permit the direct study of a hormone on isolated gene sequences in the context of specific cell types, thereby further dissecting the response obtained in the whole animal and the accompanying physiological effects of interacting hormones produced at ectopic sites.

We have found the bGH polyadenylation signal to be strong in comparison to those of other genes. Cleavage and polyadenylation of bGH message is efficient and is directed to a discrete site. We have identified a region beginning 15 to 18 nucleotides downstream of the polyadenylation site which is necessary for preservation of this accuracy. Efficient polyadenylation is one mechanism by which expression of this gene may be augmented despite its poor promoter activity. A strong polyadenylation signal and the high stability of bGH mRNA may both be the result of a high degree of secondary structure in this message.

Growth hormone mRNA sequences are very stable, and this stability may be regulated by hormones and growth conditions. Because of its long half-life, changes in the mRNA levels arise slowly and regulation at the RNA level may be inadequate to explain the fluctuations in serum GH which accompany the diurnal cycle or the application of hormonal regimes.

294

Instead, we believe that GH mRNA provides a high steady-state supply of the protein product which is then stored in secretory vesicles. Acute regulation of serum levels of growth hormone would be primarily determined by release from preformed vesicles rather than by modulation of transcription, mRNA processing, or translation. These later mechanisms are likely to be effective only for long-term regulation of gross GH levels such as during cell differentiation and development. The finding that other secreted proteins are also translated from mRNAs with long half-lives suggests a similar mode of regulation for exported gene products in general (Palmiter, 1973; Lowenhaupt and Lingrel, 1978; Guyette et al., 1979; Wiskocil et al., 1980; Brock and Shapiro, 1983).

An extension and possible generalization of this model would suggest that genes whose products must be more acutely regulated between extremes in concentrations within a short time period, such as those genes coding for key regulatory enzymes in metabolism or regulatory proteins expressed in narrow windows of a cell cycle, would be primarily regulated at the transcriptional level and thus give rise to mRNAs of short half-life. Such a regulatory scheme would enable a rapid response to multiple signals by producing a large number of new transcripts from strongly induced promoters, yet this scheme allows for an equally rapid attenuation of the response by means of mRNA degradation.

ACKNOWLEDGEMENTS

This work was supported by National Institutes of Health grants CA31810 and AM32770. D.S. was supported by a National Research Service award (AM07549) from the National Institutes of Health, R.H. was a research fellow of the American Heart Association, Northeast Ohio Affiliate, and E.G. was supported by a National Science Foundation Graduate Fellowship. Data contained in Figures 1 and 3 and Tables I, II, and III are obtained in part from articles by Camper, et al. (1985, J. Biol. Chem., 260:12246), and by Woychik, et al. (1984, Proc. Natl. Acad. Sci., 81:3944), and are used with permission of the publisher.

REFERENCES

Barinaga, M., Yamonoto, G., Rivier,C., Vale, W., Evans, R., and Rosenfeld, M. G., 1983, Transcriptional regulation of growth hormone gene expression by growth hormone-releasing factor, Nature, 306:84-85.

Barta, A., Richards, R. I., Baxter, J. D., and Shine, J., 1981, Primary structure and evolution of the rat growth hormone gene, Proc. Natl. Acad. Sci. USA, 78:4867-4871.

Berget, S. M., 1984, Are U4 small nuclear ribonucleoproteins involved in polyadenylation?, Nature, 309:179-182.

Birnstiel, M. L., Busslinger, M., and Strub, K., 1985, Transcription termination and 3' processing: The end is in sight, Cell, 41:349-359.

Brock, M. L., and Shapiro, D. J., 1983, Estrogen stabilizes vitellogenin mRNA against cytoplasmic degradation, Cell, 34:207-214.

Camper, S. A., Luck, D. N., Yao, Y., Woychik, R. P., Goodwin, R. G., Lyons, R.H., and Rottman, F. M., 1984, Characterization of the bovine prolactin gene, DNA, 3:237-249.

Camper, S. A., Yao, Y. A. S., and Rottman, F. M., 1985, Hormonal regulation of the bovine prolactin promoter in rat pituitary tumor cells, J. Biol. Chem., 260:12246-12251.

Carrillo, A. J., Pool, T. B., and Sharp, Z. D., 1985, Vasoactive intestinal peptide increases prolactin messenger ribonucleic acid content in GH_3 cells, Endocrinol., 116:202-206.

Cooke, N. E., and Baxter, J. D., 1982, Structural analysis of the prolactin gene suggests a separate origin for its 5' end, Nature, 297:603-606.

Cooke, N. E., Coit, D., Weiner, R. I., Baxter, J. D., and Martial, J. D., 1980, Structure of cloned DNA complementary to rat prolactin messenger RNA, J. Biol. Chem., 255:6502-6510.

DeNoto, F. M., Moore, D. D., and Goodman, H. M., 1981, Human growth hormone DNA sequence and mRNA structure: Possible alternative splicing, Nucleic Acids Res., 9:3719-3730.

Diamond, D. J., and Goodman, H. M., 1985, Regulation of growth hormone messenger RNA synthesis by dexamethasone and triiodothyronine. Transcriptional rate and mRNA stability changes in pituitary tumor cells, J. Mol. Biol., 181:41-62.

Doehmer, J., Barinaga, M., Vale, W., Rosenfeld, M. G., Verma, I. M., and Evans, R. M., 1982, Introduction of rat growth hormone gene into mouse fibroblasts via a retroviral DNA vector: Expression and regulation, Proc. Natl. Acad. Sci. USA, 79:2268-2272.

Evans, R. M., Birnberg, N. C., and Rosenfeld, M. G., 1982, Glucocorticoid and thyroid hormones transcriptionally regulate growth hormone gene expression, Proc. Natl. Acad. Sci. USA, 79:7659-7663.

Gick, G. G., and Bancroft, C., 1985, Regulation by calcium of prolactin and growth hormone mRNA sequences in primary cultures of rat pituitary cells, J. Biol. Chem., 260:7614-7618.

Gluckman, P. D., Grumbach, M. M., and Kaplan, S. L., 1981, The neuroendocrine regulation and function of growth hormone and prolactin in the mammalian fetus, Endocrine Rev., 2:363-395.

Gorman, C. M., Merlino, G. T., Willingham, M. C., Pastan I., and Howard B., 1982, The rous sarcoma virus long terminal repeat is a strong promoter when introduced into a variety of eukaryotic cells by DNA-mediated transfection, Proc. Natl. Acad. Sci. USA, 79:6777-6781

Gubbins, E. J., Maurer, R. A., Lagrimini, M., Erwin, C. R., and Donelson, J. E., 1980, Structure of the rat prolactin gene, J. Biol. Chem., 255:8655-8662.

Guyette, W. A., Matusik, R. J., and Rosen, J. M., 1979, Prolactin-mediated transcriptional and post-transcriptional control of casein gene expression, Cell, 17:1013-1023.

Karin, M., Eberhardt, N. L., Mellon, S. H., Malich, N., Richards, R. I., Slater, E. P., Barta, A., Martial, J. A., Baxter, J. D., and Cathala, G., 1984, Expression and hormonal regulation of the rat growth hormone gene in transfected mouse L cells, DNA, 3:147-155.

Khalili, K., and Weinmann, R., 1984, Actin mRNAs in HeLa cells. Stabilization after adenovirus infection, J. Mol. Biol., 180:1007-1021.

Lewis, U. J., 1984, Variants of growth hormone and prolactin and their post-translational modifications, Ann. Rev. Physiol., 46:33-42.

Lowenhaupt, K., and Lingrel, J. B., 1978, A change in the stability of globin mRNA during the induction of murine erythroleukemia cells, Cell, 14:337-344.

Lyons, R. H., Jr., 1985, Studies on transcription of the bovine growth hormone gene, Ph.D. Thesis, Case Western Reserve University.

Maurer, R. A., 1982, Estradiol regulates the transcription of the prolactin gene., J. Biol. Chem., 257:2133-2136.

Maurer, R. A., Erwin, C. R., and Donelson, J. E., 1981, Analysis of 5' flanking sequences and intron-exon boundaries of the rat prolactin gene, J. Biol. Chem., 256:10524-10528.

McKean, D. J., and Maurer, R. A., 1978, Complete amino acid sequence of the precursor region of rat prolactin, Biochemistry, 17:5215-5219.

Miller, W. L., and Eberhardt, N. L., 1983, Structure and evolution of the growth hormone gene family, Endocrine Rev., 4:97-130.

Murdoch, G. H., Potter, E., Nicolaisen, A. K., Evans, R. M., and Rosenfeld, M. G., 1982, Epidermal growth factor rapidly stimulates prolactin gene transcription, Nature, 300:192-194.

Nevins, J. R., 1983, The pathway of eukaryotic mRNA formation, Ann. Rev. Biochem., 52:441-446.

Nilson, J. H., Fink, P. A., Virgin, J. B., Cserbak, M. T., Camper, S. A., and Rottman, F. M., 1983, Developmental expression of growth hormone and prolactin genes in the bovine pituitary, J. Biol. Chem., 258:4565-4570.

Nyborg, J. K., Nguyen, A. P., and Spindler, S. R., 1984, Relationship between thyroid and glucocorticoid hormone receptor occupancy, growth hormone gene transcription, and mRNA accumulation, J. Biol. Chem., 259:12377-12381.

Palmiter, R. D., 1973, Rate of ovalbumin messenger ribonucleic acid synthesis in the oviduct of estrogen-primed chicks, J. Biol. Chem., 248:8260-8270.

Pavlakis, G. N., Hizuka, N., Gorden, P., Seeburg, P., and Hamer, D. H., 1981, Expression of two human growth hormone genes in monkey cells infected by simian virus 40 recombinants, Proc. Natl. Acad. Sci. USA, 78:7398-7402.

Robins, D. M., Paek, I., Seeburg, P. H., and Axel, R., 1982, Regulated expression of human growth hormone genes in mouse cells, Cell, 29:623-631.

Sasavage, N. L., Smith, M., Gillam, S., Astell, C., Nilson, J. H., and Rottman, F., 1980, Use of oligodeoxynucleotide primers to determine poly(adenylic acid) adjacent sequences in messenger ribonucleic acid. 3'-terminal noncoding sequence of bovine growth hormone messenger ribonucleic acid, Biochemistry, 19:1737-1743.

Sasavage, N. L., Smith, M., Gillam, S., Woychik, R. P., and Rottman, F. M., 1982, Variation in the polyadenylation site of bovine prolactin mRNA, Proc. Natl. Acad. Sci. USA, 79:223-227.

Seeburg, P. H., 1982, The human growth hormone gene family: Nucleotide sequences show recent divergence and predict a new polypeptide hormone, DNA, 1:239-249.

Taylor, W., Collier, K. J., Weith, H. L., and Dixon, J. E., 1981, The use of a heptadeoxyribonucleotide as a specific primer for prolactin mRNA: A prediction of ambigious RNA splicing, Biochem. Biophys. Res. Commun., 102:1071-1077.

Truong, A. T., Duez, C., Belayew, A., Renard, A., Pictet, R., Bell, G. I., and Martial, J. A., 1984, Isolation and characterization of the human prolactin gene, EMBO J., 3:429-437.

Wiskocil, R., Bensky, P., Dower, W., Goldberger, R. F., Gordon, J. I., and Deeley, R. G., 1980, Coordinate regulation of two estrogen-dependent genes in avian liver, Proc. Natl. Acad. Sci. USA, 77:4474-4478.

Woychik, R. P., Camper, S. A., Lyons, R. H., Horowitz, S., Goodwin, E. C., and Rottman, F. M., 1982, Cloning and nucleotide sequencing of the bovine growth hormone gene, Nucleic Acids. Res., 10:7197-7210.

Woychik, R. P., Lyons, R. L., Post, L., and Rottman, F. M., 1984, Requirement for the 3' flanking region of the bovine growth hormone gene for accurate polyadenylation, Proc. Natl. Acad. Sci. USA, 81:3944-3948.

Yaffe, B., and Samuels, H. H., 1984, Hormonal regulation of the growth hormone gene. Relationship of the rate of transcription to the level of nuclear thyroid hormone-receptor complexes, J. Biol. Chem., 259:6284-6291.

THE USE OF SPECIFIC cDNA PROBES TO ASSAY SERTOLI CELL FUNCTIONS

Michael D. Griswold, Michael Collard, Suzanne Hugly, and
Jodi Huggenvik

Biochemistry/Biophysics Program
Washington State University
Pullman, Washington 99163

INTRODUCTION

Spermatogenesis is the complex process whereby relatively
undifferentiated germinal cells undergo biochemical and morphological
modifications to form spermatozoa. Sertoli cells are the somatic
epithelial components of the seminiferous tubules which play fundamental
and mandatory roles in this process (for review, see Fawcett, 1975). The
close physical association of Sertoli cells with each germinal cell and
the organization of this association into a cyclic pattern have been
described in detail (Clermont and Perey 1957; Clermont, 1972). The
characterization of Sertoli cells as the "nurse" cells of the testis was
based originally on the morphological cellular relationships within the
testis. This concept of Sertoli cells functioning in a "support" or
"regulatory" role has been confirmed by both biochemical and endocrine
studies. Current evidence suggests that the Sertoli cells are the
testicular target cells for the action of testosterone and FSH, which are
the hormones most directly involved in the regulation of spermatogenesis
(for review, see Fritz, 1978). Therefore, it is possible to envision a
scenario whereby the regulation of spermatogenesis is a result of the
biochemical properties of Sertoli cells.

The identification and characterization of the specialized products
of Sertoli cells should provide insights into the manner in which this
regulatory function is accomplished. In addition, the identification of
unique products provides an assay whereby the response of the Sertoli
cells to hormones or other stimulatory or inhibitory agents can be

301

ascertained. One of the possible mechanisms by which Sertoli cells can influence germinal cells is via the secretion of proteins and glycoproteins which serve as signals or transport vehicles. The first such glycoprotein which was identified as a Sertoli cell specific secretion product was androgen binding protein (ABP) (Vernon et al., 1974; French and Ritzen, 1973; Steinberger et al., 1975; Fritz et al., 1976). This protein is secreted by Sertoli cells in cell culture and the extent of its synthesis is regulated by FSH, testosterone, and vitamin A (Louis and Fritz, 1977; Karl and Griswold, 1980). The function of ABP is not clear but is likely related to the capability of the protein to bind and transport androgens to the epididymis (French and Ritzen, 1973). Since Sertoli cells respond to a number of stimuli and have a number of proposed functions in the testis, it is important to identify and characterize the complete spectrum of secreted proteins. Sertoli cells probably assume various functional modes throughout spermatogenesis during which the secretion of proteins may change temporally (Parvinen, 1982; Wright et al., 1983). The goal of this laboratory has been to identify and characterize the proteins secreted by Sertoli cells, to examine the interaction of these secretion products with the germinal cells, and to develop sensitive and specific assays for the expression of these unique gene products. To accomplish the latter, we have utilized cloned cDNA to quantify the levels of specific mRNAs in Sertoli cells.

Glycoproteins Secreted by Sertoli Cells

Sertoli cells are secretory cells which dedicate a major portion of their protein synthetic activity to the production of secreted glycoproteins (Wilson and Griswold, 1979). The spectrum of major glycoproteins secreted by Sertoli cells has been examined by polyacrylamide gel electrophoresis (Kissinger et al., 1982). Cultured Sertoli cells were incubated with ^{35}S-methionine and the radioactive secreted proteins were collected in the culture medium after several hours. The secreted proteins were subjected to the 2-dimensional polyacrylamide gel electrophoresis (PAGE) procedure and were visualized by fluorography. An example of the profile of secreted proteins seen after this type of analysis is shown in Figure 1. This procedure allows for the detection of the most abundant secreted proteins, while some proteins synthesized in very minor amounts, such as ABP, are often not visible on the fluorogram.

The secretion products shown in Figure 1 have been characterized biochemically (Skinner and Griswold, 1980; Skinner and Griswold, 1983a;

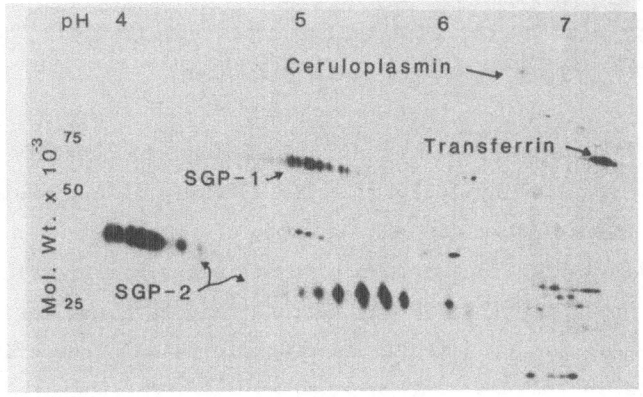

Fig. 1. Fluorogram of a two-dimensional gel electro-
pherogram of Sertoli cell secreted proteins.
Sertoli cells from 20-day old rats were placed in
tissue culture and after 3 days the medium was
changed to a medium which lacked methionine but
contained 500 μCi/ml of ^{35}S-methionine. The cells
were incubated with the isotope-containing medium
for 12 hr. The medium was collected, centrifuged
to remove debris, and subjected to electrophoresis
(O'Farrell, 1975) and fluorography (Skinner and
Griswold, 1983b).

Skinner et al., 1984; Sylvester et al., 1984). Two of the secretion
products, ceruloplasmin and transferrin, are similar to serum proteins.
The characteristics of SGP-1, SGP-2 (previously called Band 4 and
DAG-protein), ceruloplasmin, and transferrin are summarized in Table I.
These four glycoproteins comprise 60 to 90% of the total mass of protein
secreted by cultured Sertoli cells (Griswold et al., 1984). Transferrin,
SGP-1, and SGP-2 have been purified, and polyclonal antiserum has been
prepared to each of the purified proteins (Skinner et al., 1984;
Sylvester et al., 1984).

Antisera has been used to localize the secreted proteins in paraffin
sections of testis tissue (Sylvester and Griswold, 1984; Sylvester et al.,
1984). This immunohistochemical approach has contributed to the
understanding of the role of the secreted proteins in spermatogenesis.
Testicular transferrin can be localized near the acrosome region of
developing spermatids and may function to transport Fe^{+3} to specific
stages of germinal cell growth (Sylvester and Griswold 1984). This
transport system may be required in order to circumvent the blood-testis
barrier to provide a source of this important nutrient to the developing
germinal cells. Both SGP-1 and SGP-2 are secreted by the Sertoli cells
into the lumen of the seminiferous tubules and ultimately end up as a
constituent of the sperm membrane (Sylvester et al., 1984). Both SGP-1

and SGP-2 are also synthesized by the caput epididymis. The function of these sulfated glycoproteins is still not known.

A radioimmunoassay for transferrin was used to quantify the amount of this protein secreted by cultured cells (Skinner and Griswold, 1982). At least four agents regulate the amount of transferrin synthesized by cultured cells. Vitamin A and insulin stimulate transferrin synthesis to the greatest extent while FSH and testosterone have smaller effects. The radioimmunoassay for transferrin is a sensitive and convenient marker for Sertoli cells in culture, but because of the high concentration of serum transferrin in serum and lymph (4 mg/ml), the assay cannot be used to measure Sertoli cell responses in testicular tissue. This lack of an assay for _in vivo_ Sertoli cell functions was one of the reasons we initiated a program to obtain a cDNA probe to transferrin. We have recently extended our methods and rationale to the SGP-1 and SGP-2 proteins.

GENERAL APPROACH

We successfully obtained purified mRNA for transferrin, SGP-1, and SGP-2 by immunoprecipitation of polysomes using the specific polyclonal antiserum for these proteins (Palmiter, 1974). The general outline of our

Table I. Biochemical characteristics of the major glycoproteins secreted by cultured Sertoli cells.

Name	MW SDS-non reduced	MW-SDS-reduced	%CHO	Other Properties
Ceruloplasmin	130,000	130,000	N.D.[a]	Involved in Copper Transport
Transferrin	76,000	76,000	5	Involved in Iron Transport
SGP-1	70,000	70,000	25-35	Contains Sulfated Sugars
SGP-2	70,000	41,000 & 29,000	25-35	Contains Sulfated Sugars and is Made up of Covalent Dimers

Molecular weights were determined by polyacrylamide gel
electrophoresis in SDS. [a]N.D.-not determined.

methods for obtaining, cloning, and screening the cDNA is in Figure 2. We chose this approach because the transferrin, SGP-1, and SGP-2 mRNAs are present in relatively high abundance. The use of affinity purified IgG in the immunoprecipitation of the polysomes greatly enriched for the respective mRNAs (Lee et al., 1983). The final criterion for the successful cloning of a specific cDNA was the hybridization-release and translation of an immunoprecipitable precursor protein (Gurney et al., 1982).

We have obtained the purified transferrin mRNA from rat liver and the SGP-1 and SGP-2 mRNA from whole testis or from cultured Sertoli cells. The size of the mRNA and the characteristics of the precursor proteins for the transferrin, SGP-1, and SGP-2 are described in Table II. It is noteworthy that SGP-2 in the mature glycosylated and secreted form is a disulfide-linked heterodimer with subunits of 41,000 and 29,000 daltons. However, the SGP-2 mRNA codes for a single polypeptide of 44,000 daltons which is subsequently glycosylated and cleaved to achieve its mature form.

We have used several types of assays to detect and quantify specific mRNA using the cDNAs. The Northern blot analysis has been used to screen total RNA from a variety of tissues for the presence of specific mRNA (Thomas, 1980). Rat transferrin mRNA, for example, can be detected in testis and in liver tissue while a very weak signal is obtained from RNA isolated from ovaries. The specific mRNA present in Sertoli cells in a variety of situations has been quantified by nick-translation of the cDNA, by dot-blot hybridization, and also by RNA-RNA liquid phase hybridization

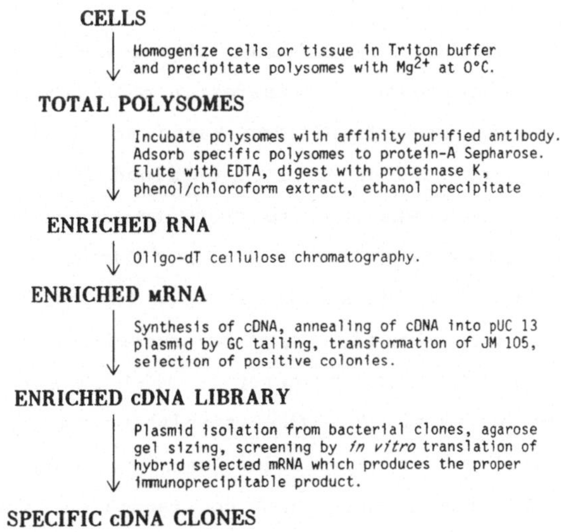

CELLS

Homogenize cells or tissue in Triton buffer and precipitate polysomes with Mg²⁺ at 0°C.

TOTAL POLYSOMES

Incubate polysomes with affinity purified antibody. Adsorb specific polysomes to protein-A Sepharose. Elute with EDTA, digest with proteinase K, phenol/chloroform extract, ethanol precipitate

ENRICHED RNA

Oligo-dT cellulose chromatography.

ENRICHED mRNA

Synthesis of cDNA, annealing of cDNA into pUC 13 plasmid by GC tailing, transformation of JM 105, selection of positive colonies.

ENRICHED cDNA LIBRARY

Plasmid isolation from bacterial clones, agarose gel sizing, screening by *in vitro* translation of hybrid selected mRNA which produces the proper immunoprecipitable product.

SPECIFIC cDNA CLONES

Fig. 2. General outline of mRNA isolation, cloning, and screening.

Table II. Characteristics of Sertoli cell mRNA and precursor proteins.

Glycoprotein	Size of mRNA	MW of in vitro Translation Product
Transferrin	2.4 KB	74,000
SGP-1	N.D.[a]	53,000
SGP-2	2.2 KB	44,000

[a]N.D.-not determined.

utilizing the SP6 plasmid system (Melton et al., 1984). The dot-blot assay can be quantitated by excising the hybridized dots and determining the radioactivity in a scintillation counter or by densitometric scans of autoradiographs of the nitrocellulose filter. The SP6 system requires that the cDNA be subcloned into the SP6 plasmid but offers a number of advantages, including increased sensitivity and lower backgrounds.

Transferrin mRNA in Cultured Sertoli Cells

The rat transferrin cDNA probe was used to quantify the amount of specific mRNA in cultured Sertoli cells (Huggenvik et al., 1984; Huggenvik et al., 1985). FSH, testosterone, insulin, and retinol are all required for maximal steady state levels of transferrin mRNA in cells cultured in the absence of serum (Table III). Nearly maximal levels of mRNA were obtained when the cells were cultured in the presence of only insulin and retinol. Retinoic acid could be substituted for retinol in all of the experiments done in culture. The results from these studies, when calculated as the amount of transferrin mRNA per mg Sertoli cell DNA and then expressed as the stimulation ratio for each hormone treatment (Table III), were very similar to data obtained from similar studies in which a radioimmunoassay was used to quantify the amount of transferrin protein secreted from cultured Sertoli cells (Skinner and Griswold, 1982). It appears, therefore, that in studies done on the cultured Sertoli cells, the amount of transferrin mRNA may be rate-limiting for transferrin secretion. When the data for transferrin mRNA is expressed as the CPM hybridized per mg of total RNA, the stimulation ratios for the hormone treatments are smaller in magnitude (data not shown). It will be necessary to measure the changes in a non-regulated "house-keeping" gene to clarify the specificity of the hormone action on transferrin mRNA.

Table III. The amount of transferrin mRNA in Sertoli cells cultured in the presence of various hormones and vitamin A.

Level of mRNA expressed as ratio of treated to control additions CPM hybridized/µg cell DNA (n = 9; ave. variation in ratio = ± 12%)

none (control)	1.00
Insulin (I)	1.66
Retinol (R)	1.42
IR	1.82
FSH (F)	1.16
Testosterone (T)	1.22
FIRT	2.03

The Sertoli cells from 20-day old rats were placed in culture in the presence of the agents listed. After 4 days of culture, the cells were removed, total RNA was extracted, and the amount of transferrin mRNA was quantified by hybridization to nick-translated cDNA. The cytoplasmic mRNA was isolated essentially as described by White and Bancroft, 1982. The RNA was applied to nitrocellulose, and treated as described by Thomas (1980). A nick-translation kit from Bethesda Research Laboratory was used to obtain the labeled cDNA. After hybridization, the dot blots of RNA were exposed to X-ray film and the radioactivity was quantified by cutting out the radioactive spots and analyzing them in a scintillation counter. In general, the hybridizations were done on 15 µg of total cellular RNA. Insulin = 5 µg/ml; FSH = 25 ng/ml; Testosterone = 0.7 µM; Retinol = 0.3 µM.

The data in Table III were obtained from Sertoli cells which had been cultured for 3 days in the presence or absence of the stimulatory agents. When the levels of transferrin secretion and transferrin mRNA are compared over a number of days, it can be seen that they both achieve maximal values at 3 and 4 days after plating (Fig. 3). In the absence of hormones and vitamin A, both transferrin secretion and transferrin mRNA levels are low and continue to decline during culture. The basis for the decline in the levels of secretion and mRNA after 4 to 6 days in culture is unknown, but this phenomenon has been reported previously and may result from sub-optimal cell culture conditions (Skinner and Griswold, 1982b).

All of the results shown in Table III and Figure 3 were obtained from Sertoli cells from 20-day old rats. The cultures were incubated at 32°C,

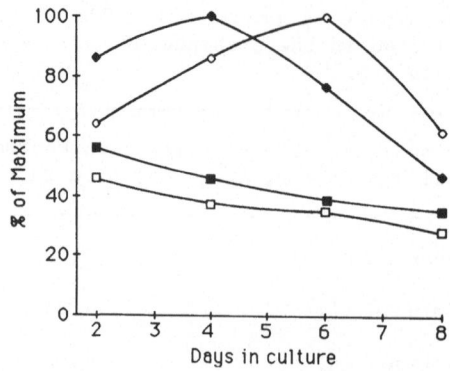

Fig. 3. Time course of transferrin mRNA accumulation
(◆,■) and transferrin protein secretion (◇,□)
in cultured Sertoli cells. The Sertoli cells were
obtained from 20-day old rats and were cultured in
the presence (◆,◇) or absence of (■,□) hormones
and vitamin A as described in Table III.

a temperature which duplicates the scrotal temperature of rats. Since
spermatogenesis is impaired at elevated testicular temperatures (this
impairment could be due to Sertoli cell dysfunction), we compared the
amount of transferrin mRNA in cells cultured at 32°C with the amount in
cells cultured at 37°C (Table IV). Cells which were maintained at 37°C in
the presence of hormones and vitamin A had the highest level of
transferrin mRNA. In other experiments we have also seen that these cells
synthesize and secrete more transferrin than do the cells which are
maintained at 32°C. However, if the cells were not treated with any

Table IV. Transferrin mRNA in cells cultured at 32°C or
37°C.

Conditions	CPM transferrin mRNA hybridized/μg Sertoli Cell DNA		
	32°	n = 3	37°
Control	31 ± 1		21 ± 2
Treated (FIRT)[a]	53 ± 2		67 ± 3
Ratio-Treated to Control	1.7		3.1

Sertoli cells from 20-day old rats were placed in culture at
the indicated temperatures in the presence or absence of
hormones and vitamin A. See Table III for hormone
concentrations. After 3 days in culture the RNA was
extracted from the Sertoli cells and the amount of
transferrin mRNA was determined by the described procedures.
[a]See Table III for abbreviations.

stimulatory agents, the level of transferrin mRNA found in cells at 37°C was consistently lower than that found in equivalent cells maintained at 32°C. The ratio of transferrin mRNA in treated cells to control cells is therefore higher when the cells are maintained at 37°C. Total protein synthesis, lactate production, and cAMP production are increased when Sertoli cells are cultured at 37°C (Hall et al., 1985). The Sertoli cells apparently do not display a decreased biosynthetic activity when cultured at 37°C; however, these studies have all been done on Sertoli cells from 25-day old rats or younger and should be repeated on cells from adult animals.

Transferrin mRNA from Sertoli Cells In Vivo

Since transferrin synthesis is apparently unique among testicular cells to the Sertoli cells, the cDNA probe can be used to measure transferrin mRNA in testis tissue. The protocol for these experiments involved homogenization of the decapsulated testis and isolation of total testicular RNA. The SP6 RNA procedure was then used to quantify the total amount of transferrin mRNA by RNA-RNA liquid hybridization. The data was expressed in terms of hybridizable mRNA per testis and hybridizable mRNA per mg tissue. Any time a measurement is made on Sertoli cells in whole testicular tissue, the problem of how to express the data arises. Expressing the data relative to the tissue protein or to total testis RNA

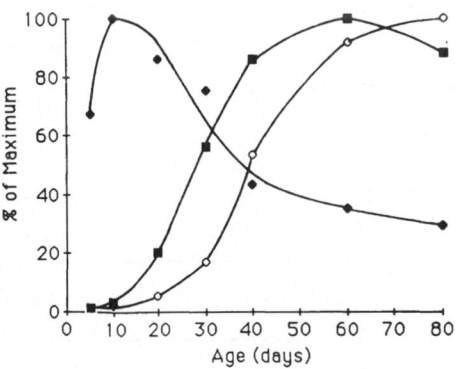

Fig. 4. Transferrin mRNA levels in the testes of rats of different ages. Testes were removed from rats of the ages shown above, the total RNA was isolated, and the transferrin mRNA was quantified by hybridization. Since this data was obtained from several different experiments, it is expressed as the percent of the maximum value. ■ = cpm transferrin mRNA/testis; ◆ = cpm transferrin mRNA/mg testis; ◇ = testis weight (max. value = 1.4 g).

is misleading because of the enormous increase in germinal cell numbers during development. Since it is well established that the Sertoli cell population is stable after 15 days of age in the rat, any data from mature rats which is expressed "per testis" is directly proportional to expressing the data as "per Sertoli cell." We first quantified the amount of transferrin mRNA in testis from rats of different ages in order to examine the temporal expression of the gene (Fig. 4). The transferrin mRNA can be detected in rats from 5 days of age, but only at a very low level. As the rat matures, the amount of transferrin mRNA per testis increases dramatically, while the transferrin mRNA per mg protein begins to decrease. Using the SP6 RNA hybridization technique we can calculate that the average Sertoli cell in the normal adult rat testis contains between 1,500 and 2,000 copies of transferrin mRNA per Sertoli cell. The number of Sertoli cells per testis was obtained from the morphometric studies of Wing and Christenson (1982). The developmental sequence of transferrin mRNA appearance and accumulation is consistent with a biological function which is consequential to spermatogenesis and in particular to spermatocytes and spermatids.

We have also examined the effect of hormone or vitamin A deprivation on the expression of the Sertoli cell transferrin mRNA in vivo. The importance of the pituitary hormones was examined in rats which were hypophysectomized at 18 days of age and allowed to regress for an additional 21 days (Table V and Fig. 5). The amount of transferrin mRNA per testis was decreased by 98% in the hypophysectomized rats and the amount of transferrin per mg of total RNA was decreased by 59%. Note that the total RNA includes RNA from any remaining germinal cells as well as the nongerminal cells of the testis.

Table V. The effect of hypophysectomy on testis weight and total testicular RNA.

Parameter	Control (n = 6)	Hypox. (n = 10)	% Change
Body Weight (g)	165 ± 5	65 ± 4	-61
Testis Weight (g)	1.3 ± .05	.09 ± .01	-93
μg RNA/ag Testis	2.4 ± 0.3	1.6 ± 0.4	-34

Rats were hypophysectomized at 18 days of age and were maintained with no hormone supplements for an additional 21 days. The testis were then removed and the total RNA was extracted.

310

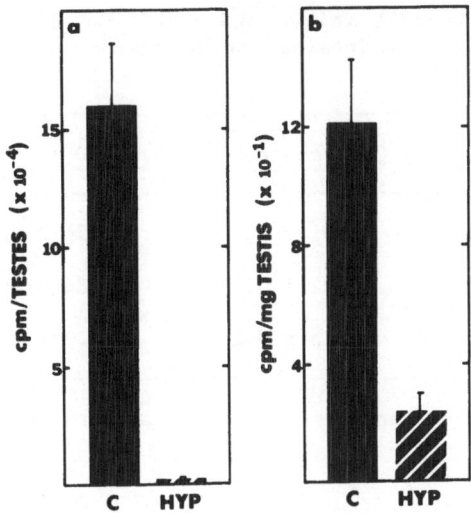

Fig. 5. Comparison of transferrin mRNA levels in testes from control and hypophysectomized rats. Experimental protocol is described in Table V. Values are reported as the sum of cpm per testis (Panel a) or cpm per mg testis tissue (Panel b). Results from the hybridization of RNA isolated from control (C) and hypophysectomized (HYP) rat testes are shown. The error bars in both Panel A and B represent the standard deviation obtained from a minimum of six determinations in duplicate hybridizations.

The data in Table V and Figure 5 clearly point out the importance of one or more of the pituitary hormones in the maintenance of the capacity of the Sertoli cells to synthesize the transferrin mRNA. In preliminary studies where we have treated the hypophysectomized rats with testosterone or FSH, it appears that FSH is the hormone which will stimulate transferrin mRNA accumulation.

In another set of experiments designed to study the regulation of the transferrin mRNA in vivo we fed rats which were initially 20 days of age a diet deficient in vitamin A (retinol). The rats were maintained on this diet for a total of nine weeks, and after that time the testes were removed and the amount of transferrin mRNA per testis was determined by hybridization to the nick-translated probe (Table VI and Fig. 6). Histological section of testes from these deficient animals showed that the seminiferous tubules contained only Sertoli cells and a few spermatogonia, but no spermatocytes or spermatids.

After nine weeks on the VAD diet, the amount of transferrin mRNA per Sertoli cell in these rats was decreased by 86%. The level of transferrin

Table VI.   Overall weight and total RNA in the testes of
            rats maintained on a vitamin A deficient (VAD)
            diet.

Parameter	Control (n = 12)	VAD (n = 10)	% Change
Body Weight (g)	261 ± 42	176 ± 32	−33
Testis Weight (g)	3.02 ± 0.45	0.84 ± 0.10	−72
μg RNA/mg Testis	2.3 ± 0.2	1.7 ± 0.3	−26

Weanling rats were maintained on a vitamin A deficient diet
or a diet supplemented with retinyl acetate for 9 weeks.

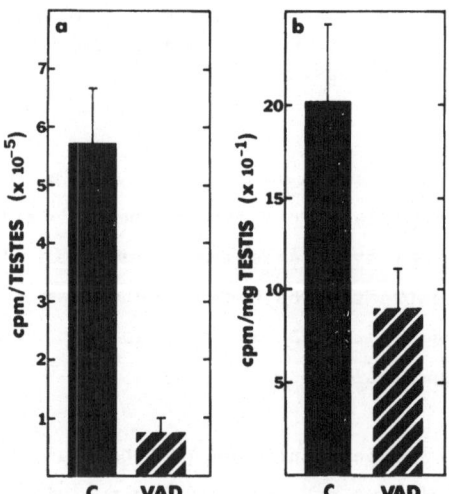

Fig. 6.   Comparison of transferrin mRNA levels in testes
           from control and vitamin A deficient rats.
           Experimental protocol was as described in Table VI.
           RNA was isolated from the testes and quantitated by
           dot-blot hybridization to nick-translated cDNA.
           Values are reported as the sum of cpm in the testes
           (Panel a) or as cpm per mg testis tissue (Panel b).
           Results from the hybridization of RNA isolated from
           control (C) and vitamin A deficient (VAD) rat
           testes are shown. The error bars in both panels a
           and b represent the standard deviation obtained
           from a minimum of eight determinations in duplicate
           hybridizations.

mRNA returned to normal values when the nine-week VAD rats were placed on a diet containing retinyl acetate for two weeks (data not shown). The action of retinol is clearly important for the expression of Sertoli cell transferrin and the interruption of the secretion of normal levels of transferrin may be one of the underlying defects which leads to aspermatogenesis in vitamin A deficiencies.

SGP-1 and SGP-2 cDNA Probes

Studies which utilize the SGP-1 and SGP-2 probes are in the initial stages. The availability of additional probes specific to Sertoli cells in the testis will be of considerable value. In addition, the SGP-1 and SGP-2 glycoproteins are produced by the epididymis, and the cDNA probes for these proteins will be useful in studies of regulation in the epididymis and perhaps other tissues (such as ovary) as well. We have utilized the SGP-2 cDNA probe in Northern blot analysis of RNA from

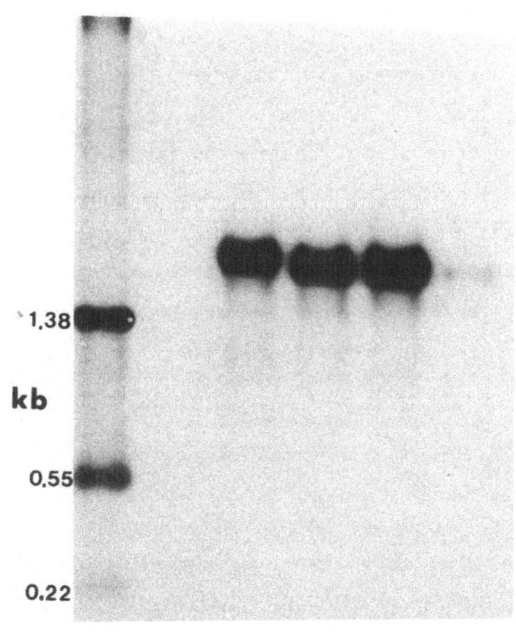

Fig. 7. Northern blot analysis of RNA from a 60-day old rat (1), from the testis of a 40-day old hypophysectomized (20 days) rat (2), from epididymides (3), and from rat ovary (4). The RNA (2 μg) was subjected to agarose gel electrophoresis, blotted to nitrocellulose, and probed with nick-translated plasmid containing the cDNA for SGP-2 protein.

cultured Sertoli cells, from testes of hypophysectomized rats, from epididymis, and from ovary (Fig. 7). RNA from each of these tissues showed a hybridization band of 2.2 KB, although the signal from the ovarian RNA was much weaker.

SUMMARY

The use of cDNA probes to measure the functional state of Sertoli cells in culture or in vivo offers a number of advantages, including sensitivity and cellular specificity. Experiments done with cultured Sertoli cells suggest that the transferrin mRNA levels are regulated by FSH, insulin, vitamin A, and testosterone. Questions relating to the specificity of the action of the hormones and vitamin A on the transferrin mRNA or on other mRNA will require the use of additional hybridization probes. It is clear that most agents which stimulate spermatogenesis also stimulate transferrin mRNA accumulation in cultured Sertoli cells. Testosterone, however, has only a small stimulatory effect on the cultured Sertoli cells.

In the in vivo measurements we have completed thus far, the level of transferrin mRNA in the testis seems to correlate well with the spermatogenic condition of the testis. In immature rats, hypophysectomized rats, and vitamin A deficient rats, spermatogenesis is abbreviated and the levels of transferrin mRNA in the Sertoli cells are correspondingly low. If this correlation holds up under additional scrutiny, assays of the transferrin mRNA levels may be effective screens for the toxic or stimulatory activities of a variety of pharmaceuticals which could affect spermatogenesis.

ACKNOWLEDGEMENTS

These studies were supported by grants HD10808, HD00263, and HD17626 from the National Institute of Child Health and Human Development.

REFERENCES

Clermont, Y., and Perey, B., 1957, Quantitative study of the cell population in the seminiferous tubules in immature rats, Amer. J. Anat., 100:241-268.

Clermont, Y., 1972, Kinetics of spermatogenesis in mammals: Seminiferous epithelium cycle and spermatogonial renewal, Physiol. Rev., 52:198-236.

Fawcett, D. W., 1975, Male reproductive system, in: "Handbook of Physiology", Vol. 5, D. W. Hamilton. and R. D. Grup, eds., American Physiological Society, Bethesda, MD, pp. 21-56.

French, F. S., and Ritzen, E. M., 1973, A high-affinity androgen binding protein in rat testis: evidence for secretion into efferent duct fluid and absorption by epididymis, Endocrinology, 93:88-95.

Fritz, I. B., 1978, Sites of action of androgens and follicle stimulating hormone on cells of the seminiferous tubule, Biochemical Actions of Hormones, V:249-277.

Fritz, I. B., Rommerts, F. F. G., Louis, B. G., and Dorrington, J. H., 1976, Regulation by FSH and dibutyryl cyclic AMP of the formation of androgen-binding protein in Sertoli cell-enriched cultures, J. of Reprod. and Fertil., 46:17-24.

Griswold, M. D., Huggenvik, J., Skinner, M. K., and Sylvester, S. R., 1984, The interactions of Sertoli cell glycoproteins with germinal cells in the testis, in: "Endocrinology," F. Labrie and L. Prioux, eds., Elsevier Science Publishers, Amsterdam, pp. 619-622.

Gurney, T., Sorenson, D. K., Gurney, E. G., and Wills, N. M., 1982, SV40 RNA: Filter hybridization for a rapid isolation and characterization of rare RNAs Analytical Biochem., 125:80-90.

Hall, P. F., Kew, D., Mita, M., 1985, The influence of temperature on the functions of cultured Sertoli cells, Endocrinology, 116:1926-1932.

Huggenvik, J., Sylvester, S. R., and Griswold, M. D., 1984, Control of transferrin mRNA synthesis in Sertoli cells, Annals of NY Acad. Sci., 438:1-7.

Huggenvik, J., Idzerda, R. L., Haywood, L., Lee, D. C., McKnight, G. S., and Griswold, M. D., 1985, Transferrin mRNA: Molecular cloning and hormonal regulation in rat Sertoli cells, submitted.

Karl, A., and Griswold, M. D., 1980, Actions of insulin and vitamin A on Sertoli cells, Biochem. J., 186:1001-1003.

Kissinger, C., Skinner, M. K., and Griswold, M. D., 1982, Analysis of Sertoli cell secreted proteins by two-dimensional gel electrophoresis, Biol. of Reprod., 27:233-240.

Lee, D. C., Carmichael, D. F., Krebs, E. G., and McKnight, G. S., 1983, Isolation of a cDNA clone for the type 1 regulatory subunit of bovine cAMP-dependent protein kinase, Proc. Natl. Acad. Sci. (USA), 80:3608-3612.

Louis, B. G., and Fritz, I. B., 1977, Stimulation by androgens of the production of androgen binding protein by Sertoli cells, Mol. Cell. Endocrin., 7:9–16.

Melton, D. A., Krieg, P. A., Rebagliati, M. R., Maniatus, T., Zinn, K., and Green, M. R., 1984, Efficient in vitro synthesis of biologically active RNA and RNA hybridization probes from plasmids containing a bacteriophage promoter, Nucleic Acids Res., 12:7035–7056.

O'Farrell, P. H., 1975, High resolution two-dimensional electrophoresis of proteins, J. Biol. Chem., 250:4007–4021.

Palmiter, R. D., 1974, Magnesium precipitation of ribonucleoprotein complexes. Expedient techniques for the isolation of undergraded polysomes and messenger ribonucleic acid, Biochemistry, 13:3606–3615.

Parvinen, M., 1982, Regulation of seminiferous epithelium, Endocrine Reviews, 3:404–417.

Skinner, M. K., and Griswold, M. D., 1980, Sertoli cells synthesize and secrete transferrin-like protein, J. Biol. Chem., 255:9523–9525.

Skinner, M. K., and Griswold, M. D., 1982, Secretion of testicular transferrin by cultured Sertoli cells is regulated by hormones and retinoids, Biol. of Reprod., 27:211–221.

Skinner, M. K., and Griswold, M. D., 1983a, Sertoli cells synthesize and secrete a ceruloplasmin-like protein, Biol. of Reprod., 28:1225–1229.

Skinner, M. K., and Griswold, M. D., 1983b, Fluorographic detection of radioactivity in polyacrylamide gels with 2,5-diphenyloxazole in acetic acid and its comparison with existing procedures, Biochem. J., 209:281–284.

Skinner, M. K., Cosand, W. L., and Griswold, M. D., 1984, Purification and characterization of testicular transferrin secreted by Sertoli cells, Biochem. J., 218:313–320.

Steinberger, A., Heindel, J. J., Lindsey, J. N., Elkington, J. S. H., Sanborn, B. M., and Steinberger, E., 1975, Isolation and culture of FSH responsive Sertoli cells, Endocrine Res. Commun., 2:261–272.

Sylvester, S. R., and Griswold, M. D., 1984, Localization of transferrin and transferrin receptors in rat testes, Biol. of Reprod., 31:195–203.

Sylvester, S. R., Skinner, M. K., and Griswold, M. D., 1984, A sulfated glycoprotein synthesized by Sertoli cells and by epididymal cells is a component of the sperm membrane, Biol. of Reprod., 31:1087–1101.

Thomas, P. S., 1980, Hybridization of denatured RNA and small DNA fragments transferred to nitrocellulose, <u>Proc. Natl. Acad. Sci. USA</u>, 77:5201–5205.

Vernon, R. J., Kopec, B., and Fritz, I. B., 1974, Observations on the binding of androgens by rat testis seminiferous tubules and testis extracts, <u>Mol. Cell. Endocrin.</u>, 1:167–175.

White, B. A., and Bancroft, F. C., 1982, Cytoplasmic dot hybridization, <u>J. Biol. Chem.</u>, 257:8569–8572.

Wilson, R. M., and Griswold, M. D., 1975, Secreted proteins from rat Sertoli cells, <u>Exp. Cell. Res.</u>, 123:127–135

Wing, T., and Christensen, A. K., 1982, Morphometric studies on rat seminiferous tubules, <u>Amer. J. of Anat.</u>, 165:13–25.

Wright, W. W., Parvinen, M., Musto, N. A., Gunsalus, G. L., Phillips, D. M., Mather, J. P., and Bardin, C. W., 1983, Identification of stage-specific proteins synthesized by rat seminiferous tubules, <u>Biol. of Reprod.</u>, 29:257–270.

MAMMALIAN GENE TRANSFER AND GENE EXPRESSION

Thomas E. Wagner, Xiao Z. Chen, and William B. Hayes

Graduate Program in Molecular and Cellular Biology
Ohio University
Athens, Ohio 45701

INTRODUCTION

It has long been the desire of molecular biologists to be able to study the regulation of expression of specific mammalian genes. With the advent of recombinant DNA technology and the availability of specific cloned genes, several new approaches to these studies have emerged. One of the most promising, which we will discuss in detail here, is the introduction of specific cloned genes into the genome of embryos and the development of adult animals from these embryos which also contain these cloned DNA sequences in their genomes. This whole animal system has many advantages. The genes may be present in a very native biological environment. If the gene is stably integrated and passed from generation to generation, a large number of similar animals would be available for study. As animals of all stages of development could be obtained, it should be possible to study specific developmental changes occurring within the integrated DNA. This information could be amplified by introducing genes with specific deletions or mutations to analyze which portions of the genes are important in genetic regulation and developmental control of these systems. The opportunities for this approach to the study of the regulation of expression of cloned genes in these whole animal systems seem extensive.

The animal used for these studies has been the laboratory mouse. This is primarily because a great deal is known about the genetic makeup of this animal, and embryological procedures for handling and culturing mouse embryos are well established. Although most published experiments, to date, have used the mouse system, there are no substantial restrictions to utilizing other laboratory or livestock species in such experiments, and recently transgenic rabbits, sheep, and swine have been produced by gene transfer (Hammer et al., 1985).

RETROVIRAL MEDIATED GENE TRANSFER

The first successful introduction of DNA into embryos was demonstrated in 1974 (Jaenisch and Mintz, 1974). SV40 DNA was micro-injected into the blastocoele cavity of mouse blastocysts. These embryos were transplanted into pseudopregnant mice. The survival rate of these embryos was similar to uninjected transplanted embryos. At one year of age, the injected mice, showing no signs of disease, were sacrificed and the DNA from various tissues analyzed for the presence of SV40 DNA. In all animals, SV40 DNA was found in some tissues examined but not all. This probably indicates that some, but not all, of the cells in the blastocyst took up the DNA. Thus, persistence of viral DNA from preimplantation stage to adult life was established, but whether this was due to stable integration or a replicating extrachromosomal entity was not established.

Following the first successful injection of DNA into embryos (Jaenisch and Mintz, 1974), little was done for several years until recombinant DNA technology made purified genes readily available and until techniques for microinjecting directly into nuclei had been perfected. However, during that time experiments using viral infection of embryos yielded some information about germline integration of foreign DNA.

In 1975 Jaenisch (Jaenisch et al., 1975) began a series of experiments using Moloney Murine Leukemia virus, M-MuLV. It was known that infection of newborn mice with M-MuLV leads to virus integration in only a few target tissues (Baluda and Drohan, 1972). To determine if M-MuLV DNA could integrate into all tissues, embryos were cultured in the presence of M-MuLV. Four 8-cell embryos with zona pellucida removed were cultured for 5 hours in the presence of M-MuLV, allowed to develop to blastocysts, and some transplanted into pseudopregnant mice. Others of the blastocysts were examined for production of virus; none was found. Thus, the blastocysts were not permissive to infection by M-MuLV. Of 15 mice born from treated embryos, one was viremic and developed leukemia. Hybridization studies of the DNA of this mouse revealed the presence of

M-MuLV DNA in non target tissue such as brain and testes, as well as target tissues such as lymph nodes and spleen. However, when hybridization studies were done on mRNA isolated from this mouse, 50 to 100 times as much M-MuLV was present in spleen, lymph nodes, and kidney than in testes and brain. The level of hybridizing M-MuLV RNA seemed to correlate with the degree of infiltration of lymphoma cells. Therefore, even though the DNA was present in all tissues, it was expressed differentially.

Male mice infected with M-MuLV at the newborn stage do not transmit the disease to their offspring (Law and Moloney, 1961), although females do through their milk. A viremic male mouse which had been infected at the 4-8 cell stage was studied (Jaenisch, 1976) to determine if germline integration had taken place. This mouse transmitted M-MuLV to 5% of his offspring. However, viremic sons, grandsons, and great-grandsons transmitted M-MuLV to 50% of their offspring. This indicates that the original mouse was a mosaic. The Mendelian inheritance pattern of the offspring indicates that exogenous foreign viral DNA has been stably integrated (become endogenous) in the germ cell line. Hybridization studies on DNA from these mice indicate one copy of M-MuLV DNA per diploid genome in all non-target tissues. However, during leukemogenesis there appeared to be an amplification of M-MuLV sequences in the tumor or lymphoid tissues to 4 copies per diploid genome. Further studies with this mouse line verified true Mendelian inheritance Patterns (Jaenisch, 1977) and also indicated no difference in phenotype in mice heterozygous or homozygous for endogenous M-MuLV DNA. This endogenous M-MuLV has been termed MOV-1 and is located on chromosome 6 (Breindl et al., 1979).

To ascertain whether the site of chromosomal integration affected expression, another male mouse infected at the 4-8 cell stage was studied (Jahner and Jaenisch, 1980). This mouse was again a mosaic, transmitting viremia to 26% of his offspring. These viremic offspring showed endogenous integration by transmitting viremia to 50% of their offspring. Southern blots on the liver of the original male showed three separate integration sites for M-MuLV DNA. Offspring showing only one integration site were examined for phenotype. Three phenotypes were found: 1) those showing early viremia, prior to 3 weeks of age (MOV-1); 2) those showing no viremia at 3 weeks but occasional activation later; and 3) those never developing viremia. Studies on nine more insertional lines (Jaenisch et al., 1981) showed these three phenotypes and one more. This fourth phenotype begins expression of M-MuLV during embryogenesis and has viral

RNA in all organs. This is in contrast to the MOV-1 phenotype where expression occurs just after birth and only in lymphoid tissues. Of the 13 different lines obtained, eight were non-expressors. The results of these two papers clearly demonstrate that for foreign retroviral DNA inserted endogenously into the germ cell line, the chromosomal position of the gene is important for its activation during development.

These mouse lines containing integrated M-MuLV viral genomes within their chromosomal complement have also proven to be of value in suggesting a new method for the identification and cloning of mammalian genes. The MOV-13 mouse line was shown to have a lethal mutation resulting from the integration of the M-MuLV sequences into the first intron of the mouse α 1(I) collagen gene (Harbers et al., 1984). The approach of inducing mutations in the mammalian genome by random retroviral integration permitting the isolation of the mutated gene using the proviral sequences as a tag has been presented and a unique M-MuLV based vector developed for this purpose (Reik et al., 1985).

The additional usefulness of lines with foreign DNA was illustrated in studies on factors involved in the expression of the M-MuLV in these lines. It had previously been shown that in lines that expressed the inserted M-MuLV genome, a somatic amplification of the endogenous genome had occurred in only those tissues expressing the gene (Jaenisch, 1979). Using the restriction enzyme HhaI which cuts the sequence 5'GCGC 3' only if the internal C is not methylated, it was shown that the endogenous M-MuLV DNA was highly methylated in all strains (Stuhlmann et al., 1981). However, the somatically acquired copies were hypomethylated. Since M-MuLV RNA is seen only in those tissues that have somatically acquired copies of M-MuLV (Jaenisch, 1979), it can be concluded that methylation plays a role in retroviral activation and expression during development of eukaryotes. Further studies of de novo methylation of the introduced M-MuLV genomes in the mouse embryo show that efficient de novo methylation occurs after the introduction of the virus into pre-implantation but not into post-implantation embryos (Jahner et al., 1982).

Animal viruses have been used to mediate the transfer of DNA into cells. DNA fragments can be inserted into the viral genome in place of certain viral sequences, and the functions of these viral sequences can be complimented by a virus competent for these gene products (helper virus). Normally, viral particles containing both the chimeric virus and the helper virus are produced. This mixed virus population can be used to infect cells with high efficiency. Murine retroviruses are attractive vectors for the transduction of genes into mammalian cells as well as into embryos and whole animals. Double stranded DNA sequences of interest can be inserted between a 5' and 3' long terminal repeat (LTR) of a provirus. The only retroviral sequences that must be manipulated are the LTRs and packaging sequences located near the 5' LTR (Mann et al., 1983). Cells can then be transfected with this chimeric proviral construct along with a helper virus and a mixed virus population consisting of helper virus and chimeric virus can be detected. If a helper virus containing deleted packaging sequences is used, this virus will not be packaged, leaving a stock consisting only of chimeric virus (Mann et al., 1983).

Retroviral vectors have been constructed to transduce either the herpes simplex virus TK gene (Tabin et al., 1982; Wei et al., 1981) or the Eco-gpt gene (Perkins et al., 1983; Mann et al., 1983). In all cases, chimeric virus could be recovered after transfection with helper virus, and this virus stock could transmit the selectable phenotypes to cells after infection at high efficiency. The advantages of retroviral vectors for the introduction of genes into cells is clear. The efficiency of transfer of the genes to cells after infection is virtually 100%. Retroviruses can package a wide variety of RNA genomes (3.4 to 9.8 kb) and can infect a number of cells types (Goldfarb et al., 1981; Goff et al., 1981). The retroviral DNA is efficiently inserted into the host chromosome (Goldfarb et al., 1981) and it is possible to produce helper free virus stocks by co-infecting with packaging-deficient retroviral mutants (Mann et al., 1983).

The use of murine retroviral vectors to mediate gene transfer in whole animals has been exploited to introduce a selectable gene into different tissues of mice (Stuhlmann et al., 1984). Cell lines producing recombinant replication-deficient murine sarcoma virus (MSV-gpt virus) were microinjected directly into post-implantation mouse embryos and the resulting live born mice contained MSV-gpt sequences in a variety of tissues including thymus, spleen, lung, kidney, and brain showing that cloned genes may be introduced into animals by retroviral vectors. More recently, van der Putten et al. (1985) have demonstrated high efficiency gene transfer into the mouse germ line by infection of pre-implantation mouse embryos with a replication-competent recombinant retrovirus. This study clearly shows that retroviral vectors may be used to establish single copy transgenic animals with exogenous functional genes inserted within the integrated proviral sequences. These workers have inserted the dihydrofolate reductase gene into the germ line of mice and demonstrated the expression of this gene within these animals.

TOTIPOTENT EMBRYONIC CELL MEDIATED GENE TRANSFER

Totipotent pre-implantation mammalian embryonic cells which have been propagated in culture have the capability of reestablishing a partial contribution to a developing pre-implantation embryo into which they are transplanted (Stewart and Mintz, 1981). The resulting fetus and offspring is usually a mosaic with contributions from both the recipient embryonic cells, and the cultured embryonic cell lines may be genetically transformed by DNA mediated co-transformation with a selectable marker so that virtually any cloned gene may be introduced into the chromosomal complement of the totipotent embryo derived cell lines. The pioneering studies in this area were carried out using teratocarcinoma cells (Mintz and Cronmiller, 1981). Although teratocarcinoma cells may be injected into the blastocoele cavity of mouse blastocysts to generate mosaic mice, only one report, describing one mouse, has reported observing germline contribution from these totipotent cell lines (Mintz and Cronmiller, 1981). To date, only one has reported both genetic transformation of teratocarcinoma cells and their introduction into the germ line of mosaic transgenic mice resulting from the introduction of these transformed cells into mouse blastocysts. Recently, cell lines derived directly from cultured, outgrown mouse blastocysts have been established as continuous cell lines known as EK cell lines (Evans and Kaufman, 1981). These EK cell lines have similar properties to the teratocarcinoma cells in that they also produce mosaic contributions to mice resulting from blastocysts into which these cells have been introduced. In the case of the EK cell lines, a substantial percentage of the resulting chimeric offspring are germline chimeras (Bradley et al., 1984) and the producing of such germline chimeras from genetically transformed EK cell lines has been reported very recently (Evans, 1984). Therefore, the EK cells offer a unique approach to the generation of animal lines into which specific genes have been introduced permanently into their germ line.

GENE TRANSFER MEDIATED BY DIRECT DNA PRONUCLEAR MICROINJECTION

Numerous mouse lines containing different added cloned genes in their chromosomal complement and termed "transgenic" have been produced by the direct microinjection of recombinant DNA into the pronuclei of fertilized eggs which developed into liveborn transgenic mice. By far the majority of experimental animal lines containing integrated foreign DNA in their germ line as a result of experimental introduction of DNA have been produced by fertilized egg pronuclear injection. Using a concept and methodology developed at the beginning of this decade (Gordon et al., 1980; Wagner et al., 1981a), virtually any DNA molecule may be introduced into the chromosomal complement of the zygote and, therefore, each of the cells of the resulting transgenic animal. Since the development of this base technology, mouse lines have been established containing cloned transgenes coding for β-globin from several species (Wagner et al., 1981a; Costantini and Lacy, 1981; Stewart et al., 1982; Lacy et al., 1983; Chada et al., 1985; Townes et al., 1985), herpes simplex virus thymidine kinase (Wagner et al., 1981a; Brinster et al., 1981), growth hormone from several species (Palmiter et al., 1982; Palmiter et al., 1983; Selden et al., 1985; Wagner et al., 1984; Hammer et al., 1985), chicken transferrin (McKnight et al., 1983), SV40 viral t-antigen (Brinster et al., 1984; Hanahan, 1985), immunoglobins μ and κ (Grosschedl et al., 1984; Storb et al., 1984; Ritchie et al., 1984), rat elastase (Swift et al., 1984; Ornitz et al., 1985), the cellular oncogene myc product (Stewart et al., 1984), rat myosin light chain (Shani, 1985), and mouse alphafetoprotein (Krumlauf et al., 1985).

The site of introduction of these DNA sequences was the pronuclei of the early one-cell fertilized mouse egg. One-cell embryos were used in these experiments to attempt to insure representation of the exogenous added genes in all cells of the developing organism. The large pronuclear regions of the fertilized egg also provided an easy target for micro-injection. In addition to these reasons for pronuclear microinjection, Wagner et al. (1981a) suggest that the early male pronucleus may provide a specialized nuclear environment for the incorporation of DNA sequences and for their inclusion into a functional chromosomal region. Although very little is presently known about molecular events within the developing pronuclei of mammalian species, biochemical studies of these events have begun to be investigated in the sea urchin. It has been suggested that sperm chromatin dispersion and male chromosomal gene expression may be manifestations of changes in the nucleoprotein content of the paternally derived chromatin within the developing male pronucleus. Kunkle et al. (1978) have shown that soon after fertilization of the sea urchin egg, the male pronuclear chromatin acquires proteins, probably from the egg, of molecular weights greater than 80,000 D and a nuclear protein composition similar to that of the female pronucleus. These workers postulate that

325

such changes in male pronucleus composition may allow the paternal genome to participate in RNA synthesis (Longo and Kunkle, 1977). Also, Laskey et al. (1977) have isolated a group of enzymes from the frog oocyte which organize cellular histones into nucleosomal chromatin units and which may function in structuring the sperm chromatin into transcriptionally functional chromosomal units during early male pronuclear development. Both of these observations suggest that extensive "rebuilding" of functional chromosomes occurs within the early male pronucleus following removal of protamines from the sperm DNA during sperm decondensation. Similar early molecular events within the mouse male pronucleus might assist in assuring the appropriate nucleoprotein structure for the microinjected and integrated exogenous DNA sequences placed in the early male pronucleus. Also, the active DNA repair and ligation enzyme activity in the fertilized egg probably contributes in large measure to the efficiency of DNA integration in pronuclear chromosomes. Clearly, the efficiency of DNA integration is dramatically enhanced in fertilized egg nuclei over the observed rate in DNA mediated transformation or nuclear microinjection in cultured cells by a factor of between 1,000 and 10,000 (Palmiter and Brinster, 1985). Although the rationale for early male pronuclear microinjection for DNA transformation of the mammalian egg may be supported by some molecular data in non-mammalian systems and probable analogous molecular and enzymatic processes in the mammal, no definitive proof has been presented which clearly delineates the molecular mechanisms which result in the highly efficient DNA integration process which occurs following introduction of DNA into egg pronuclei.

A comparison of the three methods which have been employed to produce transgenic animals for study suggests that direct pronuclear injection of fertilized eggs is at present the most effective, efficient, and generally useful method. Using direct egg microinjection, virtually any DNA sequence of any length may be introduced into the mouse germ line with an average efficiency of about 25% (Brinster et al., 1985). Most genes introduced in this fashion function to produce their protein gene product in a highly native manner, suggesting the integration process does not alter gene function. Both retroviral vector mediated and totipotent embryonic cell mediated gene transfer, at best, produce or would produce mosaic founder animals due to DNA integration at later stages of the development process or in only some of the later stage embryonic cells. Extensive and time consuming outbreeding of these founder lines would be required in order to establish lines hemizygous for an added genetic sequence. Also, retroviral mediated gene transfer is restricted by the length of DNA which may be inserted between the LTRs of the vector construction and the activity and fidelity of the gene regulatory and promoter activities in the vicinity of the strong promoter activity of the flanking retroviral LTRs. Because of these advantages, almost all of the experimentally generated transgenic animal lines have been produced by DNA injection into the pronucleus of fertilized eggs which developed into the founder animals of these lines.

DEVELOPMENT OF THE TRANSGENIC ANIMAL SYSTEM FOR THE STUDY OF GENE FUNCTION

Analysis of the first transgenic mice produced by egg pronuclear microinjection centered on determination of the number, integrity, stability of integration, and germline transmission of the introduced genes present in the transgenic mouse genomes. The first successful incorporation of foreign DNA into mice by egg pronuclear injection was reported by Gordon et al. (1980). In this study, in both mice carrying the foreign DNA sequences [herpes virus thymidine kinase (HSV-TK) gene together with SV40 sequences], gross rearrangement of the introduced DNA was observed by Southern hybridization analysis of the transgenic animal DNA. However, in all subsequent studies (Palmiter and Brinster, 1985), only occasional instances of the rearrangement of added DNA has been observed. No apparent explanation for the rearrangement in the Gordon et al. (1980) experiment has been offered or suggested by subsequent studies of transgenic animals, but it seems that DNA sequence rearrangement is not a general problem with transgenic animal production by egg pronuclear microinjection.

In addition to evaluating the integrity of the introduced sequences, researchers first studying transgenic mice sought to determine whether the added genetic sequences were stably integrated into the host chromosomes or were extrachromosomal. The initial indicating that pronuclear injected DNA became stably integrated into the host chromosomes was provided by Mintz and colleagues (Wagner et al., 1981b) studying transgenic mice containing an HSV-TK/human β-globin gene construction. In these studies, Southern blot hybridization analysis of undigested DNA from mice developed from injected fertilized eggs showed the added sequences to migrate with mouse DNA in excess of 50 kb in length. Further studies (Brinster et al., 1981; Palmiter and Brinster, 1985; Brinster et al., 1985) clearly indicate that essentially all transgenic mice resulting from fertilized egg pronuclear DNA injection integrate the added DNA at one, presumably random, site and that when more than one copy of the added gene is present, these multiple copies are arranged in a tandem head-to-tail array of several to several hundred copies of the gene at the single integration site as a consequence of a homologous recombination event between the injected DNA molecules prior to the integration event. The evidence for this conclusion is data from Southern blot hybridization analysis of transgenic mouse DNA which shows ratios of intensity of the introduced hybridizing DNA to hybridizing junction fragments between the integrated DNA and native mouse sequences adjoining the inserted DNA to be inversely proportional to the copy number of the integrated DNA. This is similar to what is seen when tissue culture cells incorporate DNA from DNA-mediated transformation experiments (Perucho et al., 1980; Robbins et al., 1981), but in contrast to the single gene incorporation seen in retroviral mediated DNA integration in embryos (Breindl et al., 1979).

An obvious consequence of the integrated status of transgenes in germ cells of transgenic animals would be germline transmission of the added genetic sequences. Germline transgene transmission was first demonstrated by Wagner et al. (1981a) and Costantini and Lacy (1981) in mice transgenic for the rabbit β-globin gene. Costantini and Lacy (1981) found that four male mice containing injected globin genes transmitted the gene to a portion of their offspring as judged by Southern hybridization analysis of the DNA from these F_1 animals. Wagner et al. (1981a) found the presence of the transgene product, rabbit β-globin, in the blood cells of the offspring of founder transgenic mice harboring the rabbit β-globin gene. Similar results during the same time period by Brinster et al. (1981) with mice transgenic for the metallothionein/HSV-TK fusion gene and Gordon and Ruddle (1981) with mice transgenic for the human interferon gene firmly established germline transmission of transgenes in transgenic mouse lines. Extensive further analysis of numerous transgenic mouse lines have confirmed true Mendelian genetic transmission of transgenes (Brinster et al., 1985).

Demonstrated stable germline integration of intact added foreign genes into the mouse genome allowed analysis of foreign gene expression in founder and decedent mice. Although Wagner and co-workers (Wagner et al., 1981a) showed high levels of expression of the rabbit globin gene in transgenic mice using immunodiffusion techniques to demonstrate the presence of protein product in the blood of these mice which cross-reacted with mouse anti-rabbit globin antisera, other early transgenic mouse experiments did not show transgene expression of genes with their natural promoter sequences [HSV-TK gene, (Gordon et al., 1980); rabbit β-globin gene flanked by lambda phage DNA sequences, (Costantini and Lacy, 1981); and human leukocyte interferon gene, (Gordon and Ruddle, 1981)]. This difficulty in expressing certain genes in transgenic animals led to an approach in which an alternative, strong promoter was fused to the coding region of these genes. In particular, the promoter of the mouse metallothionein I gene was fused to the HSV-TK structural gene (Brinster et al., 1981), the rat growth hormone coding sequences (Palmiter et al., 1982a), and the human growth hormone coding sequences (Palmiter et al., 1983), resulting in the expression of these fusion genes in transgenic animals generally resembling the expression levels natural for the endogenous metallothionein I gene in mice. The results of the alternative strong promoter fusion gene experiments were dramatic, especially when the expressed coding sequences were the growth hormone gene. Mice transgenic for the metallothionein I/rat or human growth hormone gene contained between 100 ng/ml and 100 μg/ml of the foreign growth hormone protein in their serum and grew 100% faster than otherwise isogenic control animals (Palmiter et al., 1982a, 1983; Wagner and Jochle, 1985).

Table I. Exogenous genes expressed in transgenic mice.

GENE	PROMOTER/REGULATORY ELEMENT	REFERENCE
Rabbit β-globin	Homologous	Wagner et al., 1981a
Thymidine kinase (HSV)	Homologous	Wagner et al., 1981b
Thymidine Kinase (HSV)	Metallothionein I	Brinster et al., 1981
Rat growth hormone	Metallothionein I	Palmiter et al., 1982a
Human growth hormone	Metallothionein I	Palmiter et al., 1983
Chicken transferrin	Homologous	McKnight et al., 1983
T-antigen (SV40)	Homologous	Brinster et al., 1984
T-antigen (SV40)	Metallothionein I	Messing et al., 1985
Mouse μ light chain (Ig)	Homologous	Storb et al., 1984
Mouse κ heavy chain (Ig)	Homologous	Grosschedl et al., 1984
Rat elastase	Homologous	Swift et al., 1984
myc	MMTV	Stewart et al., 1984
Rat myosin light chain	Homologous	Shani et al., 1985
Alphafetoprotein	Homologous	Krumlauf et al., 1985
Human β-globin	Homologous	Townes et al., 1985
Human β-globin	Mouse β-globin	Chada et al., 1985
T-antigen	Elastase I	Messing et al., 1985
T-antigen	Insulin	Hanahan, 1985

With the realization that non-eukaryotic DNA flanking sequences (including plasmid and bacteria phage sequences) in DNA fragments introduced into transgenic animals inhibited the genetic expression of transgenes, a wide range of genes with and without promoter replacement with the metallothionein I promoter have been introduced into transgenic mouse lines and express their protein products at essentially native levels (Selden et al., 1985; Wagner et al., 1984; McKnight et al., 1983; Brinster et al., 1984; Grosschedl et al., 1984; Ritchie et al., 1984; Swift et al., 1984; Stewart et al., 1984; Hanahan, 1985; Storb et al., 1984; Ornitz et al., 1985; Shani, 1985; Krumlauf et al., 1985). A list of genes introduced into transgenic mouse lines which express their transgene product is shown in Table I.

REGULATION OF GENETIC EXPRESSION IN TRANSGENIC ANIMALS

Although the mechanisms of gene regulation are central to the study of molecular biology, little was discovered during the first decade of available recombinant DNA technology regarding the molecular processes by which the transcriptional expression of genetic coding sequences is regulated and orchestrated within complex higher eukaryotic species. While the structure of genes, including the intron/exon organization of the structural gene, the identity of specific 3' flanking mRNA processing signals, 5' flanking signals for mRNA capping, and common Goldberg-Hogness ("TATA boxes") sequences, were identified, neither the identity nor the mode of action of those elements which control the timing and level of transcription of complex genes was elucidated. Much of the reason for the inability to identify and study genetic regulatory sequences was a lack of appropriately native test systems for the study of cloned gene expression. Although useful information regarding regulatory genetic elements should be obtainable by transfection of genes or deleted genes into cell lines in culture, often appropriate cell lines which mimic the cells of native living tissue were not available, and those which were available provided false and confusing information. Conversely, genetic sequences introduced into transgenic experimental animals reside in a highly native environment and are exposed to the complete development process. For this reason, the development of the transgenic animal system for the study of gene function has greatly aided investigation of genetic regulatory elements controlling the expression of mammalian genes. From their inception, transgenic animal experiments have proven invaluable in identifying mammalian genetic regulatory elements and elucidating their mode of action.

An important aspect of genetic regulation in mammals and other higher eukaryotic species is tissue-specific expression (i.e. production of the protein gene product exclusively, or highly preferentially, within the cells of one or more specific tissue types). A substantial number of cloned genes have now been expressed in a highly tissue-specific manner in transgenic animals. A compilation of transgenes in transgenic animals which display tissue specific expression is presented in Table II.

The first fully functional transgene in mice, the rabbit β-globin gene (Wagner et al., 1981a), showed a high degree of tissue specificity. These globin gene transgenic mice not only expressed high levels of the foreign globin in transgenic mice, but Northern hybridization studies of mRNA from erythroid bone marrow cells, liver cells, and brain cells showed a greater than 500-fold increase in hybridizing message in the erythroid tissue over that observed in non-erythroid tissues (Wagner et al., 1982). Although similar early studies by Costantini and Lacy (1981) failed to demonstrate significant expression of the rabbit globin gene in transgenic mice and indicated no erythroid cell specific expression (Lacy et al., 1983), later experiments by these workers did show significant expression levels and erythroid tissue specificity of globin genes in transgenic mice (Chada et al., 1985). In their initial experiments, Costantini and Lacy (1981) introduced a rabbit β-globin sequence into mouse fertilized eggs which contained large flanking sequences of lambda phage DNA, and only after introducing β-globin genetic sequences devoid of these prokaryotic sequences did they observe substantial expression of the genes with erythroid tissue-specific expression (Chada et al., 1985). That the

Table II. Tissue-specific gene expression.

GENE	CELL TYPE SPECIFICITY	REFERENCE
Rabbit β-globin	Erythroid	Wagner et al., 1981a
Human β-globin	Erythroid	Townes et al., 1985
Mouse/human β-globin	Erythroid	Chada et al., 1985
Rat elastase I	Pancreatic	Swift et al., 1984
Rat myosin light chain	Skeletal muscle	Shani, 1985
Mouse alphafetoprotein	Yolk sac & liver	Krumlauf et al., 1985
Mouse κ light chain (Ig)	B cells	Storb et al., 1984
Mouse μ heavy chain (Ig)	B and T cells	Grosschedl et al., 1984
Insulin/T-antigen	Pancreatic cells	Hanahan, 1985
Rat elastase I/T-antigen	Pancreatic	Messing et al., 1985

regulatory sequences required to impart erythroid specific expression to the β-globin gene are within the sequences introduced into transgenic mice has been further confirmed and elucidated by the recent studies of Townes et al. (1985). In these studies both a complete human β-globin gene with extensive 5' flanking sequences and a truncated gene with 5' flanking deletions to -48 both expressed in appropriated erythroid tissues, suggesting that elements involved in tissue-specific regulated expression of β-globin reside within the structural gene or 3' to it.

In addition to the globin gene system, tissue specificity has been demonstrated for a substantial percentage of the genes which have been introduced into transgenic mice. Perhaps the most dramatic example of selective tissue expression is in mice transgenic for the rat elastase gene which show expression levels of elastase in acinar cells of the pancreas more than 10,000-fold greater than in any other tissue examined (Swift et al., 1984; Ornitz et al., 1985). Localization of the cis-acting elements responsible for this remarkable tissue-specific regulation has been accomplished using a fusion gene constructed from the 5' flanking promoter/regulatory region of the rat elastase gene and the structural gene coding for human growth hormone (Ornitz et al., 1985). By introducing 5' deletions of the elastase promoter/regulatory sequences of this fusion gene, these workers have identified the site of the elastase tissue-specific regulatory element to a 213 bp region contiguous with the promoter for the elastase gene. This experiment also clearly demonstrates the ability to use tissue-specific regulatory elements from one gene to direct the expression of another gene in a specific tissue where the latter gene is not normally present in order to study the resulting physiological or developmental consequences of this aberrant expression.

The immunoglobin genes μ and κ have also been shown to be expressed in a tissue-specific fashion in lymphoid cells. Immunoglobin μ genes are expressed equally well in both B and T cells, and κ light chain genes are only expressed in B cells while neither are expressed to any observable extent in non-lymphoid cells (Storb et al., 1984; Grosschedl et al., 1984; Rusconi and Kohler, 1985). Interesting studies of the tissue-specific regulatory elements in the rat myosin light chain gene (Shani, 1985) and the developmentally important mouse alphafetoprotein gene (Krumlauf et al., 1985) are also underway.

Transgenic animal studies of tissue-specific gene regulation have not been performed out of context with other studies of gene regulation.

Using other, often viral, experimental systems, DNA sequences associated with the specific enhancement of the transcription of both viral and mammalian genes, termed "enhancer sequences," have been identitied (Khoury and Gruss, 1983; (Church et al., 1985; Queen and Baltimore, 1983; Banerji et al., 1983; Ephrussi et al., 1985).

Analysis of genes transcribed by polymerase II has revealed, among other promoter/regulatory elements, elements that appear to increase transcriptional efficiency in a manner relatively independent to their position and orientation with respect to a nearby gene (Khoury and Gruss, 1983). The first of these enhancer elements were discovered in animal viral genes, initially as sequences 5' to the SV40 viral structural gene and later in many other viral systems including retroviruses (Temin, 1982). The SV40 enhancer, a 72 bp tandem repeat of SV40 DNA, was first identified as a cis-acting element located more than 100 bp 5' of the cap site of the early viral genes in SV40 (Benoist and Chambon, 1981; Gruss et al., 1981). Deletion of this element dramatically reduced the t-antigen gene expression in SV40 virus mutants; and linkage of the SV40 enhancer to heterologous genes, including the herpes simplex thymidine kinase gene, the mouse metallothionein gene, and the p21 transforming gene from Harvey murine sarcoma virus, resulted in increased transcriptional activity of, at least, several orders of magnitude (Khoury and Gruss, 1983). Also, enhancer elements have been identified in cellular genes including the immunoglobin heavy chain genes (Banerji et al., 1983; Queen and Baltimore, 1983). The immunoglobin enhancer element is naturally found in the intron joining the constant region of the gene to the variable region, but movement of this element either 5' or 3' of the structural gene allows retention of enhancer activity (Gillies et al., 1983). The immunoglobin enhancers, as is the case with viral enhancers as well, may also be inverted in orientation and function equally as well (Banerji et al., 1983; Khoury and Gruss, 1983). The enhancers in the immunoglobin gene system provide tissue-specific activation of somatically rearranged immunoglobin genes and possibly cause abnormal expression of other genes such as c-myc that become translocated to their domain of influence (Banerji et al., 1983).

Recently, Church et al. (1985) have demonstrated that the tissue-specific character of immunoglobin enhancers is due to the interaction of lymphocyte-specific cellular proteins with enhancer sequences, rendering the enhancers operative only on lymphoid tissue where appropriate, active, enhancer/cell protein complexes can form.

333

It seems highly likely that the tissue-specific regulatory elements identified with the globin, elastase, myosin, alphafetoprotein genes in transgenic mouse experiments, like the immunoglobin gene tissue-specific regulatory elements, will be shown to be enhancer elements which function in a similar manner to those already identified and characterized for viral systems.

Regulation responsive to other signals than tissue-specific factors has also been demonstrated for introduced genes in transgenic animals. The pioneering studies by Brinster and co-workers (Brinster et al., 1981; Palmiter et al., 1982a, b; Palmiter et al.,1983) studying the expression of mouse metallothionein I fusion genes in transgenic mice clearly demonstrated the regulation of coding sequences positioned 3' from the metallothionein I promoter/regulatory sequences by heavy metals which generally resembled the heavy metal inducibility of the endogenous mouse metallothionein I gene. Levels of thymidine kinase activity in metallothionein I/HSV-TK transgenic mice (Brinster et al., 1981) and growth hormone in metallothionein I/growth hormone transgenic mice (Palmiter et al., 1982a; 1983) was elevated as much as 10-fold when excess cadmium or zinc were injected or fed to these transgenic mice.

However, the utility of fusion genes in transgenic experiments is limited by the expression parameters of the fusion promoter. This drawback might be overcome if the expression of the fusion gene could be modulated by including additional regulatory sequences in the injected DNA construction. For example, the expression of the fusion gene might be modified to more closely resemble that of the corresponding endogenous gene or its expression could be drastically altered from that of the endogenous gene in order to perturb a physiological or developmental system. Recently, it has been shown that a binding site for the glucocorticoid receptor is present within the first intron of the human growth hormone structural gene (Moore et al., 1985). Since the mouse metallothionein I/human growth hormone fusion gene contains this putative glucocorticoid regulation site in addition to a strong promoter, Selden et al. (1985) undertook an experiment to demonstrate if transgenic mice containing this fusion gene would express human growth hormone at differing levels depending upon the level of added glucocorticoid within the transgenic animals. The results of this experiment clearly showed a 2- to 6-fold increase in circulating human growth hormone level and a 4-fold increase in human growth hormone liver mRNA following treatment of the transgenic mice with either injected dexamethasone or orally administered triamcinolone.

Through the use of regulatory sequences such as the 15 bp gluco-corticoid regulation sequence in the human growth hormone gene or those tissue-specific enhancer elements identified for the elastase gene, the immunoglobin genes and other tissue specific transgenes, specifically designed genes may be constructed to alter gene expression in an experimentally useful manner. Experiments underway in many laboratories using such complex genetic constructions in transgenic animals offer a bright future for the use of transgenic animals to study a wide range of interesting biological problems.

ONCOGENE EXPRESSION AND TUMORIGENESIS IN TRANSGENIC MICE

Transgenic mice harboring specific cloned oncogenes or transforming genes provide the opportunity to study the role of these genes in the process of tumorigenesis. Transgenic mice containing DNA coding for the early transcribed region of the SV40 virus routinely develop choroid plexus papillomas within the first six months of birth (Brinster et al., 1984). Both the presence of SV40 mRNA and T-antigen were detected in these tumors but not in other tissues of the transgenic mouse. Further, deletions within the SV40 early region DNA in the small t-antigen or in the large T-antigen coding regions provided useful data identifying the specific gene product responsible for tumorigenesis in these transgenic lines. Whereas, truncated SV40 sequences deleted in the small t-antigen coding region yielded transgenic mice which showed tumor development essentially identical with those transgenic lines containing the complete early coding region, SV40 early region sequences deleted in the large T-antigen coding region produced transgenic mice which showed no signs of tumorigenesis (Brinster et al., 1984). Also, deletion of the SV40 72 bp enhancer sequence greatly reduced the incidence of choroid plexus tumors (Palmiter and Brinster, 1985). The results of these studies clearly demonstrate that expression of SV40 large T-antigen results in tumorigenesis and that the site and level of expression of this transforming gene are directed by specifically identifiable regulatory elements associated with the SV40 gene (i.e. the 72 bp, tandem repeat, enhancer sequences).

An important concept emerged from the seminal studies of the Brinster and Palmiter group on the role of large T-antigen, and by inference, any oncogene in tumorigenesis. This hypothesis suggests that the specific location and type of tumor formation is the direct result of tissue-specific regulatory elements expressing significant levels of oncogene product in a tissue-specific fashion and inducing tumor formation in the regulatory element targeted tissue. To support this hypothesis, transgenic mice were produced containing the large T-antigen structural gene fused to several heterologous promoters with different tissue specificities. SV40 T-antigen structural genes were fused to the metallothionein I, elastase I, and insulin promoter/enhancer elements and transgenic lines generated containing each of these fusion genes (Palmiter and Brinster, 1985; Hanahan, 1985). Those transgenic mice with the metallothionein I/T-antigen fusion gene developed hepatocarcinomas and insulinomas and a high incidence of hind limb paralysis resulting from abnormal myelination of peripheral axons (Messing et al., 1985). The transgenic animals with elastase I and insulin promoter/enhancer elements showed high incidence of pancreatic adenomas and insulinomas, respectively (Palmiter and Brinster, 1985).

Of particular importance to establishing the molecular basis for tumorigenesis in the mammal is the study of Stewart et al. (1984) in which the hormonally inducible mouse mammary tumor virus promoter/enhancer sequences (Ringold, 1983) were fused to the structural region of the cellular oncogene myc in order to regulate expression of the myc gene by prolactogenic stimuli and glucocorticoids. Transgenic mice carrying this unique construct produced myc oncogene product in the mammary tissue during lactation, resulting in spontaneous, tissue-specific (mammary), adenocarcinomas. Although myc oncogene tissue-specific expression was not absolute, the presence of the mouse mammary tumor virus promoter/enhancer sequences directed oncogene production and tumorigenesis to the mammary gland and tumors did not develop in other tissues. Since myc oncogene has never before been found associated with mammary tumors, it is clear that myc, like T-antigen, can induce tumorigenesis in abnormal target tissues when its expression is controlled by a heterologous regulatory element.

Transgenic experimental animals have already contributed greatly to the established knowledge regarding the role of oncogenes in tumori-genesis, and it seems likely that this unique whole animal gene analysis system will continue to be one of the most important tools in the study of oncogenesis.

336

INSERTIONAL MUTAGENESIS IN TRANSGENIC ANIMALS

Transgenic experimental animals, whether produced by retroviral insertional vectors or by direct fertilized egg microinjection, contain inserted exogenous DNA sequences in their chromosomal complement. Unless these exogenous inserts are positioned within repetitive DNA sequences or other non-coding or non-regulatory sequences, their presence must disrupt the function of the gene into which they have randomly incorporated. Identification of the presence of such insertional mutagenesis requires inbreeding the offspring from founder animals since in the founder transgenic animal only one of its two copies of the affected gene are disrupted.

This type of induced insertional mutagenesis has now been observed in high percentage of transgenic mice lines where inbreeding towards homozygous lines has been attempted. Jaenisch et al. (1983) have identified several insertional mutants, which are developmentally impaired, by inbreeding MOV mouse lines. The most well characterized of these mutants is the MOV-13 line where insertion of M-MuLV sequences disrupted the α 1(I) collagen gene resulting in developmental lethality (Schnieke et al., 1983; Harbers et al., 1984). Wagner et al. (1983) also found small litter sizes and an inability to inbreed to homozygous status in a human growth hormone transgenic mouse line as the apparent result of a lethal mutation in a developmentally important genetic loci. In another particularly interesting example of insertional mutagenesis, a transgenic line transmitted the exogenous added sequence only through females. Although the male transgenic mice contained the foreign DNA and were fertile, none of their offspring were found to contain the transgene (Palmiter et al., 1984). These findings were interpreted by the authors of this report to indicate that the inserted DNA disrupted a gene that is normally transcribed during spermatogenesis, when sperm are haploid (Palmiter and Brinster, 1985).

Perhaps the most interesting insertional mutant identified to date in a transgenic mouse line has been characterized by Leder and co-workers (1985) in transgenic mice containing the MMTV/myc fusion gene described earlier (Stewart et al., 1984). The mice resulting from inbreeding of the offspring from a single mouse transgenic for the MMTV-myc gene showed high incidence of syndactyly, a developmental defamation in fore and hind limbs showing fused and deleted digits. Finding this particular insertional mutant was especially fortuitous, since spontaneous mutants with the same characteristics have been identified within the breeding colony at Jackson Laboratory. Using the MMTV-myc marker sequences to probe a genomic library produced from this transgenic mouse line, the disrupted gene has been cloned (Leder, 1985). These recently completed experiments represent a feat accomplished for the first time in a mammalian system: the discovery and cloning of a developmentally important gene with an identified function (limb development) but for which no product protein or DNA sequence was known. This accomplishment shows the power and promise of the technique of induced insertional mutagenesis in transgenic experimental animals.

A very recent report (Smithies et al., 1985) describing methods to enhance the probability of DNA insertion by homologous recombination in mammalian cell DNA mediated transformation suggests that future improvements in these methods may allow transgenic experiments which combine insertional mutagenesis and gene introduction (replacement). If vectors which target the integration of injected sequences to homologous sequences within the chromosomes of the transgenic hose animal become available, it may be possible in the future to insert specific sequence modified genes into the native homologous gene within experimental animal lines. Such an experimental procedure would be far more powerful than existing transgenic technology, since it would result in gene replacement, not just gene additions, and it would allow examination of the physio- logical and developmental consequences of altering genetic regulatory and structural sequences.

Although only a few experiments have been carried out using trans- genic insertional mutagenesis, a remarkably high percentage (20% to 30%) of the transgenic mouse lines inbred in order to search for insertional mutations have yielded such mutations. It seems inevitable that the next several years will witness a dramatic increase in the number of transgenic insertional mutants produced for use in the study of gene function.

FUTURE EXPERIMENTAL POSSIBILITIES PRESENTED BY THE TRANSGENIC ANIMAL SYSTEM

In addition to the increased use of induced insertional mutagenesis in the whole animal via transgenic animals to study gene function, gene replacement, and especially the process of early embryonic development in the mammal many other new and expanded existing uses of transgenic animals will play an increasingly important role in the field of molecular biology for decades to come.

Although the cellular oncogenes are thought to play an important role in early mammalian development, the nature of that role remains unknown. Transgenic animals carrying either cellular oncotransgenes under the regulation of selected promoter/enhancer/regulators, or induced insertional mutant oncogenes promise to provide an experimental system to study this role as well as the role of oncogenes in tumorigenesis. The study of cellular oncogenes and gene function in general may be greatly aided by several, yet untested, potential new approaches using transgenic animals. Two of these possibilities, antisense transgenes and the use of unique synthetic regulatory elements, are worthy of note.

Due to the anti-parallel, duplex structure of the DNA molecule, only the transcript of one of the DNA strands will serve as an appropriate coding message for the translation of the protein product of a gene. The transcript of the other DNA strand, although useless in coding for appropriated protein, if used as a coding strand would code for a mRNA which is complementary to the appropriate message. By inverting the DNA sequences, cDNA, in a recombinant expression vector, such "antisense" mRNA may be produced. Several experiments in both prokaryote and eukaryote systems have shown that such antisense RNA inhibits the production of gene products from the native message (Mizuno et al., 1984; Simons and Kleckner, 1983; Coleman et al., 1984; Izant and Weintraub, 1984; Zamecnik et al., 1978; Rosenberg et al., 1984). The mechanism by which antisense RNA inhibits gene function is not fully known. It may act in the nucleus by inhibiting nascent poly(a)+RNA, leading to its degradation or preventing its export (Rosenberg et al., 1985), or by directly blocking translation of the mRNA by hybridization in the cytoplasm (Mizuno et al., 1984; Simons and Kleckner, 1983; Coleman et al., 1984).

Rosenberg et al. (1985) have demonstrated the use of antisense RNA in studying gene function on drosophila embryonic development. Antisense RNA complementary to the Kruppel gene message was introduced into developing drosophila embryos. Kruppel is a gene locus where mutants result in foreshortened embryos lacking all thoracic and several abdominal segments (Nusslein-Volhard and Wieschaus, 1980). Kruppel antisense RNA, synthesized in vitro from inverted Kruppel cDNA, was introduced into developing embryos containing the completely native Kruppel gene and

339

foreshortened Kruppel mutant flies resulted. This study clearly shows the action of antisense RNA to inhibit specific gene function in situ in a eukaryotic system.

This action of antisense RNA suggests the use of specific orientationally inverted genetic constructions, coding for the antisense RNA complementary to the message from genes of general or developmental interest, in transgenic animal lines to inhibit the action of the complementary native gene. Such antisense transgenic animals would offer a more efficient alternative to random insertional mutagenesis in the study of the role of known genetic sequences in the development or maintenance physiology of mammals.

Other developing technologies in addition to antisense RNA are directly applicable to the transgenic animal system. With the identification and characterization of genetic regulatory elements of minimal sequence length, such as the glucocorticoid receptor binding sites in several genes (Moore et al.,1985; Scheidereit et al., 1983; Karin et al., 1984), enhancer elements (Khoury and Gruss, 1983), and other genetic regulatory elements rapidly being identified, the possibility arises for synthetic modification of these sequences altering either their function, sensitivity, or response specificity.

Although the structural relationship between genetic effector molecules and DNA genetic regulation sequences must be quite specific and complex, recent data suggests that these relationships may be modified to elicit different genetic responses. Probably the best examples of such modifications are found in steroid hormonal regulation of gene function. Already, a wide range of steroid analogues are available which show different binding characteristics to receptor proteins (Tate et al., 1984; Rochefort et al., 1983) altering the genetic response resulting from the binding of these steroid/receptor complexes to their DNA binding sequences (Katzenellenbogen et al., 1983; Tate et al., 1984; Wakeling et al., 1984). Also, alteration of the binding affinity and specificity of DNA toward genetic effector and transcription processing proteins by synthetic sequence alteration of DNA binding sequences is well established in prokaryotic systems (Caruthers et al., 1980; Caruthers et al., 1982). Using analogous approaches, it may be possible to synthetically alter mammalian genetic regulation sequences in order to produce sequences which would serve as enhancers or activators of gene expression in response to specific synthetic small molecules. The use of such synthetic regulatory elements in recombinant genetic constructions introduced into transgenic animals might allow controlled expression of selected genes by external induction and make possible studies of the effects of native genes or antisense genes on mammalian physiology or development which, if expressed under native or existing regulatory elements, might restrict the viability of the transgenic animal.

Clearly, with existing transgenic methodology and the potential future addition of targeted gene insertion and gene replacement, antisense transgenes, and synthetically regulated transgenes, the ultimately native transgenic animal experimental system will play a dominating role in the future of experimental mammalian molecular biology.

In addition to the central role of transgenic animals in experimental molecular biology, gene transfer provides enormous opportunities in domestic animal genetics. Hammer et al. (1985) have demonstrated the application of gene transfer techniques to livestock animals, and extensive work is presently in progress to use this technology to enhance the production capacity of agricultural animals. Gene transfer bolds the possibility of allowing greater increases in genetic improvement in livestock production traits such as growth, feed efficiency, disease resistance, and reproductive efficiency in a single gestation period than presently possible using classical genetic selection procedures over several decades. The incorporation of transgenic animal studies into animal science research has already begun and represents a technology revolution in this important agricultural field.

ACKNOWLEDGEMENTS

This work was supported by grant HD09042 from the National Institute of Child Health and Human Development.

REFERENCES

Baluda, M. A., and Drohan, W. M., 1972, Distribution of deoxyribonucleic acid complementary to the ribonucleic acid of avian myelobastosis virus in tissues of normal and tumor-bearing chickens, J. Virol., 10:1002-1009.

Banerji, J., Olson, L., and Schaffner, W., 1983, A lymphocyte-specific cellular enhancer is located downstream of the joining region in immunoglobulin heavy chain genes, Cell, 33:729-740.

Benoist, C., and Chambon, P., 1981, In vivo sequence requirements of the SV40 early promotor region, Nature, 290(5804):304-310.

Bradley, A., Evans, M., Kaufman, M. H., and Robertson, E., 1984, Formation of germ-line chimeras from embryo-derived teratocarcinoma cell lines, Nature, 309:255-256.

Breindl, M., Doehner, J., Willecke, K., Dausman, J., and Jaenisch, R., 1979, Germline integration of moloney leukemia virus: Identification of the chromosomal integration site, Proc. Natl. Acad. Sci. USA, 76:1938-1948.

Brinster, R. L., Chen, H. Y., Messing, A., Van Dyke, T., Levine, A. J., and Palmiter, R. D., 1984, Transgenic mice harboring SV40 T-antigen genes develop characteristic brain tumors, Cell, 37:367-379.

Brinster, R. L., Chen, H. Y., Trumbauer, M., Senear, A., Warren, R., and Palmiter, R. D. 1981, Somatic expression of herpes thymidine kinase in mice following injection of a fusion gene into eggs, Cell 27:223-231.

Brinster, R. L., Chen, H. Y., Trumbauer, M., Yagle, M. K., and Palmiter, R. D., 1985, Factors affecting the efficiency of introducing foreign DNA into mice by microinjecting eggs, Proc. Natl. Acad. Sci. USA, 82:4438-4442.

Caruthers, M. H., Beaucage, S. L., Efcavitch, J. W., Fisher, E. F., Matteucci, M. D., and Stabinsky, Y., 1980, New chemical methods for synthesizing polynucleotides, Nucleic Acids Symp. Ser., 7:215-223.

Caruthers, M. H., Beaucage, S. L., Efcavitch, J. W., Fisher, E. F., Goldman, R. A., deHaseth, P. L., Mandecki, W., Matteucci, M. D., Rosendahl, M. S., and Stabinsky, Y., 1982, Chemical synthesis and biological studies on mutated gene-control regions, Cold Spring Harbor Symp. Quant. Biol., 47(part 1):411-418.

Chada, K., Magram, J., Raphael, K., Radice G., Lacy, E., and Costantini, F., 1985, Specific expression of a foreign beta-globin gene in erythroid cells in transgenic mice, Nature, 314:377-380.

Church, G. M., Ephrussi, A., Gilbert, W., and Tonegawa, S., 1985, Cell-type-specific contacts to Ig enhancers in nuclei, Nature, 313:798-801.

Coleman, J., Green, P. J., and Inouye, M., 1984, The use of RNAs complementary to specific mRNAs to regulate the expression of individual bacterial genes, Cell, 37:429-436.

Costantini, F., and Lacy, E., 1981, Introduction of a rabbit β-globin gene into the mouse germ line, Nature, 294:92-94.

Ephrussi, A., Church, G. M., Tonegawa, S., and Gilbert, W., 1985, B lineage--specific interactions of an immunoglobulin enhancer with cellular factors in vivo, Science, 227:134-140.

Evans, M., 1984, personal communication.

Evans, M. J., and Kaufman, M. H., 1981, Establishment in culture of pluripotential cells from mouse embryos, Nature, 292:154-156.

Gillies, S. D., Morrison, S. L., Oi, V. T., and Tonegawa, S., 1983, A tissue-specific transcription enhancer element is located in the major intron of a rearranged immunoglobulin heavy chain gene, Cell, 33:717-723.

Goff, S. P., Tabin, C. J., Wang, J. Y-J., Weinberg, R. A., and Baltimore, B., 1981, Transfection of fibroblasts by cloned Abelson murine leukemia virus DNA and recovery of transmissible virus by recombination with helper virus, J. Virol., 41:271:285.

Goldfarb, M. P., and Weinberg, R. A., 1981, Structure of the provirus within NIH 3T3 cells transfected with Harvey sarcoma virus DNA, J. Virol., 38:125-135.

Gordon, J. W., and Ruddle, F. H., 1981, Integration and stable germline transmission of genes injected into mouse pronuclei, Science, 214:1244-1246.

Gordon, J. W., Scangos, G. A., Plotkin, D. J., Barbosa, J. A., and Ruddle, F. H., 1980, Genetic transformation of mouse embryos by micro-injection of purified DNA, Proc. Natl. Acad. Sci. USA, 77(12):7380-7384.

Grosschedl, R., Weaver, D., Baltimore, D., and Costantini, F., 1984, Introduction of a μ immunoglobulin gene into the mouse germ line: specific expression in lymphoid cells and synthesis of functional antibody, Cell, 38:647-658.

Gruss, P., Dhar, R., and Khoury, G., 1981, Simian virus 40 tandem repeated sequences as an element of the early promoter, Proc. Natl. Acad. Sci. USA, 78:943-947.

Hammer, R. E., Palmiter, R. D., and Brinster, R. L., 1985, Partial correction of murine hereditary growth disorder by germline incorporation of a new gene, Nature, 311:65-67.

Hanahan, D., 1985, Heritable formation of pancreatic beta-cell tumors in transgenic mice expressing recombinant insulin/simian virus 40 oncogenes, Nature, 315:115-122.

Harbers, K., Keuhn, M., Delius, H., and Jaenisch, R., 1984, Insertion of retrovirus into the first intron of 1(I) collagen gene leads to embryonic lethal mutation in mice, Proc. Natl. Acad. Sci. USA, 81:1504-1508.

Izant, J. G., and Weintraub, H., 1984, Inhibition of thymidine kinase gene expression by antisense RNA: A molecular approach to genetic analysis, Cell, 36:1007-1015.

Jaenisch, R., Harbers, K., Schnieke, A., Lohler, J., Chumakov, I., Jahner, D., Grotkopp. D., and Hoffman, E., 1983, Germline integration of Moloney murine leukemia virus at the Mov13 locus leads to recessive lethal mutation and early embryonic death, Cell, 32:209-216.

Jaenisch, R., 1976, Germ line integration and Mendelian transmission of the exogenous Moloney leukemia virus, Proc. Natl. Acad. Sci. USA, 73:1260:1264.

Jaenisch, R., 1977, Germ line integration of Moloney leukemia virus: Effect of homozygosity at the M-MuLV locus, Cell, 12:691-696.

Jaenisch, R., 1979, Moloney leukemia virus gene expression and amplification in preleukemic and leukemic BALB/Mo mice, Virology, 93:80-90.

Jaenisch, R., Fan, H. C., and Crocker, R., 1975, Infection of preimplantation mouse embryos and of newborn mice with leukemia virus: Tissue distribution, Proc. Natl. Acad. Sci. USA, 72:4008-4012.

Jaenisch, R., Jahner, D., Nobis, P., Simon, I., Lohler, J., Harbers, K, and Grotkopp, D., 1981, Chromosomal position and activation of retroviral genomes inserted into the germ line of mice, Cell, 24:519-529.

Jaenisch, R., and Mintz, B., 1974, Simian Virus 40 DNA sequences in DNA of healthy adult mice derived from preimplantation blastocysts injected with viral DNA, Proc. Natl. Acad. Sci. USA, 71:1250-1254.

Jahner, D., and Jaenisch, R., 1980, Integration of Moloney leukemia virus into the germ line of mice: Correlation between site of integration and virus activation, Nature, 287:456-458.

Jahner, D., Stuhlmann, H., Stewart, C. L., Harbers, K., Lohler, J., Simon, I., and Jaenisch, R., 1982, De novo methylation and expression of retroviral genomes during mouse embryogenesis, Nature, 298:623-628.

Karin, M., Haslinger, A., Holtgreve, H., Richards, R., Krauter, P., Westphal, H., and Beato, M., 1984, Characterization of DNA sequences through which cadmium and glucocorticoid hormones induce human metallothionein-IIA gene, Nature, 308:513-519.

Katzenellenbogen, B. S., Miller, M. A., Eckert, R. L., and Sudo, K., 1983, Antiestrogen pharmacology and mechanism of action, J. Steroid Biochemistry, 19(1A):59-68.

Khoury, G, and Gruss, P., 1983, Enhancer elements, Cell, 33:313-314.

Krumlauf, R., Hammer, R. E., Tilghman, S. M., Brinster, R. L., 1985, Development of α-fetalprotein genes in transgenic mice, Mol. Cell. Biol., 5:1639-1648.

Kunkle, M., Longo, F. J., and Magun, B. E., 1978, Nuclear protein changes in the maternally and paternally derived chromatin at fertilization, J. Exp. Zool., 203:371-380.

Lacy, E., Roberts, S., Evans, E. P., Burtenshaw, M. D., and Costantini, F. D., 1983, A foreign beta-globin gene in transgenic mice: Integration at abnormal chromosomal positions and expression in inappropriate tissues, Cell, 34:343-358.

Laskey, R. A., Honda, B. M., Mills, A. D., Morris, N. R., Wyllie, A. H., Mertz, J. E., DeRoberts, E. M., and Gurdon, J. D., 1977, Chromatin assembly and transcription in eggs and oocytes of Xenopus laevis, Cold Spring Harbour Symp. Quant. Biol., 42:171-178.

Law, L. W., and Moloney, J. B., 1961, Studies of congenital transmission of a leukemia virus in mice, Proc. Soc. Exp. Biol. and Med., 108:715-723.

Leder, P., 1985, personal communication.

Longo, F. J., and Kunkle, M., 1977, Synthesis of RNA by male pronuclei of fertilized sea urchin eggs, J. Exp. Zool., 201:431-438.

Mann, R. C., Mulligan, R. C., and Baltimore, D., 1983, Construction of a retrovirus packaging mutant and its use to produce helper-free defective retrovirus, Cell, 33:153-159.

Martin, G. R., 1981, Isolation of a pluripotent cell line from early mouse embryos cultured in medium conditioned by teritocarcinoma stem cells, Proc. Natl. Acad. Sci. USA, 78:7634-7638.

McKnight, G. S., Hammer, R. E., Kuenzel, E. A., and Brinster, R. L., 1983, Expression of the chicken transferrin gene in transgenic mice, Cell, 34:335-341.

Messing, A., Chen, H. Y., Palmiter, R. D., Brinster, R. L., 1985, Peripheral neuropathies, hepatocellular, carcinomas, and islet cell adenomas in transgenic mice, Nature, 316:461-463.

Mintz, B., and Cronmiller, C., 1981, Nett-1: A karyotypical normal in vitro line of developmentally totipotent mouse teritocarcinoma cells, Somatic Cells Genet., 7:489-505.

Mizuno, T., Chou, M-Y, and Inouye, M., 1984, A unique mechanism regulating gene expression: Translational inhibition by a complementary RNA transcript (micRNA), Proc. Natl. Acad. Sci. USA, 81:1966-1970.

Moore, D. D., Marks, A. R., Buckley, D. I., Kapler, G., Payvar, F., and Goodman, H. M., 1985, The first intron of the human growth hormone gene contains a binding site for glucocorticoid receptor, Proc.

Natl. Acad. Sci. USA, 82(3):699-702.

Nusslein-Volhard, C., and Wieschaus, E., 1980, Stations affecting segment number and polarity in drosophila, Nature, 287:795-801.

Ornitz, D. M., Palmiter, R. D., Hammer, R. E., Brinster, R. L., Swift, G. H., and MacDonald, R. J., 1985, Specific expression of an elastase-human growth hormone fusion gene in pancreatic acinar cells of transgenic mice, Nature, 313(6003):600-602.

Palmiter, R. D., and Brinster, R. L., 1985, Transgenic Mice, Cell, 41:343-345.

Palmiter, R. D., Brinster, R. L., Hammer, R. E., Trumbauer, M. E., Rosenfeld, M. G., Birnberg, N. C., and Evans, R. M., 1982a, Dramatic growth of mice that develop from eggs microinjected with metallothionein-growth hormone fusion genes, Nature, 300:611-615.

Palmiter, R. D., Norstedt, G., Gelinas, R. E., Hammer, R. E., and Brinster, R. L., 1983, Metallothionein-human GH fusion genes stimulate growth of mice, Science, 222:809-814.

Palmiter, R. D., Chen, H. Y., and Brinster, R. L., 1982b, Differential regulation of metallothionein thymidine kinase fusion genes in transgenic mice and their offspring, Cell, 29:701-710.

Palmiter, R. D., Wilkie, T. M., Chen, H. Y., and Brinster, R. L., 1984, Transmission distortion and mosaicism in an unusual transgenic mouse pedigree, Cell, 36:869-877.

Perkins, A. S., Kirschmeier, P. T., Gattoni-Celli, S., and Weinstein, I. B., 1983, Design of a retrovirus-derived vector for expression and transduction of exogenous genes in mammalian cells, Mol. Cell Biol., 3:1123:1132.

Perucho, M., Hanahan, D., and Wigler, M., 1980, Genetic and physical linkage of exogenous sequences in transformed cells, Cell, 22:309-317.

Queen, C., and Baltimore, D., 1983, Immunoglobulin gene transcription is activated by downstream sequence elements, Cell, 33:741-750.

Reik, W., Weiher, H., and Jaenisch, R., 1985, Replication-competent Moloney murine leukemia virus carrying a bacterial suppressor tRNA gene: Selective cloning of proviral and flanking host sequences, Proc. Natl. Acad. Sci. USA, 82:1141-1145.

Ringold, G. M., 1983, Regulation of mouse mammary tumor virus gene expression by glucocorticoid hormones, Curr. Topics Microbiol. Immunol., 106:79-103.

Ritchie, K. A., Brinster, R. L., and Storb, U., 1984, Allelic exclusion and control of endogenous immunoglobulin gene rearrangement in kappa transgenic mice, Nature, 312:517-520.

Robbins, D. M., Ripley, S., Hendeson, A., and Axel, R., 1981, Transforming
DNA integrates into the host chromosome, Cell, 23:29-39.

Rochefort, H., Borgna, J. L., and Evans, F., 1983, Cellular and molecular
mechanism of action of antiestrogens, J. Steroid Biochem.,
19:69-74.

Rosenberg, U. B., Preiss, A., Seifer, E., Jackle, H., and Knipple, D. C.,
1985, Production of phenocopies by Kruppel antisense RNA injection
into Drosophila embryos, Nature, 313:703-706.

Rusconi, S., and Kohler, G., 1985, Transmission and expression of a
specific pair of rearranged immunoglobin µ and κ genes in a
transgenic mouse line, Nature, 314:330-334.

Scheidereit, C., Geisse, S., Westphal, H. M., Beato, M., 1983, The
glucocorticoid receptor binds to defined nucleotide sequences near
the promoter of mouse mammary tumor virus, Nature, 304:749-752.

Schneike, A., Harbers, K., and Jaenisch, R., 1983, Embryonic lethal
mutation in mice induced by retrovirus insertion into the α1(I)
collagen gene, Nature, 304:315-320.

Selden, R. F., Wagner, T. E., Yun, J. S., Moore, D. D., and Goodman, H.,
1985, Glucocorticoid regulation of human growth hormone expression
in transgenic mice, submitted.

Shani, M., 1985, Tissue-specific expression of rat myosin light-chain 2
gene in transgenic mice, Nature, 314(6008):283-286.

Simons, R. W., and Kleckner, N., 1983, Translational control of IS 10
transposition, Cell, 34:683-691.

Smithies, O., Gregg, R. G., Boggs, S. S., Koralewski, M. A., and
Kucherlapati, R. S., 1985, Insertion of DNA sequences into the
human chromosomal β-globin locus by homologous recombination,
Nature, 317:230-234.

Stewart, T. A., and Mintz, B., 1981, Successive generations of mice
produced from an established culture line of euploid
teratocarcinoma cells, Proc. Natl. Acad. Sci. USA, 78:6314-6318.

Stewart, T. A., Pattengale, P. K., and Leder, P., 1984, Spontaneous
mammary adenocarcinomas in transgenic mice that carry and express
MTV/myc fusion genes, Cell, 38:627-637.

Stewart, T. A , Wagner, E. F., and Mintz, B., 1982, Human β-globin gene
sequences injected into mouse eggs, retained in adults, and
transmitted to progeny, Science, 217:1046-1048.

Storb, U., O'Brien, R. L., McMullen, M. D., Gollahon, K. A., and Brinster,
R. L., 1984, High expression of cloned immunoglobulin kappa gene in
transgenic mice is restricted to B lymphocytes, Nature,
310(5974):238-241.

Stuhlmann, H., Cone, R., Mulligan, R. C., and Jaenisch, R., 1984, Introduction of a selectable gene into different animal tissue by a retrovirus recombinant vector, Proc. Natl. Acad. Sci. USA, 81:7151–7155.

Stuhlmann, H., Jahner, D., and Jaenisch, R., 1981, Infectivity and methylation of retroviral genomes is correlated with expression in the animal, Cell, 26:221–232.

Swift, G. H., Hammer, R. E., MacDonald, R. J., and Brinster, R. L., 1984, Tissue-specific expression of the rat pancreatic elastase I gene in transgenic mice, Cell, 38:639–646.

Tabin, C. J., Hoffman, J. W., Goff, S. P., and Weinberg, R. A., 1982, Adaptation of a retrovirus as a eucaryotic vector transmitting the herpes simplex virus thymidine kinase gene, Mol. Cell Biol., 2:426–436.

Tate, A. C., Greene, G. L., DeSoombre, E. R., Jensen, E. V., and Jordan, V. C., 1984, Differences between estrogen- and antiestrogen-estrogen receptor complexes from human breast tumors identified with an antibody raised against the estrogen receptor, Cancer Research, 44:1012–1018.

Temin, H. M., 1982, Function of the retrovirus long terminal repeat, Cell, 28(1):3–5.

Townes, T., Lingrel, L., Chen, H., Brinster, R., and Palmiter, R., 1985, Erythroid specific expression of human β-globin genes in transgenic mice, Eur. Mol. Biol. Org. J., 4:1715–1724.

van der Putten, H., Botteri, F. M., Miller, A. D., Rosenfeld, M. G., Fan, H., Evans, R. M., and Verma, I. M., 1985, Efficient insertion of genes into the mouse germ line via retroviral vectors, Proc. Natl. Acad. Sci. USA, 82:6148–6152.

Wagner, E. F., Covarrabias, L., Stewart, T. A., and Mintz, B., 1983, Prenatal lethalities in mice homozygous from human growth hormone gene sequences integrated in the germ line, Cell, 35:647–655.

Wagner, E., Stewart, T., and Mintz, B., 1981b, The human β-globin gene and a functional viral thymidine kinase gene in developing mice, Proc. Natl. Acad. Sci. USA, 78:5016–5020.

Wagner, T., Hoppe, P., Jollick, J., Scholl, D., Hodinka, R., and Gault, J., 1981a, Microinjection of a rabbit β-globin gene into zygotes and its subsequent expression in adult mice and their offspring, Proc. Natl. Acad. Sci. USA, 78:6376–6380.

Wagner, T. E., Murray, F. A., Minhas, B., and Kraemer, D. C., 1984, The possibility of transgenic livestock, Theriogenology, 21:29–44.

Wagner, T. E., Van Blerkom, J., and Jollick, J. D., 1982, Gene chimeric mammals: Mice developed from zygotes microinjected with rabbit β-globin genes producing rabbit globin, DNA, 1:165–166.

Wagner, T. E., and Jochle, W., 1985, Recombinant Gene Transfer in Animals: The potential for improving growth in livestock, in: "Control and Manipulation of Animal Growth," P. J. Buttery, ed., Butterworth Press, London (in press).

Wakeling, A. E., Valcaccia, B., Newbolilt, E., and Green L. R., 1984, Non-steroidal antiestrogens--receptor binding and biological response in rat uterus, rat mammary carcinoma and human breast cancer cells, J. Steroid Biochemistry, 20:111–120.

Wei, C., Gibson, M., Spear, P. G., and Scolnick, E. M., 1981, Construction and isolation of a transmissible retrovirus containing the src gene of Harvey murine sarcoma virus and the thymidine kinase gene of herpes simplex virus type 1, J. Virol., 39:935–944.

Zamecnik, P. C., and Stephenson, M. L., 1978, Inhibition of Rous sarcoma virus replication and cell transformation by a specific oligodeoxy-nucleotide, Proc. Natl. Acad. Sci. USA, 75:280–284.

REVERSIBLE DISSOCIATION OF CHICK

OVIDUCT PROGESTERONE RECEPTOR SUBUNITS

Wayne W. Grody, William T. Schrader, and Bert W. O'Malley

Department of Cell Biology
Baylor College of Medicine
Houston, Texas 77030

INTRODUCTION

Previous studies in our laboratory have shown that chick oviduct cytosol contains two 4S progesterone-binding components of different molecular weights, designated A and B, which exhibit high affinity for DNA and chromatin, respectively (Schrader et al., 1972). Both are single polypeptide chains with kinetically identical hormone binding sites (Schrader and O'Malley, 1972). Their molecular weights are 79,000 and 108,000, respectively (Coty et al., 1979; Kuhn et al., 1975). We have postulated that the native cytoplasmic receptor form, sedimenting at 6-8S in low-salt sucrose gradients and containing equal amounts of A and B (Schrader et al., 1975), is a dimer or higher aggregate consisting of one part each of the A and B proteins as subunits.

While data derived from sucrose gradient sedimentation (Schrader et al., 1975), ion-exchange chromatography (Schrader et al., 1976), and chemical cross-linking studies (Birnbaumer et al., 1979) have been consistent with this model, attempts to test the hypothesis definitively by reconstitution of A-B dimers or aggregates have been unsuccessful, since the standard treatments known to yield the 4S species (e.g., heat, 0.3 M salt, dialysis) are all apparently irreversible (Schrader and O'Malley, 1972). This failure to achieve subunit recombination was presumed to be due to partial denaturation of critical site(s) on the A or B proteins during aggregate dissociation and monomer isolation.

351

Evidence from a variety of studies led us to believe that amino groups on the two proteins might be involved at this crucial site(s) for aggregate assembly. First, the aggregate receptor does not bind to phosphocellulose, while the A and B monomers both bind (Coty et al., 1979), suggesting that there are fewer exposed basic residues on the aggregate than on either monomer. Second, A and B can be chemically cross-linked together in the aggregated state by dimethyl-suberimidate (Birnbaumer et al., 1979), an agent that reacts with amino groups, indicating that amino groups are present on the receptor combining surfaces. Finally, the DNA binding property of the A protein is masked when it is complexed in the aggregate, again suggesting occlusion of basic regions.

To probe these interactions further, we have employed an agent, pyridoxal 5'-phosphate (PALP), which reacts specifically with amino groups of proteins. We have found that PALP not only induces receptor aggregate disassembly but does so in a reversible manner, allowing aggregate reconstitution and analysis.

EFFECT OF PALP ON RECEPTOR DNA BINDING

The 6-8S native progesterone receptor species in chick oviduct cytosol is unable to bind to DNA-cellulose (Schrader et al., 1977). Ammonium sulfate fractionation liberates two 4S components, A and B, of which the former exhibits affinity for DNA (Schrader et al., 1972). Litwack and co-workers (Cake et al., 1978) have reported that PALP markedly inhibits the binding of rat liver glucocorticoid receptor to DNA-cellulose. For our DNA binding studies, [^{3}H]progesterone-labeled cytosol prepared in Buffer A (see legend to Fig. 1) was precipitated by adding 50% (v/v) of saturated $(NH_4)_2SO_4$ solution, and the pellet was resuspended in the original cytosol volume of borate buffer (20 mM Na borate, pH 8.0). Two ml aliquots of this solution were incubated for 90 min at 4°C with various concentrations of PALP, and assayed for DNA-cellulose binding as described in Figure 1.

As illustrated in Figure 1, PALP inhibits binding of progesterone receptor to DNA in a concentration-dependent manner. It has been suggested (Cake et al., 1978) that PALP interacts with proteins by forming a Schiff base with ε-amino groups of lysine residues. This metastable interaction can be transformed to a covalent bond by subsequent reduction

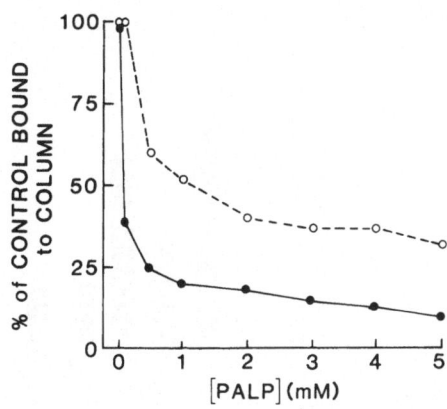

Fig. 1. Inhibition of receptor binding to DNA-cellulose by
PALP. Labeled receptor-hormone complexes were
prepared from oviducts of estrogenized immature
chicks, using Buffer A (10 mM Tris-HCl, 1 mM
Na_2EDTA, and 10 mM 1-thioglycerol) as described
elsewhere (Schrader, 1975; Coty et al., 1979).
After labeling with [^{3}H]progesterone (S.A. = 50
Ci/mmol), the receptors were precipitated at 50%
saturation of ammonium sulfate and redissolved in
the original cytosol volume of borate buffer (20 mM
Na borate, pH 8.0). PALP stock solution (100 mM in
H_2O, neutralized to pH 7.5-9.0) was added to the
receptor samples. Some samples were then treated
with $NaBH_4$, made up as a 100 mM stock solution in
H_2O. The $NaBH_4$ was added to yield a final
concentration of 5 mM. DNA-cellulose columns
(2 ml) were prepared as described previously
(Schrader, 1975). After 90 min incubation (4°C),
receptor samples (2 ml) were applied to the
DNA-cellulose columns and washed with 7 ml of
Buffer A. Radioactivity in the pooled wash allowed
calculation of the percent of the input hormone-
receptor complexes retained on the column. Samples
with (●) and without (○) the $NaBH_4$ reduction step
were compared.

with $NaBH_4$. As expected, this results in even more marked inhibition of
receptor DNA binding, reaching about 90% at 5 mM PALP (Fig. 1, closed
circles).

REVERSIBILITY OF PALP EFFECT ON DNA BINDING

Since the initial binding of PALP to proteins is non-covalent, the
reagent can be removed by the addition, in excess, of an exogenous source
of primary amino groups. This was done by adding ethanolamine at a final
concentration of 50 mM to the PALP-treated receptor samples, and
incubating in the cold for an additional 90 min. Figure 2 shows that the

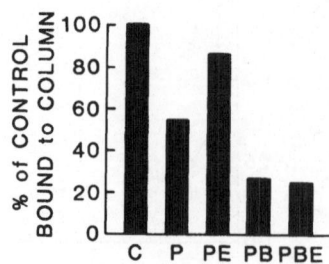

Fig. 2. Reversibility of the PALP effect on DNA-cellulose binding by receptor. [^{3}H]progesterone-receptor complexes were prepared by ammonium sulfate precipitation as described in Figure 1. Samples were treated with the following additions before assay of DNA-cellulose binding as in Figure 1. C: untreated control sample; P: 5 mM PALP; PE: 5 mM PALP followed by 50 mM ethanolamine (prepared as 1.0 M aqueous stock solution, pH 8.0); PB: 5 mM PALP followed by 5 mM NaBH$_4$; PBE: 5 mM PALP followed by 5 mM NaBH$_4$ and then by 50 mM ethanolamine.

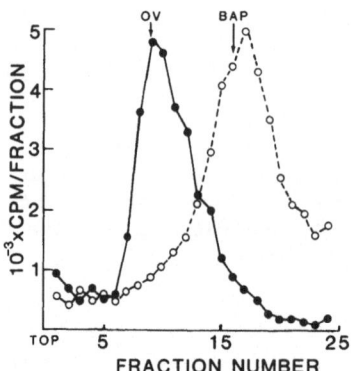

Fig. 3. Dissociation of receptor aggregates by PALP. [^{3}H]progesterone-receptor complexes in cytosol were prepared in borate buffer. Aliquots (200 μl) were layered atop 5 ml sucrose gradients (5% to 20% w/v) prepared in borate buffer and centrifuged as described in the text. Gradients were prepared with (●) or without (○) 6 mM PALP. Arrows indicate sedimentation positions of [^{14}C] standards: OV, ovalbumin (3.67S); BAP, bacterial alkaline phosphatase (6.3S).

addition of ethanolamine reverses the inhibition of DNA binding, provided the Schiff base was not first reduced with NaBH$_4$.

EFFECT OF PALP ON RECEPTOR QUATERNARY STRUCTURE

The effect of PALP treatment on receptor sedimentation behavior is illustrated in Figure 3. Centrifugation of cytosolic receptor in sucrose gradients containing 5 mM PALP causes dissociation of 6-8S receptor aggregates to 4S monomers. This is essentially the profile seen when receptors are run in gradients containing 0.3 M KCl (data not shown), but

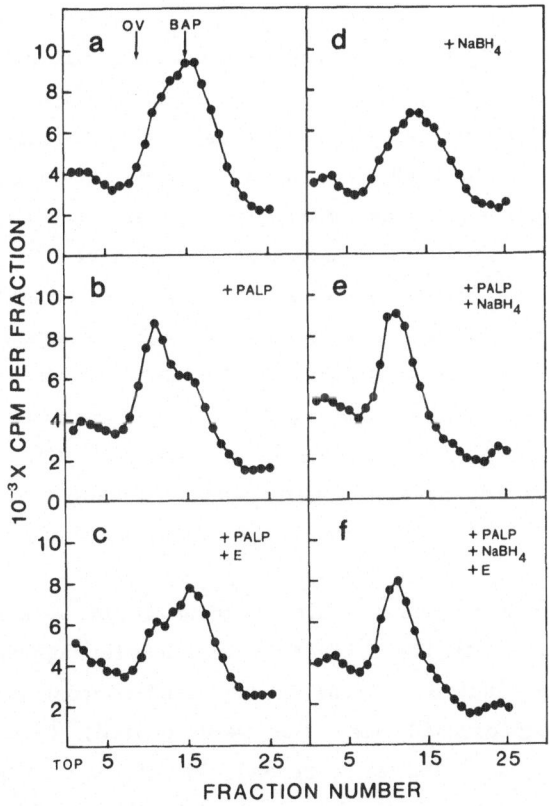

Fig. 4. Reversibility of receptor dissociation by PALP. Aliquots of labeled cytosol receptor in borate buffer were incubated at 4°C with the following additives and then analyzed by sedimentation velocity as described in Figure 3. (a) untreated control; (b) 5 mM PALP; (c) 5 mM PALP followed by 50 mM ethanolamine; (d) 5mM NaBH$_4$; (e) 5 mM PALP followed by 5 mM NaBH$_4$; (f) 5 mM PALP followed by 5 mM NaBH$_4$ and then 50 mM ethanolamine.

the low concentration of PALP used in these experiments is not consistent with a simple ionic strength effect. If receptors are exposed to 5 mM PALP for just 90 min and then run in gradients not containing PALP, the dissociation seen is substantial, but not complete (Fig. 4, panel b). However, we have found that use of a 100 mM PALP stock made up at higher pH (pH 9) will result in 80 to 90% dissociation after only 90 min (data not shown).

RECONSTITUTION OF RECEPTOR AGGREGATES

The finding that the Schiff base linkage between PALP and receptor is reversible, at least as monitored by the DNA-cellulose binding experiments described above, suggested to us that the dissociation of receptor aggregates by PALP might also be reversible. Our initial investigation of this possibility is depicted in Figure 4. [^{3}H]progesterone-labeled cytosol, containing exclusively 6-8S receptor aggregates (panel a), was incubated with 5 mM PALP for 90 min, and an aliquot removed for sedimentation analysis. As shown in panel b, this resulted in dissociation of 60 to 70% of the receptor aggregates to 4S monomers. To the remainder of the sample, ethanolamine was added to 50 mM, and after an additional 90 min incubation, aliquots were again removed for analysis on sucrose gradients. Panel c shows that the ethanolamine treatment caused reduction of the 4S shoulder and regeneration of receptor aggregate forms. However, if the PALP-treated sample was reduced with NaBH$_4$, complete dissociation occurred (panel e), which was no longer reversible by adding ethanolamine (panel f). NaBH$_4$ by itself was without effect (panel d).

We next wanted to see if this reaction could be accomplished using isolated 4S monomers in the absence of any heavier aggregates. In these experiments, peak fractions of the 4S region of sucrose gradients on which PALP-treated receptor had been run were pooled, treated with 50 mM ethanolamine, and analyzed on a second set of sucrose gradients. Both sets of gradients were run in the Beckman VTi65 vertical tube rotor to shorten the total time required for the experiment, and the second set of gradients contained 10% glycerol to facilitate proper layering of the pooled 4S sample. As shown in Figure 5, we found that ethanolamine treatment of the 4S pool obtained by PALP dissociation caused recomplexing of the monomers into the heavy (10S) aggregate forms typically seen in the vertical tube rotor (panel b); in the presence of PALP, only 6-8S aggregate forms and 4S monomers are seen (panel a).

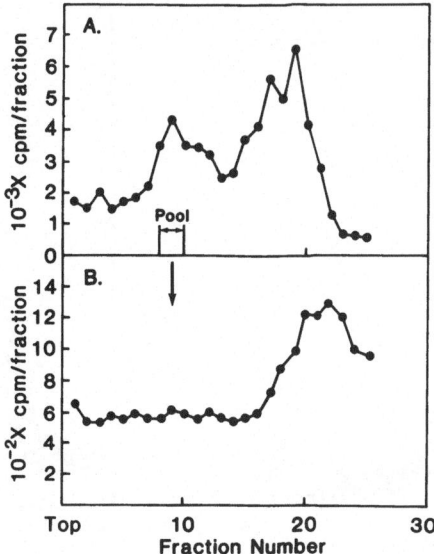

Fig. 5. Reconstitution of receptor aggregates from pooled
4S monomers. Cytosol was treated with 5 mM PALP
for 90 min, and run on sucrose gradients in the
Beckman VTi65 vertical tube rotor (fractions 8-10)
were pooled (panel a) and run on a second set of
gradients containing 10% glycerol (panel b). The
glycerol gradients were centrifuged for 105 min at
65,000 rpm.

In both types of reconstitution experiments described thus far, the
4S substrates were not characterized or purified in any way prior to
reassociation. To test our A-B dimer hypothesis more definitively, it was
necessary to begin with separate preparations of the A and B proteins, and
determine whether they could recombine to form larger aggregates when
mixed together. In addition, it was equally important to test whether the
individual A and B proteins were able to self-aggregate to form
homodimers. For this study, cytosolic receptors were dissociated, and A
and B subunits were isolated as described in Figure 6. To verify that A
and B prepared in this way are competent to reform aggregates with their
complementary monomers in PALP-treated cytosol, the experiments shown in
Figure 6 were performed. [^{3}H]progesterone-labeled A and B, prepared as
described, were incubated individually with an equal volume of
PALP-treated cytosol which had been pre-incubated with 10^{-8} M cold
progesterone. The reaction products were then analyzed on sucrose
gradients containing 10 mM Na_2MoO_4, which we have previously found useful
for stabilizing native receptor aggregate forms on sucrose gradients
(Grody et al., 1980).

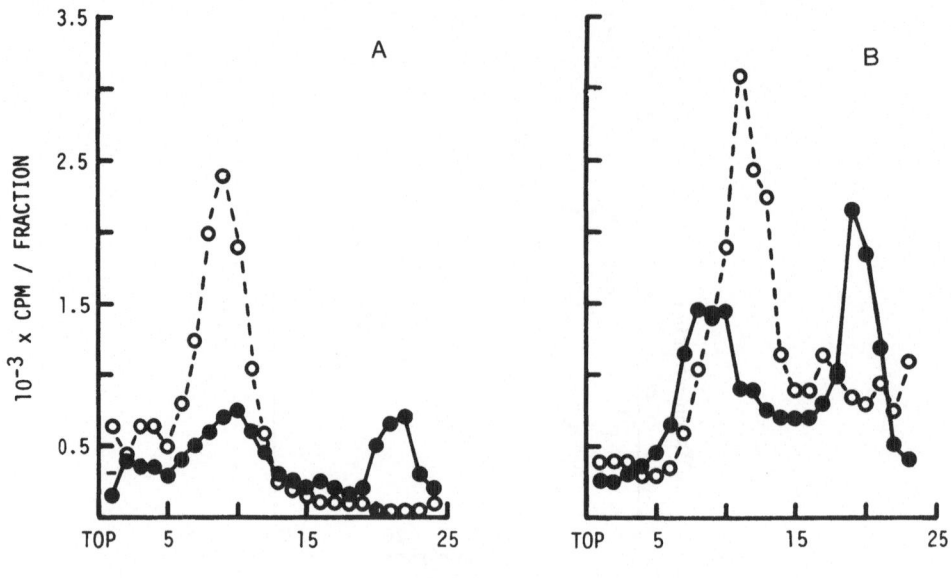

Fig. 6. Recombination of labeled receptor subunits with unlabeled cytosol which had been preincubated with 10^{-8} M non-radioactive progesterone. Receptor A (left panel) was prepared from cytosol (in borate buffer) by treating the cytosol for 90 min at 4°C with 5 mM PALP. Then the sample was brought to 0.135 M KCl and applied to a 2 ml DEAE-cellulose column. Receptor A passes through the column but receptor B is retained (Schrader and O'Malley, 1972). The pooled drop-through fraction was the source of receptor A. Receptor B was prepared from cytosol (in Buffer A), precipitated at 50% saturation of ammonium sulfate and redissolved in borate buffer containing 0.1 M KCl. This sample was applied to a 2 ml DNA-cellulose column; receptor B passed through this column but receptor A was retained. Portions of both preparations were analyzed for contamination by the other subunit by phosphocellulose chromatography (Schrader et al., 1976). Receptor samples were then analyzed by sedimentation on sucrose gradients (see Fig. 3) containing 10 mM Na_2MoO_4, either alone (o) or after 90 min incubation with 50 mM ethanolamine and unlabeled cytosol pre-treated for 90 min with 5 mM PALP (●). Sedimentation values were: fraction 10, 4S; fraction 20, 8S.

As shown in Figure 6, the A and B preparations by themselves ran exclusively as 4S monomers; but when added to cytosol, about 50% of the labeled monomer in each prep was able to recombine with elements in the cytosol to give sharp 8-9S aggregate peaks. Since the cytosol receptors had been blocked with cold progesterone, this result cannot be explained merely by transfer of labeled hormone from the A and B preparations to some residual aggregates in the cytosol which had not been dissociated by PALP. Furthermore, an analogous experiment in which all receptor components had first been dissociated only by $(NH_4)_2SO_4$ precipitation, and not by PALP, gave no detectable aggregate reconstitution (results not shown).

We next wanted to see if these partially purified A and B receptor pools could be recombined directly with each other to yield aggregates. In this experiment, receptors A and B were incubated for 90 min at 4°C, individually or mixed together at a molar ratio of 1:1, in the presence or absence of 50 mM ethanolamine. Then 200 µl of each incubation mixture was analyzed on sucrose gradients. The results, shown in Figure 7, indicate that there was no apparent self-aggregation of the individual A or B monomers, whether or not ethanolamine was present (left and center panels). Both cases gave a peak of free (dissociated) hormone at the top of the gradient, and a single receptor monomer peak at 4S. The small residual aggregate peak sedimenting in fraction 20 of panel B was not augmented or reduced by these manipulations. This peak consists of aggregates which co-eluted with receptor B subunit during its isolation. Thus, no A-A or B-B dimers or aggregates were produced. Similarly, in the absence of ethanolamine, the A and B receptors sedimented primarily as a mixture of 4S monomers even when mixed together in equal amounts (Fig. 7, right panel, open circles). But when ethanolamine was added to consume the PALP, the 4S peak was substantially reduced, and a concomitant restoration of receptor aggregates occurred (closed circles). The reaction is at best 50% efficient, when corrected for the presence of free progesterone in the profiles, and often was much less efficient than the analogous reactions performed in the presence of crude cytosol (i.e., the type of experiment illustrated in Fig. 6). It is possible, therefore, that aggregation is facilitated by some cytosolic factor which is variably lost during partial purification of A and B.

TITRATION OF A AND B IN THE RECONSTITUTED AGGREGATES

The studies just described demonstrate that receptors A and B, prepared according to our protocol, can be recombined into aggregates upon removal of the PALP used for their initial dissociation. The finding that both monomers A and B are required for the reconstitution reaction to occur is consistent with our A-B dimer hypothesis. However, it still remained important to see whether these monomers were recombining in a fixed, orderly fashion, as opposed to nonspecific aggregation.

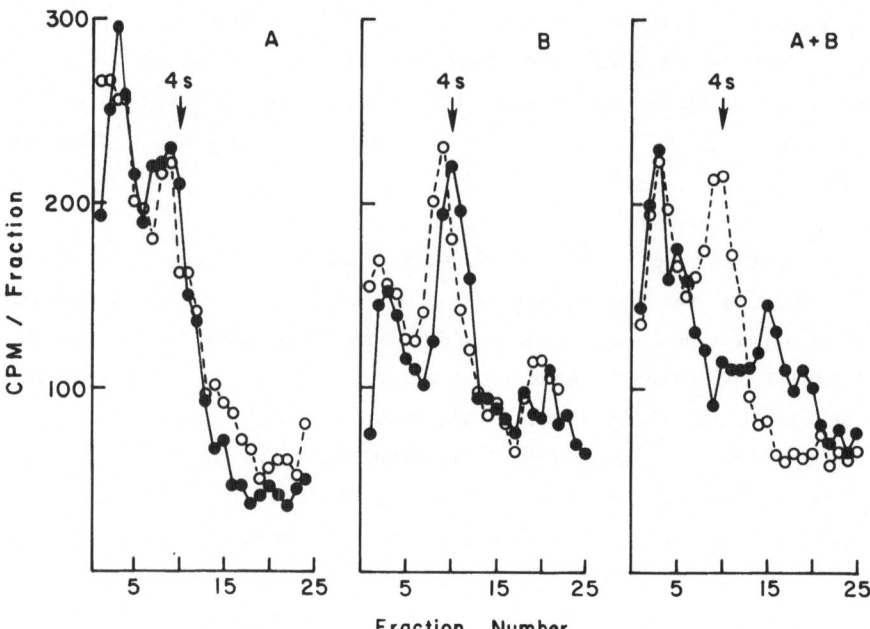

Fig. 7. Reconstitution of receptor aggregates of receptor A and B monomers. Receptors A and B were prepared as in Figure 6, and incubated for 90 min at 4°C individually or mixed in a 1:1 molar ratio in the presence (●) or absence (○) of 50 mM ethanolamine. The mixtures were then assayed by sedimentation velocity (see Fig. 3). Panel A, receptor A only; panel B, receptor B only; Panel A + B, 1:1 mixture of receptors A and B.

First it was necessary to develop a reliable assay system for quantitation of A and B in a mixed population of the two species. While DEAE-cellulose chromatography provides good resolution of the A and B proteins, it could not be used because intact A-B complexes elute at the same position as B (Schrader et al., 1975). Thus, any reconstituted aggregates remaining undissociated by salt before application to the column would cause a falsely elevated estimation of the amount of B protein in the sample.

We therefore chose to use phosphocellulose chromatography, which does not give quite as good separation of the A and B peaks, but has the advantage that intact A-B complexes do not bind at all (Schrader et al., 1975). Using control mixtures of A and B at known input ratios, and in the absence of any reaggregation, we developed an assay to optimize resolution of A and B on phosphocellulose. Basically, this involved elution with a shallow KCl gradient as described in Figure 8. Even under these conditions, the phosphocellulose elution profile is not always a

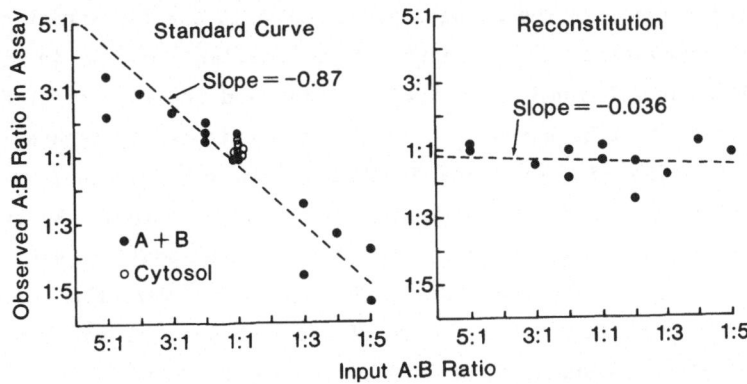

Fig. 8. Stoichiometry of receptor subunit assembly. Relative amounts of receptors A and B were measured in standard and experimental samples by phosphocellulose chromatography, on small (2 ml) phosphocellulose (Whatman P-11) columns. Aggregates were dissociated by KCl treatment as described in the text, and eluted from phosphocellulose using a 40 x 50 ml, 0.15 to 0.325 M KCl gradient (in Buffer A), collecting 1 ml fractions. A and B were quantified by summing the cpm in their respective elution peaks. In cases where separation of the two peaks was less than ideal, KCl concentrations in the eluted fractions were measured using a conductivity meter, and the following cut-off points for the two peaks were used, as determined by control chromatography of known A + B mixtures: B, 0.19 to 0.22 M KCl; A, 0.225 to 0.26 M KCl.

precise reflection of the input A:B ratio. However, the results of many such assays yielded a nearly linear correlation with the known input ratio. We used this approach to construct a standard curve for the phosphocellulose assay (Fig. 8, left panel). The validity of the assay was tested by analyzing the A:B content of crude cytosol, known from earlier experiments to contain equal amounts of both species (Schrader and O'Malley, 1972). The open circles in the left panel of Figure 8 show that the assay yields the expected ratio (1:1) of A:B from cytosol.

We then performed the same type of analysis on A and B found in the reconstituted aggregates. [³H]progesterone-labeled A or B, prepared as described, were added at various molar excesses to labeled cytosol which had been pre-treated with PALP (PALP stock solution used here was made up at pH 9.0, which in companion studies was shown to facilitate 80 to 90% receptor dissociation in 90 min). The reaction was performed at input ratios of 1:1 (cytosol only) and with added excesses of either A or B to produce input ratios ranging from 5:1 to 1:5. The experiment documented in Figure 6 had already established that the added A and B monomers were able to recombine with the endogenous monomers in the PALP-treated cytosol to yield aggregates. This experiment is similar to that in Figure 6, except that the A:B input ratios were varied and the PALP-treated cytosol was labeled so that both monomers in the reconstituted aggregates could be quantitated. The mixtures were incubated with ethanolamine, and run on 36 ml preparative sucrose gradients in a Ti45 rotor. Reconstituted aggregates were separated from unreacted A and B by pooling peak fractions of the 6S region of the gradients. The aggregates were then dissociated by treatment with 0.5 M KCl for at least 8 hours, before loading onto phosphocellulose columns. This long period of KCl treatment is necessary to ensure dissociation of the reconstituted aggregates, since partially purified receptor complexes appear to be markedly more stable to salt (and dialysis) than are the native cytosolic aggregates (see "Discussion").

The results of a series of phosphocellulose analyses of the A:B content of reconstituted aggregates are plotted in Figure 8 (right panel). As can be seen, the curve generated from these data is quite different from the standard curve generated from phosphocellulose assays of non-recombining mixtures of A and B (Fig. 8, left panel). Unlike the standard curve, which reflects an approximately linear correlation between input A:B ratio and assayed A:B ratio, the reconstitution curve shows little or no such correlation. Regardless of the input A:B ratio, the assayed A:B ratio in the reconstituted aggregates is constant, approximately 1:1.

CHARACTERIZATION OF RECONSTITUTED AGGREGATES

Finally, it was important to determine, by some criteria other than sucrose gradient sedimentation, whether the reconstituted aggregates behave similarly to the native receptor. In Table I, the DNA and phosphocellulose binding properties of the reconstituted aggregates are presented. A reconstitution reaction was performed by adding labeled A to cold cytosol, as in the experiment depicted in Figure 6. The reaction mixture was run on preparative sucrose gradients, and the reconstituted aggregate peaks pooled. Receptors in these pools were assayed for DNA-cellulose and phosphocellulose binding and compared to the binding behavior of native receptor aggregates in crude cytosol and to native aggregates pooled from sucrose gradients. Receptor A monomers isolated by ammonium sulfate precipitation and DEAE-cellulose were used as controls. As shown in Table I, the behavior of the reconstituted aggregates on

Table I. Binding characteristics of reconstituted aggregates.

| | Percent of Input [3H] Receptor Bound to Column | |
Receptor Source	DNA Cellulose[d]	Phosphocellulose[e]
[3H] A monomers[a]	84	85
Crude cytosol aggregates	11	21
Cytosol aggregate pool[b]	20	44
Reconstituted aggregate pool[c]	23	10

[a][3H]receptor A monomers were prepared by passing resuspended cytosolic $(NH_4)_2SO_4$ precipitates through DEAE-cellulose at 0.125 M KCl.

[b]Labeled cytosolic receptor aggregates were isolated by running crude cytosol on low-salt sucrose gradients, and pooling the 6-8S peaks.

[c]Reconstitution reaction was performed by incubating [3H]receptor A, prepared as described in Figure 6, with PALP-treated cold cytosol as described in the legend of Figure 6. Reconstituted aggregates were isolated by pooling the 6-8S peaks of sucrose gradients.

[d]DNA-cellulose binding assays were performed as described in Figure 1. Drop-through volumes were checked for the presence of free hormone by the DEAE-slurry assay (Vedeckis et al., 1980), and typically contained < 20%. Results are averages of several experiments.

[e]Phosphocellulose binding assays were performed identically to the DNA-cellulose binding assays.

DNA-cellulose and phosphocellulose is identical to that of the native aggregates. In both cases, the labeled A receptors are unable to bind quantitatively to the columns when in the aggregated state. Moreover, the A subunits in the reconstituted aggregates must be complexed with B subunits in a configuration resembling that of the native cytosolic aggregates, since in both cases the DNA-binding and phosphocellulose-binding domains of the proteins are apparently occluded. Indeed, the reconstituted aggregates exhibit even less binding to phosphocellulose than native aggregates pooled from sucrose gradients, probably because the native complexes partially dissociate during sedimentation, while the more stable reconstituted aggregates remain intact. The inhibition of DNA binding cannot be ascribed to the presence of PALP, since the agent remains at the top of the preparative sucrose gradients. Also, assay of drop-through volumes for free hormone by the DEAE-cellulose assay (Vedeckis et al., 1980) revealed that dissociation of [^{3}H]progesterone from receptor cannot account for the large drop-through percentages.

EFFECT OF OTHER PROBES OF DNA BINDING

The finding that PALP both blocks the receptor's DNA binding ability and dissociates receptor aggregates suggested to us that these two effects might be mediated at or near the same site on the A subunit. If this is so, this would imply that other agents besides PALP which are known to inhibit DNA binding, presumably by interacting with this site, should also dissociate the A-B aggregates as PALP does.

In a preliminary investigation of this hypothesis, we have studied Cibacron blue F3GA, a sulfonated polyaromatic dye which is believed to interact with the dinucleotide fold of DNA-binding proteins (Thompson and Stellwagen, 1976), and has been reported to block binding of mouse uterine estradiol receptor to DNA (Kumar et al., 1979). By the same assay procedures used in the investigation of PALP reported here, we have found that Cibacron blue indeed blocks the receptor A protein's DNA binding ability and dissociates receptor aggregates when added at a concentration of 0.5 mM. However, it is unlikely that all these agents are acting at precisely the same site or in exactly the same way, for they do not give a mutually consistent pattern of competition for binding to two other matrices, Cibacron blue-agarose and phosphocellulose. The results of many such binding assays are summarized in Table II.

DISCUSSION

The results presented here constitute the first successful attempt to regenerate chick oviduct progesterone receptor dimers from their individual subunits. This has been an important goal of this laboratory ever since the initial characterization of the A and B receptor proteins, at which time many manipulations were tried (Schrader and O'Malley, 1972). Though A and B are known to be present in equal amounts in the native 6-8S aggregate species (Schrader et al., 1979), and can even be cross-linked chemically (Birnbaumer et al., 1979), generation of 4S monomers by all previously known methods (heat, salt, etc.) has not been reversible.

Table II. Inhibition of receptor A binding to resins by various agents.

Reagent Added[a]	Percent of Control Bound to Column[b]		
	DNA Cellulose	Cibacron Blue-Agarose	Phosphocellulose
None	100	100	100
5 mM PALP	38	49	37
200 μM PALP	13	23	11
1 mg/ml Hep	6	45	N.D.[c]
50 mM ATP	43	75	42
500 μg/ml oligo(dA)$_{40-60}$	23	42	35

[a]Competitors were incubated with [^{3}H]receptor A monomers, prepared as described in footnote a of Table I, for 90 min at 4°C and binding assays performed as described for the DNA binding assay in Figure 1. CB, Cibacron blue F3GA; Hep, heparin.

[b]Binding was assayed using 2 ml columns of each matrix. Percent binding of treated receptor samples was compared to binding of untreated (control) samples.

[c]N.D., not done because of direct interaction of heparin with phosphocellulose resin.

This consistent failure to achieve subunit recombination has in the past been ascribed to loss or destruction of some low molecular weight "coupling factor" during monomer preparation (Schrader and O'Malley, 1972). In view of the present results, it is no longer necessary to invoke such an explanation. It now appears more likely that the previous dissociation methods irreversibly altered some site on the receptor molecule itself which is essential for maintaining the dimer structure. However, we cannot unequivocally rule out participation of additional co-factors at this time since the receptor subunits used in these experiments were not purified to homogeneity.

In the experiments reported here, we have dissociated receptors by a novel method--treatment with millimolar concentrations of PALP--which apparently leaves the critical site intact, allowing subsequent aggregate reconstitution if PALP is removed. Interestingly, this treatment also differs from the other dissociation methods in that it does not result in receptor "activation" as defined by the ability to bind to DNA).

Two observations lead us to believe that this crucial site may be at or near the receptor's DNA-binding domain. First is the finding, reported here, that PALP blocks the receptor A subunit's ability to bind to DNA cellulose. The same effect has been seen by Litwack and co-workers in the rat liver glucocorticoid receptor system (Cake et al., 1978), by Muldoon and Cidlowski in the rat uterine estrogen receptor system (Muldoon and Cidlowski, 1980), and in the chick oviduct progesterone receptor system by Nishigori and Toft (Nishigori and Toft, 1979), who also noted inhibition of receptor binding to ATP-Sepharose. Secondly, in our successful recombination of partially purified A and B monomers, described above and illustrated in Figure 7, only the A protein had been prepared by PALP treatment; the B protein had been obtained following a routine dissociation procedure, ammonium sulfate precipitation (necessitated because dissociation by PALP would have caused both A and B to drop through the DNA-cellulose column, preventing isolation of B). This suggests that the critical binding site for interaction with PALP lies on the A protein, which is the one possessing the high affinity DNA binding site (Schrader and O'Malley, 1972; Schrader et al., 1977).

If this is true, it would assign to the A protein's DNA-binding site a previously unsuspected function, that of binding to a complementary site on the B protein in order to maintain the dimer complex. We already know that this site on the A protein is occluded or otherwise masked by B when the two are complexed together (Coty et al., 1979). This is supported by our finding that other agents besides PALP which are known to inhibit DNA binding, such as Cibacron blue (Kumar et al., 1979), also dissociate the A–B dimer as PALP does. In fact, we have recently obtained evidence that DNA itself can promote aggregate disassembly (data not shown). Still, the data in Table II argue against the idea that all of these agents act upon exactly the same receptor sites. This is not surprising, however, for the chemical structures of these inhibitors are so different that one would not expect all of them to interact with the same amino acid residue(s). It remains to be seen whether these agents, like PALP, dissociate receptors in a manner that is also reversible.

Our observation that the reconstitution reaction requires the presence of both A and B, and that both proteins are found in equimolar amounts in the reconstituted aggregates (regardless of the input ratio),

366

supports our hypothesis that the native 6S form of the progesterone receptor is a dimer consisting of one each of the A and B proteins as subunits. This is entirely consistent with our earlier analyses using other methods.

We have also shown that the reconstituted dimers behave functionally as the native receptor dimers, at least with regard to phosphocellulose and DNA binding. We are presently in the process of testing whether they resemble the native complexes by other criteria such as chromatin binding and antigenic determinants.

The only difference we have noticed so far between the reconstituted aggregates and the native aggregates in crude cytosol is the pronounced stability of the former to dissociation by salt and dialysis. In related studies in our laboratory, we have noticed that the native aggregates, when partially purified about 500-fold by passage over several ion-exchange resins, are also more stable than they behave in crude cytosol (unpublished data). This suggests that there is loss of some cytosolic aggregate-destabilizing factor during preparation of the monomers used in the reconstitution reaction, and during the native aggregate purification procedures. Indeed, we have found that salt-dependent dissociation of reconstituted aggregates for the phosphocellulose assay can be facilitated by adding back a small volume of crude cytosol treated only with cold progesterone (data not shown).

The composition of the larger cytosolic complexes ($\geq$ 8S) seen on sucrose gradients remains unknown. We have always assumed that they represented some type of higher-order aggregation of the 6S dimers, possibly as an A_2B_2 tetramer. Dougherty and Toft (1982) have observed a relatively salt-stable 8-9S progesterone receptor series in the presence of molybdate and phosphate buffer. We, too, have observed stabilization of higher aggregate forms by both molybdate (Grody et al., 1980) and phosphate (unpublished data). However, we have been hesitant to conclude that these aggregates have any in vivo significance as opposed to non-specific in vitro aggregation. Dougherty and Toft have been able to resolve these aggregates into two 8S species by DEAE-cellulose chromatography: one appears to contain only A, or an A precursor, and the other, only B.

Their observation of apparent A-A and B-B dimers under special conditions can be reconciled with our data if one postulates a 9S tetramer

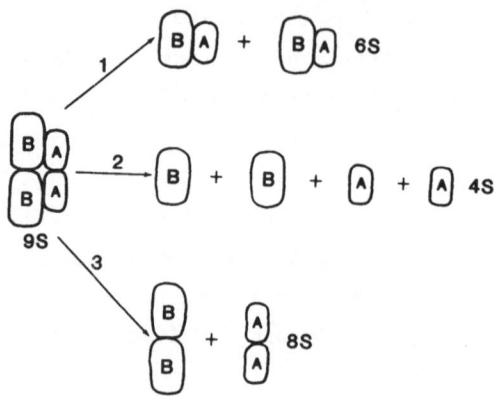

Fig. 9. Postulated structure of heavy (≥ 8S) cytosolic
        receptor aggregates, showing possible breakdown
        pathways and sedimentation values on sucrose
        gradients: (1) spontaneous breakdown in Tris
        buffer (Buffer A); (2) salt-dependent dissociation
        in Tris buffer; (3) salt-dependent dissociation in
        phosphate molybdate buffer.

of the form diagrammed in Figure 9. Such a structure would contain
equimolar amounts of the A and B proteins, and might break down
spontaneously (in the absence of molybdate and phosphate) to yield two 6S
A-B dimers. While 0.3 M salt in Tris buffer will completely dissociate
both the 9S and 6S structures to 4S monomers, molybdate and phosphate
might act by stabilizing the bonds between homologous subunits, yielding
the 8S A-A and B-B dimers. We have investigated previously the relative
stability of receptor aggregates in cytosol to dissociation by time of
dialysis (Schrader and O'Malley, 1978) and found that the large receptor
aggregates (8-10S) are much less stable than the 6S A-B dimer. Thus,
alternate pathways of dissociation of the largest complex are already
known to exist experimentally.

We are currently addressing these possibilities in an effort to
establish the true in vivo nature of the native receptor. We already have
reason to believe that it may be of a size larger than a 6S dimer, albeit
relatively less stable in vitro. We have found (Compton et al., 1984)
that sucrose gradient sedimentation of native cytosolic receptors in
Buffer A (without molybdate or phosphate) in a Beckman VIi65 vertical tube
rotor gives a single sharp receptor peak at 10S. Possibly the much
shorter centrifugation time required with the vertical tube rotor (shorter
by a factor of 10) allows us to see this unstable native species before
its breakdown to the 6-8S forms seen on longer centrifugations.

SUMMARY

We report here the reversible dissociation of chick oviduct progesterone receptor subunits by pyridoxal 5'-phosphate. This agent has been reported to inhibit binding of steroid hormone receptors to DNA and nuclei (Cake et al., 1978).

We have found that pyridoxal 5'-phosphate inhibits binding of chick oviduct progesterone receptors to DNA-cellulose, and also dissociates 6-8S cytosolic receptor aggregates to 4S monomers. Both of these effects are shown to be reversible if pyridoxal phosphate is removed, allowing in vitro reconstitution of receptor aggregates. Fidelity of reconstitution has been assessed by testing the reconstituted aggregate for binding to DNA-cellulose, phosphocellulose, and by studies using sedimentation velocity measurements. By these three criteria, the reconstituted product is indistinguishable from the native cytosol complex from which the monomers were derived. The reconstitution reaction shows an absolute requirement for the presence of both receptor monomers A and B. Titration experiments show a molar ratio of 1:1 for A and B in the reconstituted aggregates. These reconstitution studies confirm our hypothesis (originally based upon dissociation experiments) that native receptor aggregates are composed of the A and B proteins as subunits.

ACKNOWLEDGEMENTS

This investigation was supported in part by USPHS postdoctoral fellowship AM06137 to W.W.G., and by USPHS grant HD07857.

REFERENCES

Birnbaumer, M. E., Schrader, W. T., and O'Malley, B. W., 1979, Chemical cross-linking of the chick oviduct progesterone-receptor subunits by using a reversible bifunctional cross-linking agent, Biochem. J., 181:201-213.

Cake, M. H., DiSorbo, D. M., and Litwack, G., 1978, Effect of pyridoxal phosphate on the DNA binding site of activated hepatic glucocorticoid receptor, J. Biol. Chem., 253:4886-4891.

Compton, J. G., Schrader, W. T., and O'Malley, B. W., 1984, Progesterone receptor binding to DNA: Studies by sedimentation velocity methods, J. Steroid Biochem., 20:89-94.

Coty, W. A., Schrader, W. T., and O'Malley, B. W., 1979, Purification and characterization of the chick oviduct progesterone receptor A subunit, J. Steroid Biochem., 10:1-12.

Dougherty, J. J., and Toft, D. O., 1982, Characterization of two 8S forms of chick oviduct progesterone receptor, J. Biol. Chem., 257:3113-3119.

Grody, W. W., Compton, J. G., Schrader, W. T., and O'Malley, B. W., 1980, Inactivation of chick oviduct progesterone receptors, J. Steroid Biochem., 12:115-120.

Kuhn, R. W., Schrader, W. T., Smith R. G., and O'Malley, B. W., 1975, Progesterone binding components of chick oviduct: X. Purification by affinity chromatography, J. Biol. Chem., 250:4220-4228.

Kumar, S. A., Beach, T. A., and Dickerman, H. W., 1979, Effect of Cibacron blue F3GA on oligo-nucleotide binding site of estradiol-receptor complexes of mouse uterine cytosol, Proc. Natl. Acad Sci. USA, 76: 2119-2203.

Muldoon, T. G., and Cidlowski, J. A., 1980, Specific modification of rat uterine estrogen receptor by pyridoxal 5'-phosphate, J. Biol. Chem., 255:3100-3107.

Nishigori, H., and Toft, D., 1979, Chemical modification of the avian progesterone receptor by pyridoxal 5'-phosphate, J. Biol. Chem., 254:9155-9161.

Schrader, W. T., 1975, Methods for extraction and quantification of receptors, Meth. Enzymol., 36:187-211.

Schrader, W. T., Coty, W. A., Smith, R. G., and O'Malley, B. W., 1977, Purification and properties of progesterone receptors from chick oviduct, Ann. N.Y. Acad. Sci., 286-:64-80.

Schrader, W. T., Heuer, S. S., and O'Malley, B. W., 1975, Progesterone receptors of chick oviduct: Identification of 6S receptor dimers, Biol. Reprod., 12:134-142.

Schrader, W. T., and O'Malley, B. W., 1972, Progesterone-binding components of chick oviduct: IV. Characterization of purified subunits, J. Biol. Chem., 247-51-59.

Schrader, W. T., and O'Malley, B. W., 1978, Molecular structure and analysis of progesterone receptors, in: "Receptors and Hormone Action," vol. 2, B. W. O'Malley, and L. Birnbaumer, eds., Academic Press, New York, pp. 189-224.

Schrader, W. T., Schwartz, R. J., Kuhn, R. W., Buller, R. E., and O'Malley, B. W., 1976, Progesterone-receptor mediation of chromatin RNA transcription in a cell-free system, J. Tox. Envir. Health Suppl., 1:77-96.

Schrader, W. T., Toft, D. O., and O'Malley, B. W., 1972, Progesterone-binding protein of chick oviduct: VI. Interaction of purified progesterone-receptor components with nuclear constituents, J. Biol. Chem., 247:2401-2407.

Thompson, S. T., and Stellwagen, E., 1976, Binding of Cibacron blue F3GA to proteins containing the dinucleotide fold, Proc. Natl. Acad. Sci. USA, 73: 361-365.

Vedeckis, W. V., Freeman, M. R., Schrader, W. T., and O'Malley. B. W., 1980, Progesterone-binding components of chick oviduct: Partial purification and characterization of a calcium-activated protease which hydrolyzes the progesterone receptor, Biochemistry, 19:335-343.

CONTRIBUTORS

Om P. Bahl, Ph.D.
    Professor of Biological Sciences
    Department of Biological Sciences
    State University of New York
    Cooke Hall 347
    Buffalo, New York 14260

Irving Boime, Ph.D.
    Professor of Pharmacology and
        Obstetrics and Gynecology
    Department of Pharmacology
    Washington University
        School of Medicine
    660 S. Euclid Avenue
    St. Louis, Missouri 63110

William W. Chin, M.D.
    Associate Professor of Medicine
    Harvard Medical School
    Joslin Diabetes Center
    1 Joslin Place
    Boston, Massachusetts 02215

Michael P. Czech, Ph.D.
    Professor and Chairman
    Department of Biochemistry
    University of Massachusetts
        Medical Center
    550 Lake Avenue N.
    Worcester, Massachusetts 01605

Dharam S. Dhindsa, D.V.M., Ph.D.
    Executive Secretary/Referral Officer
    Reproductive Biology Study Section
    Division of Research Grants
    National Institutes of Health
    Westwood Building, Room 307
    Bethesda, Maryland 20205

David L. Garbers, Ph.D.
    Professor of Pharmacology
        and Physiology
    Department of Pharmacology
    Howard Hughes Medical Institute
    Vanderbilt University
        School of Medicine
    Light Hall, Room 751
    Nashville, Tennessee 37232

Michael D. Griswold, Ph.D.
    Professor of Biochemistry
    Department of Biochemistry
      and Biophysics
    Washington State University
    627 Fulmer Hall
    Pullman, Washington 99163

Mark T. Groudine, M.D., Ph.D.
    Associate Member
    Fred Hutchinson Cancer
      Research Center
    1124 Columbia Street
    Seattle, Washington 98104

David W. Hamilton, Ph.D.
    Professor and Chairman
    Department of Anatomy
    4-135 Jackson Hall
    University of Minnesota
    321 Church Street, S.E.
    Minneapolis, Minnesota 55455

Vincent T. Marchesi, M.D., Ph.D.
    Professor and Chairman
    Department of Pathology
    Brady Memorial Laboratory
    Yale University School of Medicine
    New Haven, Connecticut 06510

Michael G. O'Rand, Ph.D.
    Associate Professor of Anatomy
    Department of Anatomy
    Laboratory for Cell Biology
    University of North Carolina
    Chapel Hill, North Carolina 27514

Barry I. Posner, M.D.
    Professor of Medicine
    Director, Division of Endocrinology
    Royal Victoria Hospital
    McGill University
    687 Pine Avenue West
    Montreal, Quebec, Canada H3A 1A1

Fritz M. Rottman, Ph.D.
    Professor and Chairman
    Department of Molecular Biology
      and Microbiology
    Case Western Reserve University
      School of Medicine
    Cleveland, Ohio 44106

Harry Schachter, M.D., Ph.D.
    Professor and Chairman
    Department of Biochemistry
    University of Toronto
    555 University Avenue
    Toronto, Ontario, Canada M5G 1X8

William T. Schrader, Ph.D.
    Professor of Cell Biology
    Department of Cell Biology
    Baylor College of Medicine
    1200 Moursund Avenue
    Houston, Texas 77030

Thomas E. Wagner, Ph.D.
    Professor and Chairman
    Molecular and Cell Biology Program
    Ohio University
    Athens, Ohio 45701

Bruce D. Weintraub, M.D.
    Chief of the Molecular, Cellular and
      Nutritional Endocrinology Branch (NIADDK)
    National Institutes of Health
    Building 10, Room 8D14
    Bethesda, Maryland 20205

Equine choriogonadotropin, 32–46
  oligosaccharides
    characterization of, 35
    preparation of, 32–35
Estrogen, 245
Exoglycosidases
  in determination of sugar se-
      quences and anomeric
      linkages, 17–18
  oligosaccharides and, 31, 36–40

Flavobacterium meningosepticum, 8
Follicle stimulating hormone, *see*
      Follitropin
  structural similarity to thyroid
      stimulating hormone, 87
Follitropin, 22
  function, 245
  inhibition of, 248
  production, 245
Fucoidin, inhibition of sperm
      binding to zona pellucida,
      134
Fucose, microheterogeneity, 20

Gene
  α-subunit, 251
  c-myc, transcription, 227
  chymotrypsin, 272
  cloned, nucleotide sequence com-
      parisons, 282
  de novo methylation, 223
  deoxyribonuclease I and, 205
  elastase, 332
  encoding, 249
  flanking regions, 285
  fusion, in transgenic experiments,
      334
  gag, 237
  globin, 206, 208
  growth hormone, bovine, 281
  immunoglobulin, 272
  insulin, 272
  placental lactogen, 269
  prolactin, bovine, 281
  subunit
    chorionic gonadotropin, 268,
        269–271
    chromosomal localization, 253–258
    isolation of, 251
  transcriptional rates, 282
Gene expression
  abnormal, 333
  chromatin structure and, 209–212
  differential, carbohydrate micro-
      heterogeneity and, 57–58
  during embryogenesis, 207
  enhancers, 340
  exogenous, in transgenic mice, 329
  immunoglobulin, 332
  molecular aspects of, 205

Gene expression
  regulation of, 330–335
  templating information for, 226
  tissue specificity, 272, 331
Gene replacement, 338
Gene transcription, 249
  cycloheximide and, 229–230
  lutropin, 259
Gene transfer
  mediated by DNA pronuclear
      microinjection, 324–327
  retroviral mediated, 320–323
  totipotent embryonic cell
      mediated, 324
Genetic polymorphism, 55–57
Germline transgene transmission, 328
Glutamine, as component of
      receptor, 112
N-glycanase, 8
Glycophorin A, 113
O-Glycosidic carbohydrate units,
      structure and location, 20–22
Glycoprotein
  animal, 2
  glycopeptide bonds, 2
  hormone
    carbohydrates of, 11–51
    glycosylation and biosynthesis
        of, 246
    tissue specificity, 246
  lateral mobility in the bilayer,
      132
  microheterogeneity, 4–7
  molecular weight, 123
  Sertoli cell secretion of, 302
  sperm membrane, 121–129
  synthesis, control at substrate
      level, 64–79
  in zona pellucida, 133
Golgi fractions, 190
  insulin accumulation in, 187
Gonadotropins, 245–265, *see also*
      Follitropin, Lutropin
Gonadotropin-releasing-hormone, 247
Growth hormone
  alternative messenger RNA splicing,
      291
  bovine, 281
    precursor transcript, 290
  human, 267
  messenger RNA, 292
Guanylate cyclase, 151
  molecular weight, 151
  resact and, 156

Histone, as substrate for phosphoryla-
      tion of insulin receptor, 167
Human choriogonadotropin
  alpha, 11
  beta, 11

378